THE ADAPTIVE DESIGN
OF THE HUMAN PSYCHE

THE ADAPTIVE DESIGN
OF THE
HUMAN PSYCHE
Psychoanalysis, Evolutionary
Biology, and the Therapeutic Process

Malcolm Owen Slavin
Daniel Kriegman

THE GUILFORD PRESS
New York   London

© 1992 The Guilford Press
A Division of Guilford Publications, Inc.
72 Spring Street, New York, NY 10012

All rights reserved

No part of this book may be reproduced, stored in retrieval system, or transmitted, in any form or by any means, electronic, mechanical, photocopying, microfilming, recording, or otherwise, without written permission from the publisher.

Printed in the United States of America

This book is printed on acid-free paper

Last digit is print number: 9 8 7 6 5 4 3 2 1

Library of Congress Cataloging-in-Publication Data

Slavin, Malcolm Owen.
    The adaptive design of the human psyche : psychoanalysis. evolutionary biology, and the therapeutic process / Malcolm Owen Slavin, Daniel Kriegman.
      p. cm.
    Includes bibliographical references and index.
    ISBN 0-89862-795-8
    1. Psychoanalysis—Philosophy. 2. Sociobiology. 3. Adjustment (Psychology) 4. Genetic psychology. 5. Psychotherapist and patient. I. Kriegman, Daniel H., 1951– . II. Title.
    [DNLM: 1. Evolution. 2. Psychoanalytic Theory. WM 460 S6305a]
    RC506.S577 1992
    150.19'5—dc20
    DNLM/DLC
    for Library of Congress   92-49856
                                                                              CIP

*For Joyce and Ginny,
our children,
our patients,
and their patience*

Preface

Readers of this book should, perhaps, be aware of some of its history. Twenty five years ago, one of us, Malcolm Slavin, and Robert Trivers wrote a series of children's books on animal behavior and evolution. In trying to convey the concept of evolved adaptation at a fifth grade level, the authors became engaged in a debate and quest that has not abated. A set of vexing questions arose about whether psychoanalytic notions about the seemingly irrational, conflict-filled nature of the human mind could be reconciled with the Darwinian search for the fundamentally adaptive designs that govern all living creatures. By the mid-1970s, the other of us, Daniel Kriegman, found himself to be afflicted, as it were, with the same questions—especially with the observation that in the clinical setting we had one of our most powerful demonstrations of how ancient, evolved imperatives actually operate at the core of an empathic therapeutic relationship. Apparently, once one engages with these questions, they do not readily subside. Caveat lector. These questions (we hope) can absorb and possess you, too.

Robert Trivers continued, throughout the 1970s, to create a body of seminal evolutionary biological theory that served as the critical first planks across the chasm that divides our thinking about the inner workings of the mind and the world of nature. His work, and the incisive, challenging nature of his mind, has fundamentally altered our basic outlook. We are most grateful for his influence and for his willingness to contribute a Foreword to our book.

By the late 1970s, Jonathan Slavin—who is not (genetically) related to Malcom Slavin—became enlisted in the quest for an evolutionary psychoanalysis and has been an invaluable critic and supporter of our work, as well as a contributor to the application of the evolutionary perspective in soon to be published papers of his own. His close, astute reading of our manuscript (as well as earlier papers) had an impact on the book's final shape.

Throughout the late 1980s, Karen Norberg, David Edelstein, and then Kalman Glantz—participants in the Evolution and Psychoanalysis Study Group—debated energetically with us as the book was beginning to take form. More recently, Don Greif has joined the quest, generously sharing his keen grasp of philosophical issues and superb judgment in the final shaping of the text.

Other colleagues have supported our effort to think about and present our work in ways that have given us opportunities to learn from and better communicate evolutionary ways of thinking to psychoanalytic and social scientific audiences. We are grateful to Barry Anechiarico, Jim Barron, Michael Basch, Beatrice Beebe, Don Burke, Anne Carter, Marc Fried, Arnold Goldberg, Alan Lloyd, Stephen Mitchell, Arnold Modell, Amy Morrison, Randy Nesse, Paul and Anna Ornstein, Joe Schwartz, Robert Stolorow, and Judy Teicholz.

Sharon Panulla, our editor at The Guilford Press, has been an encouraging, constructive guide throughout the entire publishing process; and Suzanne Little made some valuable editorial suggestions about the organization of the text. Rowena Howells, Managing Editor, steadfastly saw us through what must have seemed like more major, eleventh-hour changes than has been known in the history of publishing; and Curt Tow, Art Director, generously lent his design talents to what may have seemed like two of the most opinionated authors.

* * *

There is no adequate way to express what my wife, Joyce Klein, has contributed to the lengthy, arduous project that culminates in this book. I discovered long ago that if a thought does not make sense to her, it does not make sense. Fortunately, her patience and love come a close second to her judgment. Our five-and-a-half-year-old son, Sam, told me a while ago that this book was "too much about the universe, but you can make people like it if you work on it a lot more." The suggestion to work further was quite altruistic advice since both he and Anna, our daughter, have had to endure that "lot more work" throughout the major part of their lives. My parents and my brother, Dennis, have also endured the deprivations of the writing and helped immensely by taking up the kinship burdens when dad was engaged in "more work." They have imparted, in the most heartfelt sense, the evolutionary understanding that, literally, others live within us and we, in others. So, too, has the memory of my other "brother," Dennis Cherlin, whose belief, warmth, and sustaining friendship lives always in my heart.—M.O.S.

* * *

Recently I was told about an absurd statistical "fact": in 4,000 years or so, at the current rate of reproduction, the aggregate mass of human flesh and bones will exceed the mass of the universe. Absurd because of the nature of making a prediction about humanity 4,000 years into the future. And absurd, of course, because all of the matter in the universe would need to be converted into human flesh. Yet, the inevitable catastrophic collision between human nature and reality—a collision that overpopulation and pollution suggests is already occurring—indicates that unless we change our ways, ugly human-generated horrors are as inevitable a part of our future as they were of our past. The small island of relative peace and safety that my parents provided for me and that, thus far, I have been able to provide for my children ultimately must be a blip in the tragic story of the human struggle for survival. With this knowledge, I either despair or turn to continue the work begun by Sigmund Freud. Not because of his particular insights. Rather, because despite the inadequacy of many of his ideas, he most clearly articulated our need to develop an understanding of the forces that define and live us—that operate upon and through us beyond our awareness—if we are ever to become masters of our own fate. To this end, I have found that new vistas—true visions—open up if we view ourselves through the lens of the theory of evolution. I hope we have done our task well enough to enable the reader to share this developing vision. I hope that as a species our capacity to envision enables us to provide our children and our children's children with a chance to live relatively horror-free lives. In this spirit, I ask my sons, Orion, Isaac, and Joshua, to forgive me for all the occasions on which I said, "I can't right now. I have to work on the book." And I thank my wife, Ginny, for being there when I was not.—D.K.

Foreword

Many have dreamt of building a bridge between Darwin and Freud (including Freud himself). Nowadays we dream of building a bridge between modern evolutionary biology and modern psychoanalytic theory. This task is considerably more daunting than one might imagine, since evolutionary biology deals with the way in which biological (i.e., genetic) traits change over long periods of time (natural selection), while psychoanalytic theory concerns itself with one very specific and intense human dilemma: how one person interacting with another may alleviate the latter's suffering and reorient his or her thinking so as to lead a more satisfying life.

Fortunately, there has been considerable progress recently on the biological side. Since the early 1960s our understanding of the evolution of social life has been revolutionized by our attention to natural selection at the level of individuals and, indeed, genes. We now have an understanding of how living creatures are expected to act toward relatives, the logic of their parent–offspring conflict, the importance of reciprocity (especially in relations between unrelated individuals), and the logic by which social deception becomes intra-psychic *self-deception*. These are exactly the topics we need developed to bridge the gap, but the distance still remains large. After all, we have no biological theory (nor much in the way of evidence) to suggest what kinds of developmental experiences may cause an organism to lead an unsatisfying life—much less, what best to *say* to such a creature to help improve the situation!

Malcolm Slavin and Daniel Kriegman have now stepped forward with a brilliantly argued book linking the two worlds of evolution and psychoanalysis. After discussing the implicit biology hidden in psychoanalytic assumptions, they review the recent biological work, with special attention to its psychological implications. Their treatment of the biology is expert; indeed, they render the biological work in a new psychological form. The gene's-eye view of evolution becomes the psyche's-eye view of its social surroundings. The genetic "inclusive fitness" is replaced by the psychological "inclusive self-interest." A series of intermediate arguments and constructs are designed to carry us step-by-step into the therapeutic situation: the ever present social mixture of cooperation and conflict (with their internal representations), the myriad ways in which development can be disturbed (especially by close relatives), the value of a helper, or therapist, in this situation, the dynamics of deceit and self-deception on both sides of the therapeutic relationship, and, finally the benefits to both parties of reciprocity and symmetry. Throughout, Slavin and Kriegman show that the evolutionary approach leads to a new view of human development which reconciles the two major narratives of current psychoanalytic thought into a single, more powerful form.

This book is sophisticated in its evolutionary thinking and rich in its therapeutic detail. It is based on more than forty years of combined experience in the practice of psychotherapy and psychoanalysis, itself increasingly informed by the synthesis the authors describe here. Let us hope that this book ushers in a new era for psychoanalysts and psychotherapists in which the central issues of their trade are seen to benefit from evolutionary insight: what is an individual's real inclusive self-interest, how can one's ability to satisfy it be adversely affected by past events, and what relationship with another will best reorient the organism in its present dilemma. Slavin and Kriegman have shown in depth and in detail how to critique and transform psychoanalytic thinking using evolutionary logic.

ROBERT TRIVERS, PH.D.
Professor of Biology
University of California (Santa Cruz)

Contents

Preface	vii
Foreword by Robert Trivers	xi
Introduction	1

PART I
The Psychoanalytic Problem and Basic Evolutionary Approach

ONE   *Competing Psychoanalytic Visions of the Human Condition: Classical and Relational Narratives*   19

TWO   *Early Darwinian Versions of the Psychoanalytic Narratives and the Challenge for Contemporary Evolutionary Theory*   33

THREE   *The Modern Evolutionary Perspective: Some Basic Considerations*   55

PART II
Contemporary Evolutionary Theory and the Relational World: The Average, Expectable (Good-Enough) Environment Reconsidered

FOUR   *Conflict and Mutuality in the Relational World: Kin and Reciprocal Altruism*   83

FIVE   Conflict and Mutuality in Development: 107
        Parent–Offspring Conflict Theory

SIX   A Revised View of the Average, Expectable   121
        Environment

SEVEN   The Paradoxical Challenge of Human   137
        Adaptation: Constructing a Self in a Biased,
        Deceptive Relational World

PART III
Intrapsychic Dynamics and the Self System as Evolved Adaptations

EIGHT   Blind Mechanisms with Built-in Adaptive   155
        Vision: Repression, Endogenous Drives, and the
        True Self

NINE   Negotiating and Re-negotiating the Self: 179
        Transference as an Evloved Capacity to
        Promote Change (With a Special Focus on
        Adolescence)

PART IV
Contemporary Evolutionary Theory and the Clinical Process: Conflict, Negotiation, and Influence in the "Good-Enough" Therapeutic Relationship

PRELUDE Clinical Discovery, Comparative Psychoanalytic   209
         Narratives, and the Evolutionary Perspective

TEN   Transference, Resistance, and the Evolved   213
         Capacity for Creative Self-Revision

ELEVEN   The Ambiguities of Empathy and the Creation   242
         of an Alliance with the Patient's Inclusive
         Self-Interest

PART V
Toward an Evolutionary Foundation for Psychoanalysis

TWELVE The Evolved Design of the Psyche and the 263
Classical–Relational Dialectic: Toward a
Rapprochement of Competing Psychoanalytic
Narratives

Appendices

APPENDIX A Has There Been Sufficient Time for the 287
Evolutionary Design of Complex
Psychological Structures through Random
Variation and Natural Selection?

APPENDIX B Freud, Lamarckism, Haeckel's Law, and 290
Modern Evolutionary Thought

APPENDIX C The Evolution of the Human Cortex and 298
Its Relation to Civilization and Guilt

APPENDIX D The Confusion of Proximal and Distal 302
Causes

APPENDIX E Specific, Functional, Affective, 305
Motivational Systems from an
Evolutionary Perspective

APPENDIX F Regressive Renegotiation in the Treatment 308
of the "Regressed" Patient

References 311

Index 326

Introduction

> I could not leave off reading it. At the same time I did not know why it was so important to me, but I see now that the main thing was that it showed that living things could be examined scientifically with a corollary that gaps in knowledge and understanding need not scare me.
> —D. W. WINNICOTT, *On Discovering Darwin's* Origin of Species[1]

> With the mention of phylogenesis ... fresh problems arise, from which one is tempted to draw cautiously back. But there is no help for it, the attempt must be made—in spite of a fear that it will lay bare the inadequacy of our whole effort.
> —S. FREUD, *The Ego and the Id*

Fear of Biology

In the world of psychoanalysis, there is now a widespread uneasiness about using biology to further an understanding of human experience. Biological concepts are seen as violating the complexity of human psychological life and even dehumanizing the therapeutic relationship. For almost two decades, this psychoanalytic wariness about biology has been reflected in the broad critique of the quasi-biological notion of "drives." There are now many who argue that essential meanings are lost when physicalistic notions (e.g., tension reduction) are used to explain the human experience of desire, an experience that invariably occurs within a complex, relational context.

Indeed, over the past several decades, psychoanalysis has taken huge strides toward freeing itself from the reductive, biologized language of the classical Freudian metapsychology. Psychology and social science as a whole have moved beyond a naively physicalistic view of the mind that would presumably confer a mantle of scientific objectivity on psychological theories. There is a greater appreciation of how our

---
[1]Quoted in Phillips (1988, p. 1).

theories of the mind are relative and contextual; how the mind itself may operate by way of narratives that are constructed through interactions with the relational environment. We are now appropriately wary of any discourse in which there is a significant role for innate universals. Many people think about this shift as though it were, ipso facto, the creation of "pure psychology" (i.e., a definitive move away from biology).

In a climate such as this, what can an unabashedly "biological" approach such as ours contribute? Our inquiry is explicitly grounded in Darwinian principles. We speak of the psyche as part of nature, imbued with a legacy from ancient human history. Can an evolutionary understanding of this "archaic heritage" speak to the needs of clinicians who are determined to remain close to their patients' present-day experience? Can an approach that speaks of inherited genetic design actually aid clinicians in their struggle to empathically perceive the complex inner meaning of the strivings that underlie their patients' pain and apparently dysfunctional behavior? Further, can an evolutionary perspective on the innate design of the mind speak to theorists who are acutely sensitive to the plasticity of the human mind, the way it is constructed and sustained through relational interactions with other individuals, symbols, and ideologies?

We believe that the answer to these questions is definitely yes. Rather than reducing or overlooking the importance of individual experience, we hope to demonstrate how modern evolutionary biology can be used in a way that actually enhances our appreciation of the role of experience—both uniquely personal as well as cultural—in the growth and development of each individual psyche. We believe that listening with the evolutionist's ear can attune us to some of the deepest, most vital meanings encoded in our patients' communications.

Our goal is to demonstrate how thinking in "evolutionary–adaptive" terms provides a way of fitting various psychoanalytic models—models that are often disparate and competing—into a larger picture that more fully embraces the many complexities of human nature. Once we begin to think about human psychodynamics as an adaptive system with a long evolutionary history, we find ourselves listening to theoretical debates in a very different way. Current views of the self, drives, affects, and therapeutic change take on altered meanings when these dimensions of human experience are viewed as parts of an overarching, evolved, and adapted psyche. In the clinical setting, we find ourselves hearing the basic adaptive strivings in our patients' experience—strivings that, in the absence of an evolutionary sensibility, may be missed entirely or, worse, simply defined as pathology or resistance. Of vital importance to the clinician, the evolutionary perspective may open new ways to understand and respond to the inevitable conflicts and paradoxes within the therapeutic relationship.

Yet we shall not simply try to persuade the reader that it is interesting and productive to consider what biology can say about the human condition. It is inescapable. All of us do it all the time. *All psychoanalytic—indeed, all psychological—models are, at some level, biological models.* That is, they are all founded on a set of basic assumptions about what, as Greenberg and Mitchell (1983) put it, is "primary" in human nature. In our view, these assumptions include positions on issues such as what is and is not innate, essential, and universal in ourselves and the world into which we are born.

In this broad sense, even the most "antibiological" of psychoanalytic models, such as Kohut's and the narrative, hermaneutic tradition of George Klein and Schafer, are only nonbiological in the narrow sense of their rejection of a particular kind of physicalistic speculation found in Freud's classical drive theory. We shall show that the models that these and every "nonbiological" theorist constructs are rife with their own sets of implicit biological assumptions about human nature. Indeed, as Diana Fuss, the feminist literary critic, has argued, "Any radical constructionism can only be built on the foundations of a hidden essentialism" (1989, p. 12). In other words, *you simply can't create a theory of human experience and subjectivity without making assumptions, however limited, about what is basic and universal about the mind as well as what is universally "average and expectable" about the relational world in which the mind develops.* In a sense our approach is dedicated to making the inevitable "hidden essentialism" more open and explicit. Loaded as the term may be, to call this level of discourse something other than biological seems to us unproductive, evasive, and self-deceptive.

But the very term "evolutionary biology" brings a flood of connotations with it, many of them highly misleading. So we shall say something about what evolutionary biology does not mean. It does not mean neurobiology, physiology, pharmacology, or any of the other current physical or biochemical approaches that are commonly equated with what is "biological." These aspects of biology deal with the physical and chemical mechanisms that may correlate with, and are presumed to underlie, experience. We shall draw, instead, from a branch of contemporary biology that is quite different.

Whereas the neurobiologist may look at psychological structures and capacities and ask the question, "*How* are they physically built into the mind?" the evolutionary biologist asks, "*Why* has such and such a structure or capacity become part of the human repertoire?" The evolutionist's questions are, thus, the functional ones: what are the consequences, the adaptive advantages of a given psychological feature or developmental process taking one particular form versus the myriad other ways it could or might have been organized and built into human functioning? From the evolutionary viewpoint, psychodynamic processes can be just as integral, just as primary biologically, for the living organism as are biochemicals and neuropathways. As Ernst Mayr, the renowned evolutionist, has frequently noted, physiochemical mechanisms are the *means* by which the organism accomplishes its adaptive goals.

Let us consider a way of talking about biological design that shall serve as a clearer, more concrete bridge between evolutionary concepts and psychoanalytic constructs. Chomsky has shown that a complex, "hard-wired" biological program is required for the immense amount of cognitive processing needed to master languages even though each language is, patently, a culturally constructed product. This evolved, universal program is part of the biological design of our mind. It was only possible, however, to recognize the need for such underlying innate linguistic structures when, after extensive study, the magnitude of the *adaptive problem* facing the child and the fluent adult language user was appreciated. Chomsky's contribution lay in good part in his recognition that simple models of learning grounded in a tabula rasa notion of the mind simply could not begin to explain the stunning success with which each new generation solved an adaptive challenge that would be insurmountable for a mind that was not prepared to meet the challenge with an innate, evolved design.

As the evolutionists Tooby and Cosmides (1990) have said (echoing the philosopher Emmanuel Kant's insight into the paradoxical need for "knowledge" that precedes experience):

> The world is itself, complex in ways that are not . . . analyzable or deducible without an enormous amount of a priori knowledge. In order to solve a task you must already know a great deal about the nature of the circumstances in which the task is embedded. (p. 12)

For evolutionists such as Ernst Mayr, it is not simply that there are complex interactions between the psychological and biological. More important, there is a patterning within the psychological realm itself—within our very way of construing and ordering the experience of the environment. This patterning has been shaped by a huge amount of evolutionary history. Our minds, as Mayr suggests, are not biological entities *because* they are connected to brains made up of neurons and biochemical processes. Our minds as psychological phenomena—constellations of feelings, perceptions, cognitions—are, themselves, biological functional designs for living as human beings within a complex human environment (Mayr, 1983b).

Extending Chomsky's notion that universal, preexisting psychological structures must underlie the mastery and use of language, we shall try to demonstrate that many aspects of our psychodynamics can be usefully understood as also having a preexisting, underlying, "deep structural" organization. Subjective constellations of thought and feeling such as those that, in different psychoanalytic traditions, we call the "self," "introjects," "transferences," "empathy," "drives," "ego," etc., may exist precisely because we have evolved flexible, responsive innate programs for generating such "structures of subjectivity." Remember, this is not to reify these psychological terms, to reduce them to some kind of physical entity. Nor is it to diminish the role of experience in shaping the individual psyche. Rather, it is to say that these subjective, psychological entities will only arise because our mind has come equipped with an evolved program that will enable us to use our experience to generate these kinds of inner dynamic forms and processes. Moreover, as we shall see, the deep structural, adaptive design of the human psyche may serve as a way of solving the staggeringly complex adaptive dilemmas universally faced by the human child and the socially interacting adult.

What are these adaptive dilemmas? How does the evolutionary perspective afford us an appreciation of their magnitude and universal difficulty? These are the central questions we shall pose in this book. At this point, however, some readers may conclude that we are about to embark on either a bewilderingly unfamiliar—or, perhaps, all too familiar and discredited—path. So we hasten to note a few basic points and make some fundamental distinctions between our approach and other approaches that may appear superficially similar to it.

Beyond Sociobiology

As the philosopher Jaques Monod noted, almost everyone has some personal conception of Darwin's theory of evolution and, quite commonly, some beliefs about what evolutionary thought may or may not have to say about human nature and the human condition. Contemporary evolutionary theory has gone far beyond Darwin's basic framework, building on it, modifying it, and extending Darwinian thinking to the mind and social behavior. These modifications have not overturned or contradicted Darwin's essential vision. Our book should interest those who would like to learn what some of these newer developments in evolutionary theory may have to say about human nature. No prior knowledge about evolutionary theory is needed to follow our thinking. While we aim to contribute to issues that go well beyond current concerns in the field of psychoanalysis, a basic grasp of psychoanalytic concepts will enable the reader to relate to the wider issues of human development and motivation that we address.

Scientific and pseudoscientific conceptions of the relationship of biology to human behavior (conceptions of human nature) have been with us for a long time. In addition to critiques of the scientific validity of such views (with which we shall deal in Chapter Three) critics such as Steven J. Gould and Richard Lewinton have attacked these perspectives for other reasons. There appear to be two major concerns. The first is that individuals in power will use theories of what is natural—of what is *in our biology*—to rationalize the status quo in an effort to even further skew the distribution of power and resources in society. As Konner (1982) notes, historically evolutionism has commonly been used in this way. Yet we hope to demonstrate that there is an equal, possibly greater, potential in contemporary social evolutionary thought. As the evolutionist Robert Trivers has noted, the emphasis on conflict, deception, and self-interested motivations in modern evolutionary theory can actually heighten our awareness of the tendency toward exploitation in human relations; even more significantly, evolutionary social theory can provide a unique analytical tool for unmasking the often subtle ideological guises in which exploitation is concealed.

The second concern is that biological conceptions of what it means to be human inevitably lead to setting arbitrary limits on what we believe humans can potentially become. This may engender a dangerously self-fulfilling prophesy: it actually limits what we strive to be. We may settle for less than we should. People then lose hope about changing themselves and their situations because, after all, the way things are is the way they were meant to be. You cannot fight city hall, especially when the governing forces are the inherent designs in nature. Such deterministic, constraining uses of evolutionary thought have, of course, been common and are always possible.

Yet, as we shall discuss, human plasticity, our relative freedom from preprogrammed reflexive patterns, is itself a *biological* feature of the human psyche. As we shall see, there is good reason to believe that the very capacity to use culture to construct our identities—the capacity that, indeed, frees us from certain types of "biological" constraints—is itself a built-in feature of our mind. The need and capacity to socially construct ourselves evolved over eons of evolutionary change precisely because of the adaptive benefit it conferred on our ancestors. We believe that a more complete view of what is universal in human nature may actually enable us to make more effective use of our ability to actualize our unique, individual selves—to free ourselves from archaic and unnecessary constraints.

A related concern is of particular importance to clinicians. When so-called biological approaches are accompanied by facile assumptions about "biochemical disorders" and the automatic use of drug treatment, crucial dimensions of patients' communications may be overlooked, essentially "defined away" by a reductive biologizing of the problem. Traumatic histories may be ignored and symptoms suppressed with no appreciation of their meanings: the often vital messages they carry about thwarted adaptive strivings. Rather than adding another rationale to the movement toward psychopharmacological interventions, we shall see how an evolutionary adaptive perspective underscores the importance of an empathic understanding of the subjective meanings in our patients' communications. Although medication is unquestionably useful in many circumstances, the evolutionary perspective alerts us to the dangers of using medication to suppress communicative symptoms. This perspective also alerts us to certain powerful motivations on the part of clinicians to avoid the exceedingly difficult task of "decentering" from our own experience and empathically entering our patients' subjective worlds. Indeed, we hope to show that an evolutionary sensibility, as it were, entails a resolute search for the pervasive—though often hidden, even socially proscribed—adaptive or functional meanings that lie behind many symptoms and other problematic aspects of our experience.[2]

In no sense do we believe that modern evolutionary theory—as some sociobiologists would claim—can itself generate a useful model of human psychology, certainly not a new or useful clinical discourse. The level of human experience captured in psychoanalytic models of the mind is essentially inaccessible to the evolutionary theorist; like a wide-angle lens, the evolutionary perspective captures the vast sweep of the context of nature and archaic history. Psychoanalytic models, like a magnifying lens, give us a very narrow slice but show us great in-

---

[2]Some evolutionists (Williams & Nesse, 1991) have begun to bring this broader, functional conception of biological patterns to bear on certain physiological conditions; and we find that a number of physical signs and symptoms may have meanings—important adaptive functions—that previously eluded medical understanding. Consider, for example, the fact that "morning sickness," the nausea commonly experienced during early pregnancy, may represent far more than a side effect of elevated pregnancy hormones; it may represent an adaptive physiological strategy, as it were, to minimize the high risk of fetal damage in the first trimester as a result of ingesting a variety of substances. Similarly, other painful symptoms—fevers, allergic reactions, trauma-induced swellings—may turn out to have substantial adaptive significance in facilitating the complex process of self-protection and healing. These larger patterns are not apparent when our biology is looked at simply as a set of physiological mechanisms rather than as an evolved system of defensive and self-protective strategies.

dividual detail. The reason the two images are rarely even juxtaposed, not to mention integrated, is that we usually do not even recognize that we are trying to comprehend aspects of the same problem. Yet, given a way to relate these radically divergent discourses—to see them as efforts to narrate and construct the same phenomena—these perspectives can potentially complement, deepen, and alter each other very powerfully. The levels of discourse interpenetrate; there is no neat way to divide them, to segregate the psychological from the biological. In short, evolutionary theory only becomes truly useful when it is integrated with a good-enough theory of individual development and subjective experience. As we hope to show, evolutionary versions of psychology that fail to capture the depth and complexity of human experience cannot, ultimately, make productive use of biological theory.

The Plan of the Book

Our book is an inquiry into human nature that makes use of evolutionary theory in a particular context: the context of an ongoing, long-standing psychoanalytic debate about what is primary about the human mind and the relational world in which the mind functions. In the first part (Chapters One, Two, and Three), we try to immerse our readers in this context, first describing central themes in the psychoanalytic debate and then introducing evolutionary questions and concepts bit by bit as they relate to our main inquiry.

Thus, in Chapter One we do not begin with anything explicitly "biological." Rather, we start with the crucial distinctions between the "relational" and "drive" paradigms in psychoanalysis as discussed by Greenberg and Mitchell (1983). Proceeding beyond the framework developed by Greenberg and Mitchell, we present a "deconstruction" of the broader, hidden assumptions about the human condition that lie beneath the multiplicity of psychoanalytic theories. Two distinct ways of telling the story of the "normal" human social environment emerge—two ways of describing the basic motives of parent and child and the relationship between individual and society. Recognizable as two distinct narrative traditions, or paradigms of human nature, these narratives are woven into the complex theoretical models and clinical principles we know as the Freudian, ego psychological, Kleinian, object relational, self psychological, and interpersonal traditions in psychoanalysis. We believe that these traditions can, indeed must, eventually be integrated into a more unified model of the mind and a broader, more effective clinical approach. However, we are convinced that an important step must be taken first. We must have a clearer appreciation of how each psychoanalytic tradition actually represents a particular kind of bias in viewing the human condition. Only by breaking things down in this way, revealing the contradictory assumptions beneath the models, do we develop a useful sense of how these traditions fit into a more comprehensive picture of human nature. The essential, contradictory insights found in different psychoanalytic models contain vital parts of a larger whole.

Rather than immediately putting the psychoanalytic narratives in an evolutionary context, we take a historical tack in Chapter Two for several reasons. By examining Freud's turn-of-the-century evolutionism before presenting a modern evolutionary approach, we introduce many basic evolutionary concepts in the context of Freud's confusion about them—confusions that we find recur continually in the thinking of many highly educated people, indeed, even in many modern critics of Freud's "biologizing" of psychoanalysis. Here, we ask, can one think about the *experiences* of our ancestors as having an effect on our own ways of experiencing the world today without introducing mystical notions of inherited memories? How can we think about the ways we may be designed to make vital use of an ancestral human legacy to complement our extraordinary capacity to imbibe our knowledge about and view of the world from our families and culture?

Our brief excursion into the history of the intellectual relationship between evolution and psychoanalysis is, thus, intensely focused on conceptual problems that are still alive and relevant to aspects of our discussion throughout the book. By following not only Freud but Hartmann, Bowlby, Erikson, Kohut, and others' attempts to relate (or detach) some version of psychoanalysis to one or another version of evolutionary biology, we hope to clearly convey that such efforts are by no means value-free quests for a pure scientific authority. Freud's belief in resorting to a larger biological authority did not include any doubt that biological authority would basically validate and extend the particular view of human nature that underlay the model he spelled out from 1900 to 1939. Similar recourse to a biological "authority" has been utilized by others in their efforts to bolster, bend, or transform Freud's version of psychoanalysis. So, too, with those who would have us exclude *explicitly* biological views while introducing their own *implicit* biology as the validation of an emphasis on one or another analytic narrative. Our own work is, of course, also prone to the selective choice and use of biology to reshape a psychoanalytic view of the human condition. Therefore, by beginning with a picture of past efforts to influence the course of psychoanalytic theorizing through biological arguments, we hope to launch our readers on a continuing effort to evaluate our own efforts to depict the evolutionary narratives that are relevant for psychoanalysis.

To accommodate a broad range of readers' backgrounds and interests, we've adopted the use of "boxes" and appendices for synoptic presentations of issues that entail a degree of detail or technical complexity in psychoanalysis or biology that is not completely integral to the flow of our central theme. The boxes are used for interesting sidelights and a few frequent questions that have been posed to us when we present the evolutionary viewpoint. The appendices have been reserved for more technical discussions. Our goal in using boxes and appendices is to give the reader greater control over how to approach the material in this book. Some readers will find this material a valuable elaboration or counterpoint to the text. Some may prefer to stay with the flow, omitting or returning to this additional material at another time.

Chapter Three takes the reader further into contemporary evolutionary theory than Freud or any other psychoanalytic thinker has gone. There we shall elaborate on our definition of the central terms "adaptation" and "psychological deep structure": *adaptive psychological deep structures—like the self, or introjects—are those universal, shared ways of processing experience that have been favored by "natural selection." That is, they have been shaped and retained as universal human capacities because, over millions of years, they allowed individuals to negotiate more successfully (though often unconsciously) the incredible complexities of the human social environment.*

Woven throughout Chapter Three is a continuing discussion of the pitfalls of engaging in such discourse about presumed universal human capacities. Yet, as we show, the notion that the child comes equipped with some type of universal deep structure that fits the exigencies of the relational world is not by any means a new one. Such a notion is implicit in the metaphors found in every effort at psychoanalytic model building from Freud's id and ego, to Klein's "phantasies" and "introjects," to Kohut's "selfobject seeking self," to Sullivan's "security seeking self," or Winnicott's "true self." All of these "deep structures" are ways of dealing with the particular challenges that a given theorist assumes to be the primary, relatively invariant, and universal features of the relational environment. What is new in our approach lies not in the recognition that such deep structures may well be adaptive necessities. What is new is the fact that we directly acknowledge and make use of this metaphorical level of theorizing about the fit between the presumed design of the mind and the relational environment by making it more explicit. We spell out a view of the universal, relational problems for which our deep structure represents a working, evolutionary solution.

Indeed, the whole of Part II of the book (Chapters Four, Five, Six, and Seven) consists of a journey through what Hartmann called "the average expectable environment" with the evolutionary biologist as our guide. Our journey reveals a 4-million-year-old landscape only glimpsed in unconnected pieces by Freud, Hartmann, and the classical psychoanalysts as well as by Bowlby, Winnicott, Kohut, Sullivan, and Fromm in the competing relational psychoanalytic tradition: a landscape filled with the wrenching dilemmas that human children have had to face throughout our species' history in dealing with their own inherently conflicting aims as well as "normal" parental motivations; a landscape containing paradoxes buried in our very capacity to internalize parental and cultural identities; a landscape demarcated by underlying boundaries between genetic self and other that are, in reality, virtually as fluid and overlapping as a psychotic fantasy might have it.

A basic metaphor emerges that will carry us through the rest of the book: (1) since the time of our earliest hominid (prehuman) ancestors, the relational matrix in the human family presented a very complex set of adaptive problems that still exists today; (2) the individuals who best solved these relational problems were those in whom there evolved an intricate, inner, psychodynamic system that could serve as the guide to the negotiation of the conflicts and ambiguities of the relational world; (3) bit by bit, as we inherited a complex psychodynamic architecture, our species could "risk" taking on the quintessentially human plan of development in which we allow our minds to be powerfully constructed by family and culture.

In Part III, we explore the notion that psychoanalytic theory can be seen as a set of attempts to grasp pieces of this evolved, adaptive system. Psychoanalytic theories describe the separate facets in great detail, but each theory operates without access to a view of the overaching, adaptive challenges and continuing functional requirements that shape the whole. Specifically, in Chapter Eight, we examine the possible evolved functions for aspects of human experience that correspond to the central psychoanalytic notion of "repression," Winnicott's "true self," Kohut's "self," and both Freudian and Kleinian conceptions of "instinctual drives" and "innate phantasies." Put back into the evolutionary context, familiar psychodynamic processes reveal hidden adaptive dimensions such as the way repression preserves alternative developmental possibilities and Kleinian projective fantasies may serve as crucial supplements to the child's limited capacity to know the mother through inferences based simply on its own direct observations.

As we go on, we recognize the outlines of *an evolutionary narrative in which individuals are seen as active, if not always successful, strategists of their own development, synthesizers of their own "inclusive self-interest" in a world of deceptively competing and overlapping interests.* In Chapter Nine we spell out this vision of development. Transference is seen as an evolved capacity to use the past to negotiate the present and future. This central concept emerges from its traditional clinical meanings as a crucial way in which we engage and enlist others in the process by which we continuously probe our own psyches in order to revise our sense of "inclusive self-interest." *The self is conceived as organized around its innate evolved functions. It serves to record, monitor, and promote a dialectical process in which the ambiguous, deception-filled web of competing and overlapping interests in the relational world is continuously negotiated and renegotiated.* The evolutionary perspective on adaptive change is contrasted with the history of the major psychoanalytic views of adolescence—from Anna Freud through Blos to Erikson. This brief comparative psychoanalytic study uses a focus on adolescence to illustrate the ways in which the evolutionary narrative illuminates the biases in both the classical and relational psychoanalytic views of human nature while offering the basis for a new paradigm that incorporates the crucial truths embodied in each.

Chapters Ten and Eleven move on to an exploration of the clinical process through a careful comparison of the evolutionary perspective with classical Freudian, ego psychological, object relational, self psychological and interpersonalist approaches to treatment. No theoretical perspective substantially reduces the lived uncertainties of the therapeutic encounter. Yet, by the time we arrive at Chapter Ten, the evolutionary perspective has become a platform that is sufficiently broad and sufficiently apart from any particular analytic model to provide a place to stand while we allow the major analytic clinical perspectives, as Hoffman (1987) put it, to "deconstruct and dereify" each other. Using illustrative material from analytic treatment, we consider the ways in which a wide range of symptoms can be seen as our patients' expressions that in some fundamental way their approach to living exacts an intolerable cost to their self-interest. We examine how therapeutic transferences may represent ways that psychoanalytic therapies have hit upon techniques for intensifying and capitalizing on the evolved capacity for self-revision rooted in a universal, transference-based probing of the relational world.

In Chapter Eleven, we move still deeper into the implications of an evolutionarily informed therapeutic sensibility in which the analytic process is seen as a complicated intersubjective negotiation between two ancient, evolved relational entities—the psyche of patient and analyst. There is a paradox at the core of this negotiation: the relationship between therapist and patient, like that between any two individuals, is inevitably fraught with pervasive elements of conflict and potential deception. Yet, the vast weight of clinical evidence (from a variety of different analytic perspectives) argues that significant analytic change in the structure of the self is only likely to occur—to be permitted to occur by the evolved core of the patient's psyche—in the context of a relationship in which the patient experiences in a profound, reliable way that the analyst grasps and is allied with his or her interests.

Perhaps the most crucial aspect of the analytic situation lies in the process by which patient and analyst come to establish, inevitably doubt, lose, search for, and reestablish the vital sense that the analyst is sufficiently allied with the patient's core self and, as we put it, "inclusive self-interest." While such a process is crucial if the patient's innate capacity for transference-based change is to be activated, we shall see that, in an important sense, such an alliance can never be fully achieved.

Moving through and beyond the inevitable subjective distortions that occur in any analysis toward an alliance of interests is a complicated two-way negotiation process—one that carries our evolved capacity for negotiated self revision to an unusual, and in some ways unnatural, extreme. The evolutionary perspective suggests that an *inherent skepticism* exists in both patient and analyst concerning a new situation, particularly one between unrelated individuals, that offers itself as a context inviting transference and renegotiation of one's identity.

A purely "ontogenetic" perspective on object relations (in which development is understood only in the context of the individual's specific life experiences) lacks a view of the way in which we are, in effect, millions of generations older, with innately greater wisdom than could be conferred by the experiences of our lifetimes. Without an evolutionary (or "phylogenetic") perspective on relatedness and its innate, evolved meanings, our patients' intuitive knowledge of some of these meanings tends to be interpreted as resistance in a far narrower, pathologized, or overly individualized sense. We hope to demonstrate how the analytic situation elicits profound inner signals that appear to indicate resistance but that, in fact, derive from an ancient, *adaptive* legacy—a legacy that informs and interweaves tightly with personal history and, of course, pathology, but at root does not have individual, ontogenetic, or pathological origins. In a primary, clinically relevant way, the resistance prompted by the patient's innate skepticism begins not as an individual feature indicative of pathology but as a universal, adaptive aspect of the psyche.

Ultimately, one of the major ways in which therapists fail their patients (and parents similarly fail their children) is by using self deceptive strategies for protecting or enhancing their own interests while casting this as dedication to the patient's interests. In our view, such a "confusion of interests"—and the resultant loss of the absolutely vital capacity to define, know, and promote one's own interests—is the common denominator in a large range of less than good-enough, or pathogenic, environments. And such destructive self-deception that enables the analyst to pursue the enhancement of his or her own interests at the patient's expense is commonly replicated in many analytic treatments. Often such self-deception is confused with notions of "technique."

These views move us away from the idea that the analyst can ever assume a single technical stance—or even a formulated set of shifting stances—that, in itself induces the process of change. *Each treatment ultimately entails the negotiation of a specific set of mutual understandings and reciprocal changes in analyst and patient.* Thus, while recognizing the absolute centrality for the patient of the experience of having his or her perceptions recognized and validated, we believe that many of the times when this does not occur we are not dealing with empathic (or selfobject) failures in the sense of technical failures or breaks in some process that could or should be sustained. Many of these moments represent normal, continuing shifts by the analyst away from the highly difficult task of maintaining a focus on the patient's subjective viewpoint toward the more natural tendency to express the analyst's own individual bias—manifested often by the telltale signs of self-deception and deception that protect it. In these ongoing "countertransferential self-revelations," as Russell (1990) put it, the empathic clinician gives the patient the opportunity to know and correct, as it were, for the inherent deceptions entailed in the good-enough analyst's sustained, empathic inquiry. Hoffman (1983) describes this process in his references to "the patient as interpreter of the analyst."

The evolutionary perspective ultimately sustains a view of the analytic process as a profoundly reciprocal negotiation—a process of mutual influence—in which the patient's probing of the analyst's subjectivity, of the analyst's transference, is just as crucial as the child's probing efforts to ferret out and find the parent's true identity. We hope to show that this probing is an attempt to correct for the normal deceptions practiced by all family members in the good-enough environment, and probing to know the analyst (parent) is part of an attempt to better define one's own identity. The engagement of patient and analyst becomes, at some levels, a limited yet intimate mutual exchange of personal information.

Overall, Part IV, the clinical Chapters Ten and Eleven, fundamentally shifts the reader away from an emphasis on patients' "pathology" as a predominant focus: away from an understanding of symptoms, transferences, and resistance as "distortions" of reality toward the ways in which these operations of the psyche function as tactical efforts to communicate crucial adaptive information and induce a reciprocal process of awareness and influence. The evolutionary perspective has led us to a conviction about the existence of an overarching adaptive process that exists as a basic feature of the human psyche—a process that is designed to use a range of psychic means to continuously revise (or renegotiate) "inclusive self-interest" and identity.

In our final chapter, Chapter Twelve, we take the reader back to contemplate the possibility of moving decisively toward a genuine synthesis of the competing psychoanalytic visions of human nature and their clinical implications. We argue throughout the book that, like the incompatible descriptions of the elephant offered by the blind men in the parable, a new working synthesis for psychoanalysis simply cannot mix or combine incompatible models and clinical methods derived from the classical and relational narratives of human nature. Rather, a meaningful synthesis requires that the models be radically reexamined, essentially "deconstructed" into their underlying meanings, and then modified with reference to a larger, embracing reality. In this chapter, we reexamine each of the major dimensions in which the existing psychoanalytic traditions clash and sketch the outline of a synthesis that incorporates the essential features of their competing visions.

From a philosophical point of view, contemporary evolutionary biological theory embodies what is, arguably, the first major step taken in Western thought to reconceptualize the boundaries of the individual psyche. It does so in a way that expresses the intrinsically social, relational essence of the psyche without omitting those critical features that remain inherently, competitively individual in their aims. Indeed, it is in this regard that our evolutionary perspective fundamentally validates Kohut's and Winnicott's radical revision of traditional, psychoanalytic individualism: the concept of the "selfobject" and the idea that "there is no such thing as an infant, only a nursing couple." At the same time, it extends and further clarifies the universal nature of the conflict between the narrowly individual aims of the self and its innately social and relational aspects and motives.

In this fashion, our philosophical position differs markedly from both the individualistic (atomist) tradition of 18th-century British philosophy (including 19th-century utilitarian derivatives [e.g., Bentham, Mill]) and the social (collectivist) tradition of Continental thought (e.g., Rousseau, Hegel, Marx, Durkheim). *In our view, contemporary evolutionary theory represents a substantive philosophical revision of the individualist–collectivist dichotomy in Western thought.*

Based on the notion of an irreconcilable philosophical dichotomy, Greenberg and Mitchell concluded that the "deeper divergence" between psychoanalytic models is "inherently unresolvable." The evolutionary perspective, however, may offer a framework that points toward an eventual synthesis of the dialectic between the two major competing psychoanalytic narratives of human nature. In our last chapter, we return to consider the key features of such a new psychoanalytic narrative—an evolutionary psychoanalysis—grounded in these altered philosophical assumptions.

PART I

The Psychoanalytic Problem and Basic Evolutionary Approach

CHAPTER ONE

Competing Psychoanalytic Visions of the Human Condition: Classical and Relational Narratives

> Psychoanalytic models rest upon . . . irreconcilable claims concerning the human condition.
> —J. R. GREENBERG AND S. A. MITCHELL,
> *Object Relations in Psychoanalytic Theory*

In recent years, many people interested in the field of psychoanalysis have come to see psychoanalytic theoretical models and clinical approaches as tending to coalesce around two divergent paradigms. Although everyone's map of the field is different, there is a great deal of overlap between several of these broad characterizations of the psychoanalytic landscape. Greenberg and Mitchell (1983), for example, refer to "drive/structure" and "relational/structure," Modell (1984) to "one person" and "two person" psychologies, and Eagle (1984) discusses the contrasts between "instinct" versus "deficit" models. Roughly speaking, the divergent paradigms typically correspond to the classical and ego psychological traditions on one side and aspects of object–relations theory, interpersonal approaches, and self psychology on the other. Moreover, it is often argued that "the underlying premises upon which the two models are based are fundamentally incompatible" (Greenberg & Mitchell, 1983, p. 403); that they belong to "two different conceptual realms . . . two apparently irreconcilable contexts" (Modell, 1984, pp. 257-258; see also Cooper, 1991).

We believe that there is, indeed, a major division between what are essentially different psychoanalytic "world views." Although rarely articulated in their most clearly contrasting form, deep, competing intellectual currents run beneath the everyday clinical, developmental, and dynamic concepts promoted by different psychoanalytic schools or traditions. These powerful, competing currents concern basic assumptions about the *nature of the mind* as well as the fundamental *character of the relational world*.

In this volume, we try to reveal these competing currents—these views of the "psyche-in-the-world"—by elucidating the narrative structures implicit in the different accounts given by the major psychoanalytic perspectives on human psychological development. Specifically, we examine what we believe is the most fundamental issue around which the major psychoanalytic traditions clash: the complex set of assumptions about the basic relationship between the individual and the social matrix—the nature of connection, exchange, and negotiation between the individual and others, starting at the moment of conception and lasting throughout the span of each individual's life.

As we shall try to show, an implicit paradigmatic "narrative" underlies all psychoanalytic theories. Each theory recounts a presumed universal, developmental story of the psychological operations used by children and adolescents within the family and then, through a series of transformations, in the larger, adult relational world. The contrasting narratives become visible in the story lines contained within the different psychoanalytic theoretical traditions.[1]

Two Narratives of the Human Condition

Consider the following two narratives.

The Classical Narrative

In the classical view, the primary developmental task lies in the need to manage the tensions that derive from the inevitable clash between the endogenous, bodily based, driving forces within the individual and the norms and limits of ordinary social reality. Classical drive theory, Klei-

---

[1]The idea that psychoanalytic theoretical models reflect and help to construct narrative accounts of development has been promoted by theorists such as Schafer (1983) and Spence (1982) in an attempt to combat the kind of naive reifications of psychoanalytic concepts that had become a tradition in psychoanalytic theorizing. Unfortunately, from our point of view, this valuable methodical emphasis (on the socially constructed nature of theory and its root metaphors) has become equated with a certain antibiological bias on the part of many of the so-called hermeneutic theorists who promote it. Some of the valid historical reasons for this now almost reflexive antibiology will be discussed in Chapter Two. For now, suffice it to say that the fruitful insights to be gained by "deconstructing" theories into their underlying story lines shall turn out to serve as a vital bridge between the psychoanalytic field and parallel intellectual currents in the field of evolutionary biology.

nian theory, and their innumerable permutations into different versions of modern ego and developmental psychology (e.g., Freud, 1930; Freud, 1966; Hartmann, 1958; Jacobson, 1964; Mahler, Pine, & Bergman, 1975; Blos, 1979; Kernberg, 1976, 1980) basically start with this conception of an essentially innate "dividedness" and tension at the core of human nature and the human condition.

In the course of development, normal growth entails a reluctant shift away from the child's less organized, more selfish and self-centered modes of construing reality and organizing subjective experience. This shift is partially opposed by the child's innate nature. Successful maturation yields a compromise—one that is inevitably somewhat unsatisfactory to the child—between the child's self-centered motivations and the demands of the social world. These motivations and related perceptions are, in part, repressed in order not to disrupt the conscious process of subjectively organizing experience and behavior. Such disruption would conflict with the child's adaptation to the realities of the social world. On the whole, the repressive process as well as a significant part of what is repressed can be said to be disguised in a way that is intrinsically deceptive to the individual and others (Freud, 1915b).

The continued presence of, indeed fixation on, repressed contents of the psyche, albeit in deceptively disguised form, represents a continuing threat to the accommodation the child has made to external (largely parental) reality. Moreover, it is assumed in the classical tradition that there is actually an inherent tendency of the repressed to return, and while it seeks direct expression, it inevitably achieves a degree of expression in ever-changing, deceptive guises. This unbidden return of the repressed is one major form of the "repetition compulsion"—the tenacious, often painful repetition of archaic patterns that lies at the heart of the classical Freudian and Kleinian conceptions of psychopathology (Fenichel, 1945; Segal, 1964; Kriegman & Slavin, 1989)—and has been conceptually tailored to fit the altered premises of modern ego psychology. In this view, subjective reality is illusory, prone to distortion and self-deception in the form of various compulsively repetitive defenses, transferences, and resistances. Behind this deceptive facade lies a truer, objective reality. Clinically, the well-analyzed, neutral analyst is given substantial responsibility for addressing the patient's distortions and unconsciously motivated repetitions.

The Relational Narrative

In contrast, the relational model focuses on a fundamentally different type of tension. On one side lies the child's unique configuration of individual needs, identity elements, or vital experiences of self; on the other side lies a social environment that frequently is insufficiently attuned to, or invested in, the recognition and cultivation of these individual elements. Certain of the "interpersonalist" theorists (e.g., Fromm, 1941; Sullivan, 1953), psychosocial theory (Erikson, 1956), aspects of Winnicott (1965), Guntrip (1971), Fairbairn (1952), and Balint (1968), as well as the self psychologists (Kohut, 1984; Stolorow, Brandchaft, & Atwood, 1987) emphasize this relational view of the patterning of inner conflict. They represent a radically different sensibility within psychoanalysis, and they take a different view of the process of selectively altering awareness—of defensively reorganizing consciousness and meaning.

In this view, a true spontaneous or authentic self is hidden from a less than adequate environment. "Hiding" the self is necessary because the inadequate, misattuned environment fails to conform to the high degree of mutuality, convergence, and synchrony of the aims of different individuals that are assumed to comprise the "good-enough" environment (Slavin, 1990). This hiding comprises a subjective alteration of meaning and awareness that can be seen as a version of the process of "repression." Clinically, the analyst is more likely to be seen as a participant in whatever seeming distortions may occur in the patient's subjectivity. The analyst's role entails a greater recognition of the "kernel of truth" in the patient's experience than in the classical model; and the process of striving for repair and revision may procede by way of "enactments" that may *appear* simply to be symptomatic repetitions.

Repression in the Two Narratives

Each of these two models has its own version of the role and function of repression. What is critical to the difference between the classical and relational narratives is not that nonclassical models may call these alterations in awareness a split or dissociation in the self, or a "disavowing" of aspects of reality, but rather that in the relational tradition the process is seen essentially as sheltering or protecting parts of the self in order to maintain self continuity and relational ties, and/or to preserve the possibilities for future growth and development. Defenses and transferences are viewed less as a subjective facade that distorts reality and more as developmentally creative efforts to re-envision reality in an attempt to reactivate and reinitiate thwarted growth (Winnicott, 1960a; Kohut, 1984). This is opposed to the classical tradition in which egocentric, forbidden, or dangerous wishes are deceptively disguised to avoid inner conflict and then compulsively repeated (Kriegman and Slavin, 1989).

Deconstructing the Narratives: Four Basic Dimensions

Implicit in these two narratives are four dimensions, each of which embodies an aspect of the dichotomous tension between these two psychoanalytic views of the human condition. Greenberg and Mitchell (1983) provided us with a conceptual map that begins to separate out two key dimensions on which these narratives differ. These dimensions concern the *structure of the psyche* itself—the building blocks that create the mind.

Dimension 1: The Individual Mind versus the Interpersonal Field as the basic unit of analysis. The classical narrative can begin to be understood as a story in which we find the "fundamental premise that the individual mind, the psychic apparatus, is the most meaningful unit for the study of mental functioning" (Greenberg & Mitchell, 1983, p. 402). In contrast, the relational narrative portrays the psyche as deriving from (and only understandable within) the context of the interpersonal field into which we are born, and within which we must negotiate a complex web of human relationships. The essential design of the psyche is embedded in and built of relational components—or some form of internally represented relationships between the self and others.

Dimension 2: Endogenous (inner) Forces versus Relational (environmental) Experiences in the development and structuralization of the individual psyche. In the classical narrative, psychic structure is patterned and regulated by the vicissitudes of the discharge of instinctual bodily drives. In the relational narrative, psychic structure derives directly and irreducibly from the vicissitudes of interpersonal experiences in the human environment.

Moving beyond Greenberg and Mitchell's characterization of the *structure of the mind*, we discover two other narrative dimensions that represent fundamental assumptions about the *nature of relationships* between individuals, including direct implications regarding the interplay between subjective experience and "reality" and, as a corollary, the role of the past and the present in the patterns, choices, and reactions that are unconsciously repeated by individuals throughout their lives.

Dimension 3: Inherent Conflict versus Inherent Mutuality in the relational world. In the classical narrative, from the outset of life, the child's needs and way of viewing reality are seen as clashing with the norms of the environment. Significant amounts of selfishness, rivalry, competition, and aggression are assumed to be part and parcel of relationships within the family. In contrast, in the relational model family members are motivated by interpersonal needs that contain no *inherently* conflicting aims.

A corollary of this dichotomy relates to what is assumed to be the normative state of the psyche. In the classical narrative the psyche is inherently divided and, to some degree, inevitably involved in a degree of inner struggle: the very metaphor of a "tripartite structure" embodies this sense of normative dividedness. The aims of the id intrinsically diverge from and conflict with the aims of other individuals. This inherent conflict is represented in the struggle between the superego and the id. Thus, in the classical narrative the metaphor of the *divided psyche* is a logical reflection of *conflict in the interpersonal world*. The relational narrative, on the other hand, has shifted the basic metaphor to that of the "self," or the equivalently unitary notion of an "identity." Cohesion, rather than inevitable struggle, is assumed to be the normative state unless "pathological" distortions have been introduced by destructive or deficient experiences with others.[2]

Dimension 4: subjective experience as Deceptive Distortion versus Valid Communication. Classical narratives invariably stress the ways in which subjective experience conceals or misrepresents reality. Defenses, symptoms, and transferences are built around inherently self-deceptive mental operations that, when uncovered, reveal underlying, painful, objective truths. Relational narratives, on the other hand (including much of the relational/structure tradition, but especially Winnicottian and Kohutian versions), emphasize the ways in which subjective experience often represents crucial, developmentally vital truths; defenses, symptoms, and transferences are seen as inherently valid, self-protective expressions of a personal reality that has been misunderstood by others.

As a corollary, classical narratives stress how painful repetitions—including many "negative therapeutic reactions"—derive from an inherent compulsion to repeat. The compulsive nature of the repetition is given various sources: the death instinct; stubborn instinctual fixations; the recurrent attempt of the repressed to achieve expression (incomplete repression); and/or the attempt to hold onto older patterns of relatedness out of loyalty to others and/or because of enduring instinctual object ties. Relational narratives stress the ways in which individuals

---

[2] As Mitchell (1984) points out, certain versions of the relational narrative do not ignore the existence of conflict, particularly in their clinical appreciation of the tensions between "separateness and merger" and in "competing loyalties" to different objects. However, as we shall try to demonstrate, in virtually no theory that we term "relational" is a "genuine clashing" (Goldberg, 1988) of individual aims and goals given a fundamental status as an inherent feature of the basic motives of normal individuals in the good-enough environment. See Kriegman and Slavin (1990) for an extensive discussion of this issue in regard to self psychology where we show that, though Kohut (1980, 1983) emphasized his belief in conflict and did not deny its existence, he had a clear tendency toward *deemphasizing* the normative nature of conflict and toward attributing it to parental pathology.

repetitively engage in the inner maintenance of "bad objects" because a tie to a bad object is "better than none" (Fairbairn, 1952). Even a painful repetition that maintains a tie to a bad relationship serves to adaptively preserve the self so that it can continue to search for appropriate relationships for self revision and restoration (see Brandchaft, 1983; Kriegman & Slavin, 1989). As Modell (1990) notes:

> Most analysts who view the therapeutic process as primarily restorative also tend to de-emphasize the significance of transference repetition. (In fact, one cannot find the term *repetition* in Winnicott's or Kohut's indexes.) (p. 148)

Instead, the radically different current that moves beneath relational narratives is captured in the repetitive striving to actualize a curative fantasy (Ornstein, 1984), realize one's "nuclear program" (Kohut, 1984), or find expression for some version of a "true" (Winnicott, 1960a) or "authentic" (Fromm, 1941) self.

Psychoanalytic Schools and the Underlying Paradigms

Cast in terms of the essential paradigms that underlie these narratives, a "pure classical" theorist would see the mind as intrapsychically structured by the vicissitudes of endogenous motives in a world that is inherently characterized by interpersonal conflict, deception, and self-deception. In this view, reality and meaning can, nevertheless, be sorted out to a significant degree by a relatively neutral, objective observer. A "pure relational" theorist would see the mind as understandable only within a particular relational context, and interactionally structured from the vicissitudes of prior and current relationships in a world where, despite the central human striving toward authentic development within a human community, interpersonal conflicts exist as an unfortunate by-product of complexity, human limitations, and pathology. In this view, relational reality, or truth, is a social construct that has essential meaning only within the relationship in which it emerges. The observer creates reality in the act of observing within a social or intersubjective context.[3]

---

[3]One could also suggest that there are really three larger, underlying dimensions, or axes—along which one could position each psychoanalytic theorist in three-dimensional space. These axes are (1) *the nature of the mind* (Greenberg & Mitchell's [1983] drive/struc-
(continued on p. 27)

TABLE 1. Classical Versus Relational Narratives

Four dichotomies	The classical narrative	The relational narrative
1. Basic unit of analysis	The individual psyche: emphasis on intrapsychic dynamics	The interpersonal field: the individual can only be understood within an interactive context
2. Basic source of patterning (or structure) in the psyche	Vicissitudes of endogenous drives: relational ties are derivative	Vicissitudes of interpersonal interactions
3. Relationship between basic aims of self and other	Inherent clash of normal individual aims: selfishness, rivalry, competition are motivationally primary (corollary: the normal [tripartite] psyche is "divided" in a way that reflects the inevitable conflicts between the individual and others)	Emphasis on mutual, reciprocal, convergent aims: significant interpersonal clashes are due to pathology or environmental failure (corollary: the psyche [the self] is an essentially "holistic" entity that reflects the possibility of relative harmony between the individual and others)
4. Role of deception and self-deception in subjective experience	Defense and transference as distortions of reality: resistance serves to conceal truth (corollary: repetition is a means of not knowing; people defensively repeat the acting out of unconscious motivations to maintain repression, to preserve ties, or to maintain instinctual fixations)	Defense and transference as inherently valid expressions of personal reality: resistance serves to elicit recognition of individual truth for further growth (corollary: repetition is an artifact created as people repeatedly try to elicit a needed response from the relational world to reinitiate thwarted growth and revise the self; people repeat in the process of preserving and protecting a vulnerable self—e.g., by maintaining needed relational ties—so that revision can be tried again in the future)

There are clearly certain limitations in this way of mapping psychoanalytic theories. For example, the Kleinian sensibility (including the work of Kernberg) clearly fits within the classical narrative in terms of its emphasis on massive, intrinsic conflict in the relational world as well as the subjective distortions of reality in the mind of the child and the compulsive instinctual foundation of repetition (dimensions 3 and 4). Yet, in terms of the purely "mental structure" criteria used by Greenberg and Mitchell (dimensions 1 and 2), there are aspects of the Kleinian model that can certainly be seen to be consistent with the relational narrative. Kleinian theory is, after all, an object relations theory in its emphasis on innate (endogenous) relational imagery as the essential building blocks of the child's mind. Overall, however, we believe the deeper implications—especially the clinical implications—of Kleinian narratives are better illuminated by classifying them as lying in the drive, or what has been called the "one-person" tradition (Modell, 1984). The huge emphasis on the primacy of the child's embroilment with destructive–reparative conflicts, the ease with which a patient's conscious, subjective ideas are assumed to be defensive distortions, and the assumed universality and tenacity of the compulsive need to repeat all align the Kleinian model and its derivatives with the classical analytic vision of the human condition.

In a similar fashion, the early and contemporary interpersonalists (Sullivan, 1953; Fromm, 1941; Levenson, 1983; Bromberg, 1989) create accounts of the mind and the relational world that clearly belong in some respects within the relational narrative tradition. They best illustrate Greenberg and Mitchell's (1983) view of relational mental structure. Moreover, Sullivan and Fromm view life's most painful conflicts (difficulties in living) largely as functions of disruptive, misattuned (irrational and inconsistent) environments. Yet, on the last dimension (and its corollary)—the issues of subjective distortion (deception and self-deception) and compulsive repetition—interpersonalist models tilt back strongly toward the classical narrative. Sullivan's (1953) concern with "parataxic distortions" of reality and Fromm's (1941) emphasis on self-deception often end up clinically, in our view, quite similar to Hartmann's (1958) classically rooted notion that psychoanalysis can be

---

ture vs. relational structure, see dimensions 1 and 2, Table 1); (2) *the nature of the relational world* (conflict vs. mutuality, see dimensions 3 and 4, Table 1); and (3) *the nature of reality* (the objectivist vs. the constructivist view of knowledge (see Hoffman, 1991), which we discuss in Chapters Eleven and Twelve). Psychoanalytic theories tend toward one end or another of these axes. Though a single theoretical perspective may mix different ends of different axes, each theory tends to make a consistent commitment to viewing things in a fashion that is best represented by one end of each axis.

called a "science of self-deception." Only Winnicott (in some of his work when he is less of a Kleinian), Kohut, and, to an even greater degree, the intersubjectivists and social contructivists (e.g., Stolorow et al., 1987; Hoffman, 1983, 1991) broke more decisively with the classical emphasis on deception and distortion of reality, diminishing any notion of objective truth and absolute reality in favor of a more relativistic, subjectivist point of view.

Very recently, Hoffman's (1991) "social constructivist" approach gives this dimension of what we call the relational paradigm clear and consistent expression.[4] The alignment of many interpersonalists with the classical narrative (the classical vision) comes to the fore most clearly when we consider the meaning of many everyday clinical phenomena. As we discuss in Chapters Ten through Twelve, there is a clear tendency among the interpersonalists to focus on the deceptions and distortions in the patient's transference. Also, they emphasize the patient's prominent use of enacted repetitions (within transferences, in and out of the therapeutic relationship) as ways of maintaining old relational patterns and avoiding experiences that would generate growth. Indeed, though both Sullivan and Fromm partook of the relational belief in the existence of a central need for growth and development, it is, again, only Winnicott, Kohut, and the self psychologists who made a major, primary shift away from a belief in the power of a conservative instinct to self deceptively repeat the past toward a vision in which the "dread of repetition" (Ornstein, 1974) is assumed to be a stronger force than the compulsion to repeat. In this vision, the search to reinstitute thwarted growth becomes the dominant psychic striving.

Premature Resolutions of the Dialectical Tension

While the Kleinian and interpersonalist traditions give us accounts of the human condition that interweave elements of competing psychoanalytic narrative structures, several other theorists have attempted various strategies for mixing or integrating what are inevitably sensed as contradictions between these psychoanalytic visions. We do not discuss these efforts in any detail. In essence, we believe that it has been convincingly shown that most historical attempts at theoretical integration turn out to be either internally inconsistent, or heavily tilted toward

---

[4]Because Hoffman uses Greenberg & Mitchell's (1983) narrower definition of the relational paradigm (see discussion earlier in this chapter) he concludes that social constructivism is not part of the relational paradigm but is independent and crosscuts the classical-relational dichotomy.

the basic assumptions of one of these underlying paradigms (Greenberg & Mitchell, 1983). Thus, although they may appear to transcend their origins, attempts to broaden a given perspective—sufficiently to deeply *embrace* (not just allude to or assume a parallel with) the valid insights of the other—inevitably remain tied to key underlying assumptions of one of these paradigms. For example, Sandler (1962, 1981), Pine (1985), and Schafer (1976) made attempts at various types of integration and expansion that, we believe, nevertheless remain fundamentally loyal to "classical" assumptions and the classical vision. While, on the other side, Stolorow et al.'s (1987) attempt at a "pure psychology" of conflict, as well as aspects of Erikson's (1956) "psychosocial" revision of the Freudian viewpoint, ultimately tilt the narrative structure heavily toward the relational sensibility and vision of the human condition.

An acceptance of the incompatibility of these models sometimes finds expression in a "strategy of complementarity" (Kohut, 1977; Wallerstein, 1981; Gedo & Goldberg, 1973). This dualistic strategy is used to guide a tactical oscillation between the two perspectives in the clinical context. Frequently it is implied that such theoretical pragmatism, combined with an open, flexible stance in modern clinical practice, can achieve, de facto, a perspective that is tantamount to a true integration (Peterfreund, 1983; Pine, 1985). However, it is usually not possible to view our patients' transferences as attempts to (1) compulsively repeat (act out) socially unacceptable, dangerous instinctual wishes in disguised form in order to keep the repressed from entering consciousness; and (2) simultaneously view transferences as being valid expressions of subjective experience that, because of former empathic failures, were disavowed and are now in need of a mirroring, accepting response; that is, valid expressions that are primarily aimed at fuller self-revelation, integration, and growth. In our experience, individual clinicians—and sometimes whole clinical settings—reveal that their underlying thinking leans heavily toward one or the other of these narratives.

Many readers will no doubt object that few of us work in such a dichotomous way: That the "either/or" quality of our two narratives has long since been intricately integrated theoretically in the psychoanalytic literature (e.g., Loewald, 1980; Modell, 1984; Pine, 1985) as well as continually combined and blended in the rich tapestry of real clinical work. No doubt there is some truth in this objection. Yet, the overwhelming evidence from actual practice shows that when one's overall orientation is viewed as a whole, contemporary clinicians and clinical settings strongly tend to polarize around one or the other basic view of the primary nature and origin of the conflicts we must repress, and, consequently, of the very meaning of repression (and repetition). Greenberg and Mitchell (1983) have called a similar range of deep and pervasive psychoanalytic beliefs "paradigmatic loyalties," and while, moment to moment, or phase to phase, some analytic treatments may shift between the two narrative emphases, the deeper trends, the basic metaphors and narrative structure of a given treatment will tend to coalesce around one or the other of these organizing perspectives. On a still broader scale, the historical trends within psychoanalytic theory, starting with Freud's radical rejection of his early seduction theory, have moved like a pendulum from one side to the other of this basic divide (see Rappaport, 1960; Erikson, 1964; Lawner, 1985). In the final analysis, we remain in a historical era in which one or the other vision of the human condition, and its related understanding of the true meaning of our patients' communications, continues to predominate (Kriegman & Slavin, 1989, 1990; Slavin & Kriegman, 1990).

It is unquestionably true, in our view, that the classical and relational visions of the human condition partially derive from and profoundly reflect the "deeper division" in Western philosophical traditions that Greenberg and Mitchell (1983) allude to at the close of their book. They noted the way in which drive/structure theories could be traced to the highly *individualistic, atomistic tradition* of Locke and Hobbes in British political philosophy including nineteenth century utilitarian derivatives (e.g., Bentham, Mill). In this view, "[M]an cannot live outside society, but society is in a fundamental sense inimical to his very nature and precludes the possibilities for his deepest, fullest satisfactions" (p. 402). Relational/structure theories, on the other hand, are traceable to the *collectivist tradition* of continental philosophy including Rousseau, Hegel, and Marx. In this tradition, "individual man" is literally inconceivable. "The very nature of being human draws the individual into relation with others, and it is only in these relations that man becomes anything like what we regard as human" (p. 403).

There is, however, a very important implication in these clashing philosophical currents that Greenberg and Mitchell allude to but did not directly address in their discussion of psychoanalytic theories, namely, that *individualistic* philosophical traditions tend to assume that there is an inherent divergence and clash of interests between individuals. This clash of interests can only be overcome by the imposition of a social (or political) structure that will bring the crucial element of *shared* or mutually held interests into the picture so that naturally clashing individual aims will not erupt in destructively conflictual ways. The collectivist tradition, on the other hand, takes the dichotomously opposed view that individual interests are inherently a part of a larger collective entity— that is, that the sharing of overlapping interests is the primary state out of which a modicum of apparent individuality may emerge. Moreover, for Rousseau, Hegel, and Marx, the natural (original) and ultimate (future) state of relationships is one of mutuality and harmony, that is, participation in the "higher" relational order (Parsons, 1954).

These contrasting beliefs concerning the ultimate reality and meaning of individual interests, aims, and desires—as inherently divergent and competing versus inherently emanating from a group or supraindividual realm that precedes individuality and can ultimately dissolve it—are profoundly built into psychoanalytic visions. Greenberg and Mitchell's division of classical and relational theories stops short of addressing this vital issue. They limit the psychoanalytic focus to a far narrower set of issues by considering only the relational versus drive based views of the structure of *mind*. What is not addressed are the complex yet inextricably related set of issues concerning basic conceptions of the nature of the sources of, and interrelationships between, individual interests (and the whole range of subjective aims, wishes, desires, urges, impulses, etc., by which these interests find conscious and unconscious expression). It is the crucial issue of the nature and structure of the *relational world* that necessitated the third and fourth dimensions of our map (see Table 1, p. 26) of the story lines comprising classical and relational narratives. And it is these issues and the way in which contemporary evolutionary biology has crystallized our understanding of them—and may represent a genuine advance in conceptualizing them—that provided the impetus for writing this book. We believe that we can reframe some of the most pressing, central psychoanalytic concerns in a useful manner by utilizing the evolutionary viewpoint.

Despite the persisting difficulties in creating a truly integrative perspective, we question the ultimate conclusion that the deeper philosophical differences that divide the psychoanalytic paradigms are, in fact, ultimately "irreconcilable" (Greenberg & Mitchell, 1983), or must remain in the realm of unresolvable "paradoxes" (Modell, 1984). Nor do we believe that such basic conceptual issues can be adequately addressed and resolved primarily through empirical research (Eagle, 1984).

Rather, we shall try to demonstrate that, indeed, a certain kind of rapprochement of the two traditions is now possible. But it is only possible if we are able to capture and retain the deeper dialectical tension concerning the crucial question of how and why individual aims arise, how these aims relate to one's true interests, and why—both internally and interpersonally—they are observed to be in ubiquitous conflict. Any new paradigm that more closely integrates the two psychoanalytic narratives must be rooted in a set of assumptions about the mind and the human condition that do not have the same tendency to polarize in the dichotomous way that has characterized both analytic thought and, indeed, Western philosophy. Ultimately, in order to resolve the tension between the two narratives, a new paradigm must be capable of thoroughly incorporating the crucial and valid elements of both existing psychoanalytic traditions, while simultaneously providing the basis for making the necessary, fundamental alterations in some of the basic premises of each.

We believe that the foundation for understanding these paradoxical tensions within one genuinely inclusive paradigm exists in the contemporary evolutionary (phylogenetic) perspective.

It is obviously necessary to present a foundation for such a complex discussion. After we briefly review how evolutionary theory has been used historically by psychoanalytic theorists, we ask the reader to accompany us on a journey through some unfamiliar territory. We return to the issues posed in this chapter; indeed, we never actually travel very far from them. In various ways, they form a leitmotif through the entire book. We keep our focus on the implications of the evolutionary perspective for the development of an "evolutionary psychoanalysis"—a psychoanalytic theory and clinical practice that is informed by a more consistent, inclusive understanding of our nature as a species.

CHAPTER TWO

Early Darwinian Versions of the Psychoanalytic Narratives and the Challenge for Contemporary Evolutionary Theory

> The problem is not, as some critics would have it, that Freud confused psychology with biology, but that he employed... fallacious biological principles.
> —A. MODELL, *Psychoanalysis in a New Context*

> Even in the post-Freudian discussion of metapsychology within psychoanalysis, the antiquated elements in Freud's conceptions of evolutionary biology seem not to have been scrutinized... and not, in the light of new knowledge, to have been revised...
> —I. GRUBRICH-SIMITIS, *A Phylogenetic Fantasy*

Several psychoanalytic theorists used evolutionary theory to bolster and elaborate particular conceptions of the human psyche. Let us look briefly at earlier attempts to link psychoanalytic visions with Darwinian thought.

Freud's Search for the Archaic Social Determinants of Psychic Structure

In Freud's first published works (e.g., 1878) he showed that there is a continuity between the nerve cells of lower and higher animals. This finding was contrary to accepted thought and supported an evolutionary viewpoint in which no sharp distinction is drawn between humans and lower animals. As a medical student Freud's studies focused extensively on evolutionary biology (Ritvo, 1990; Jones, 1953). This decisively influenced his vision of the animal nature and roots of human psychology, as could already be observed in his early neuroanatomical writings (Kriegman, 1988). It seems clear, based on the strong Darwinian leanings of his biological mentors, his own first neurological writings, and the repeated evolutionary view of the human species stated throughout his psychological writings, that Freud was strongly influenced by a basic evolutionary scientific *Zeitgeist* within which all of his thinking occurred (Sulloway, 1979; Ritvo, 1990).

> We know that . . . the [research] of Charles Darwin . . . put an end to the presumption . . . that man is a being different from other animals or superior to them; he himself is of animal descent being more closely related to some species and more distantly to others. The acquisitions he has subsequently made have not succeeded in effacing both in his physical structure and in his mental dispositions the evidences of his parity with them. This [discovery] was . . . the *biological* blow, to human narcissism. (Freud, 1917, p. 140)

Many of Freud's vehement rejections of contributions that modified psychoanalytic theory can be traced to their failure to remain true to this evolutionary vision of the continuity between humanity and the rest of the animal kingdom. Indeed, Freud was vigilant in guarding psychoanalysis against the intrusion of ideas that would deny what he saw as our essential animal nature.[1] He believed that it was inconceivable for people to accept the painful truths of psychoanalysis unless they had accepted the "*biological* blow to human narcissism" inherent in the theory of evolution. Despite many profound changes in Freud's psychoanalytic views, his effort to maintain a consistency between psychoanalysis and evolutionary biology was firmly rooted in his most basic beliefs about life and nature. These beliefs remained unchallenged.

After Freud abandoned his *Project for a Scientific Psychology* (1895)—the keystone of his initial attempt to make psychoanalysis into a general scientific psychology (Holt, 1989)—he became considerably less oriented to a reductionistic, physicalistic form of biology and more interested in the historical and social–functional explanations possible in evolutionary thought (Holmes, 1983; Slavin, 1988). Though increasingly committed to a model of the psyche as fashioned by the vicissi-

---

[1] Freud persistently tended to equate our animal nature with our bodily nature or neurophysiology. Our neurophysiology, in turn, was equated with Freud's particular physicalistic conception of the instincts or drives. This tendency to blend classical drive theory with the far broader (and distinctly separate) vista of evolutionary thought generates a confusion over the implications of the evolutionary perspective that we discuss at several points in this book.

tudes of drives and the id, Freud also appears to have become increasingly convinced of the validity of studying the biological (i.e., adaptive) reality of psychodynamic structure and process in itself without reducing it to neurophysiology (Solms & Saling, 1986).

Freud's view of the evolutionary process was in some ways quite crude and quaint. For example, he drastically abbreviated its time scale: eons of what we now understand to have been essentially random, incremental changes were cast in terms of one-shot cataclysmic social events such as the revolt against authority in the primal horde (Freud, 1912–13). Yet, the general thrust of Freud's evolutionary thinking can be understood as a search for what modern evolutionists (Trivers, 1985; Hamilton, 1964; Mayr, 1983b; Tooby & Cosmides, 1990) would call the "social selection pressures" that shaped the intricate, inner design of the psyche.

Originally envisioned in terms of the prehistoric drama in *Totem and Taboo* (1912–1913), virtually *all of Freud's later references to the evolutionary process illustrate his groping for an understanding of the origins and functions of intrapsychic structure as responses to social or relational dilemmas*. In an important attempt to depict the superego as a *universal* structural feature of the mind (i.e., as a "deep structure" that, to some degree, precedes and shapes *individual* experience) Freud announced his conviction of its social, evolutionary origins.

> The Superego, according to our hypothesis, actually originated from the experiences that led to totemism ... it has the most abundant links with [our] phylogenetic acquisition ... [our] archaic heritage. (1923, p. 36)

Even features of the psyche that subsequent classical (drive–conflict theory) analysts became accustomed to viewing as somehow having no social or relational origins—the "endogenous" or organically rooted drives and the id—could, in principle, be traced back to the shaping effects of ancient *external* (social) experiences.

> There is ... nothing to prevent our supposing that the instincts themselves are, at least in part, precipitates of the effects of external stimulation which, in the course of phylogenesis, have brought about modifications in the living substance. (1915a, p. 120)

> In the Id, which is capable of being inherited, are harbored residues of the existences of countless egos. (1923, p. 38)

For Freud, "affective states" of all kinds were seen as having become incorporated in the mind as precipitates of primaeval traumatic experiences and when a similar situation occurs they are revived like mnemic symbols . . . we must not overlook the fact that biological necessity demands that a situation of danger should have an affective symbol so that a symbol of this kind would have to be created . . . (1926, pp. 93–94)

Freud's dramatic "primeval tragedies," as Gubrich-Simitis (1987) put it—the narratives of paternal domination, castration, exile, murder, and mourning in *Totem and Taboo* (1912–1913) and a relatively recently discovered paper, *The Phylogenetic Phantasy* (1915d)—represent his vision of the kinds of ancestral social events that necessitated the formation of innate mechanisms like guilt, signal anxiety, and the superego as ways of channeling individual behavior in socially congruent directions. Lacking an accurate conception of natural selection, not to mention a grasp of the evolution of social behavior, Freud groped for a way of conceiving the bridge between ancient interpersonal experiences and the universal underlying features of internal psychic structure designed to mediate what he assumed to be the inherent conflict between the individual and others. He intuitively sensed, but could not accurately conceptualize, a crucial understanding of the link between the experiences of our ancestors in confronting the relational dilemmas of the past and the built-in adaptive guidelines—the evolved design features of the modern human psyche.

Because Freud lacked a model that explained how social selection pressures could create a complex, internal, purposive, psychological design, he grasped at one aspect of Lamarckian theory—the notion that acquired traits can be inherited. Freud sought to portray the transformation of phylogenetic experience into enduring, inner structures in a fashion that paralleled individual, ontogenetic learning—learning that is, in one way or another, rooted in actual memory traces of events.[7]

Yet, contrary to the received wisdom from within psychoanalysis (Rappaport, 1960) and outside the field (Gould, 1987), the essential theoretical implications of Freud's sweeping Lamarckian metaphors may be greatly overemphasized. Indeed, it can be argued that Freud made use of these metaphors, the intellectual "tropes" of his day, as vessels in which to contain a far more significant current of scientific (or protoscientific) thought—as heuristic devices in the search for the evolutionary (functional) origins of universal features of the psyche.

Let us consider one of Freud's last attempts to grapple with this

---

[7]See Appendix B: "Freud, Lamarckism, Haeckel's Law, and Modern Evolutionary Thought," p. 290, for a clarification of Lamarckian versus Darwinian thinking.

CRITICAL TRENDS IN THE REACTIONS TO FREUD'S EVOLUTIONISM

There have been two major trends in *reactions* to Freud's evolutionary biological thought. From within psychoanalysis, (including Jones, 1953; Rappaport, 1960; and others) there is the view we shall call the "symbolist" tradition; from outside psychoanalysis (including biologists like Gould [1987] and intellectual historians like Sulloway [1979]) there is the view we shall call the "literalist" tradition.

The symbolist tradition broadly includes diverse thinkers like Jones and Rappaport, essentially classical analysts, or sympathizers, who basically deplored the Lamarckian and speculative nature of Freud's works such as *Totem and Taboo* and the pervasive references to phylogenetic origins.

To protect the clinical and, what they understood to be, the scientific integrity of psychoanalysis, the symbolists developed a line of interpretation (apology) in which Freud's evolutionary speculation was treated largely as a dramatization, a symbolic illustration—a metaphor in the restrictive sense of metaphor; that is, as purely illustrative—of the intrapsychic and developmental truths (about conflict, guilt, mourning, and idealization) that were embodied in the classical analytic model.

The symbolists left us, historically, with the sense that Freud's evolutionary biology should be treated largely as a metaphor in the narrow or superficial sense of "just a metaphor," rather than as a leading edge in the process of looking for, or constructing something new (Cassirer, 1962; Kuhn, 1962). They never attempted to see Freud's evolutionary biology as a set of meanings that could be "deconstructed," or analyzed to reveal underlying structural truths about the direction of his thinking in psychoanalysis.

In this way, the symbolists left psychoanalysis wide open to the literalists. Largely from outside psychoanalysis, and distant from its clinical data and concerns, the literalists emphasized the other side of the issue. They took at face value Freud's explicitly stated convictions that, in one way or another, the ultimate answers to many psychoanalytic riddles would one day come from biology. They also seized on his noted Lamarckism—the notion that characteristics acquired in the course of ontogenesis (learning) can be transmitted to subsequent generations. And they also cite Freud's misplaced affection for Haeckel's law—"ontogeny recapitulates phylogeny"—an antiquated doctrine that clearly did intrigue Freud (see Appendix B: "Freud, Lamarckism, Haeckel's Law, and Modern Evolutionary Thought" p. 290).

The literalists, Gould and Sulloway most notable among them, take this material and make what is, in effect, a rather large inferential leap. They view Freud's use of these now dated historical notions as meaning that he had a primary adherence to, or interest in, furthering these ideas themselves. In Sulloway's case this is taken to the extreme claim that Freud's real agenda was to "biologize psychoanalysis," as if the clinical and theoretical dilemmas had virtually no determinative power or reality of their own in forging the trajectory of Freud's thought. The literalists thus take the historical medium that Freud used (Lamarck and Haeckel, for example) to be the message he wanted to convey. They fail to consider the way, as Mayr (1988) repeatedly points out, in which all manner of arcane historical forms have been used, indeed sometimes must be used, as vehicles in the search to illuminate and explain extremely puzzling contemporary phenomena.

We shall take the position that Freud's evolutionary biological strivings can (and, indeed, *must*) be read as a metaphor; not as a simple "illustrative" metaphor but, rather, as possibly a *leading edge* of his thought. From the literalists, we take seriously his highly flawed quest for a biological solution—not a "reduction to biology" (as in Sulloway's notion of "biologist of the mind") or an overemphasis (as in Gould, 1987) on the ultimate relevance of his attachment to certain dated, nineteenth-century doctrines. Rather, we take Freud's biologism as a determination to refuse conventional distinctions about what is "biological" and what is "psychological," and about when our human history really started. His determination was to bridge "the gulf which other periods of human arrogance had torn too wide apart between mankind and the animals" (Freud, 1939, p. 100), the gulf between *human* nature and nature (Slavin, 1988).

topic—an attempt that happens to be one of his most simple and lucid. In parts of the extended set of evolutionary speculations at the end of *Moses and Monotheism* (1939), Freud put right out on the table, so to speak, what he was looking for beneath the evocative notions of an archaic heritage and apparently mystical notions of phylogenetic memories. Earlier, in *Analysis Terminable and Interminable* (1937) he had noted:

> We know that we must not exaggerate the difference between inherited and acquired characteristics into an antithesis . . . Nor does it imply a mystical overvaluation of heredity if we think it credible that even before the ego has come into existence, the lines of development, trends and reactions which it will later exhibit are already laid down for it. (1937, p. 240)

Freud is explicitly trying to surmount simplistic dichotomies between innate and learned characteristics—even the innate is, in a sense, the result of prehistoric, phylogenetic "learning." He implies that although we know that the ego emerges from experience, aspects of its structure and functions may, indeed must, exist prior to *individual* experience. In 1939, he continued:

> If any explanation is to be found of what are called the instincts of animals which allow them to behave from the first in a new situation in life as though it were an old and familiar one—if any explanation at all is to be found of this instinctive life of animals, it can only be that they bring the experiences of their species with them into their

new existence, that is, that they have preserved memories of what was experienced by their ancestors. The position in the human would not at bottom be different. His own archaic heritage corresponds to the instincts of animals even though it is different in its compass and contents. (p. 100)

In the quotes on the preceding pages, Freud was clearly trying to talk about the evolved structure of many crucial, socially adaptive features of the psyche—the superego, the ego, signal anxiety, and "signal affects" like guilt. Freud's concept of the "archaic heritage" does not simply refer to visceral urges or elemental impulses that, like the notion of the id, could be relegated to some quasi-organic status as if these features were somehow vestiges of an earlier bestial stage of the species. Compare the essential direction of Freud's thought with the way in which the same issues are now articulated by contemporary evolutionists:

All adaptations evolved in response to the repeating elements of past environments, and their structure reflects in detail the recurrent structure of ancestral environments. Even planning mechanisms (such as "consciousness"), which supposedly deal with novel situations, depend on ancestrally shaped categorization processes and are therefore not free of the past.... each species design functions as an instrument that has registered, weighted, and summed enormous numbers of encounters with the properties of past environments. Species are data recording instruments that have directly "observed" the conditions of the past through direct participation in ancestral environments. (Tooby & Cosmides, 1990, pp. 375, 390)

Freud assumed that certain kinds of links between the experiences of past and present generations of living organisms must exist in nature in order to prepare individuals to be capable of executing complex adaptive strategies. Such adaptive strategies can only be possible if there is some way in which the past can be used to evaluate conditions in the present environment and anticipate the future. Though Freud was dealing with complex dynamic phenomena, for example, human meaning at shifting levels of awareness, the basic issues of evolved adaptive design are quite similar to those dealt with directly by Darwinian theorists since Freud's death. Links between experiences throughout the ancestral past and the capacity to evaluate situational or contextual meanings and act in the present became commonplace observations of modern ethology in the '50's (Tinbergen, 1951; Lorenz, 1955). These observations and early descriptive studies of individual species' behavior have been incorporated into the broader, far more powerful theoretical models developed in the past several decades in evolutionary social biology (Williams, 1966; Hamilton, 1964; Mayr, 1983b; Trivers, 1971, 1974, 1985).

More recently there has emerged a realization that natural selection has probably operated to shape complex inner psychological designs—adaptive designs for evaluating the internal experiences, outer realities, and complicated interaction between the two that must be faced by all social species (Tooby & Cosmides, 1990; Symons, 1989). This emphasis on the evolution of inner, *functional designs* as distinct from the direct evolution of *behavioral patterns* per se has brought evolutionary theory a major step closer to the central psychoanalytic concern with inner dynamic organization (Slavin & Kriegman, 1988).

But we are skipping ahead. We discuss the fascinating issues at the interface of evolutionary and psychoanalytic theories starting in Chapter Three. At this point, suffice it to say there was much validity in Freud's Darwinian convictions that (1) in certain respects, the mind can be understood as an adaptive organ, the basic structure and function of which was shaped over vast evolutionary time, and (2) the metaphor of "phylogenetic memories" can be effectively read as referring to the fact that, as Tooby and Cosmides (1989) point out:

> The world is, itself, complex in ways that are not ... analyzable or deducible without an enormous amount of a priori knowledge: in order to solve a task, you must already know a great deal about the nature of the circumstances in which the task is embedded. (p. 26)

Yet Freud's understanding of evolutionary theory was extremely fragmentary and general (Ritvo, 1990). As did most intellectuals of his day, Freud lacked a grasp of the basic genetic theory of natural selection; and he died long before the recent, crucial developments in the evolutionary theory of mutuality and conflict in nature (e.g., Hamilton, 1964; Trivers, 1971, 1974) that have made possible a sophisticated analysis of motivation in social creatures such as ourselves. Overall, Freud was far from being in a position intellectually to apply evolutionary thinking in a very powerful, not to mention critical, way to the structure of psychoanalytic theory.

In many respects he did not (could not) use evolutionary theory to examine or modify his developing psychoanalytic vision of the human condition. Rather, despite the profundity of his scientific convictions of the need for consistency between analytic and evolutionary biological views, he tended to use what he knew of Darwinian theory to bolster his vision of the human condition (Slavin & Kriegman, 1992). Freud gave us an account of the relational world that in terms of the competing visions discussed in Chapter One, is largely a narrative of conflict and deception.

Indeed, one might well argue that Freud made use of the evolutionary stage as a way of depicting and substantiating certain of his earliest views of the origin of neurotic conflict. Though, by the time he wrote *Totem and Taboo* (1912–1913), he had long rejected the emphasis on familial politics and deceit that characterized the views we now call "the seduction theory" (Masson, 1984), in that primeval drama he found a way of calling our attention to the primary role of the parental environment in the etiology of infantile conflict. The notion of the "primal father" dominating and rationalizing his power in a fashion that directly contravenes the interests of his sons became a way of inculpating paternal motives in oedipal conflict in a fashion that he had left behind in the shift toward drive theory by the turn of the century. As Gubrich-Simitis (1987) noted:

> The traumatic real experience in Freud's early conception of the etiology of hysteria appears ... set back into the distant past of the prehistory of the species, that is, transposed from the ontogenetic to the phylogenetic dimension. (p. 99)

Of course, in the comfortably distanced setting of human prehistory, Freud does not challenge the motives of contemporary fathers. Indeed, it is striking that while he speculates freely about an "archaic heritage" of parricidal fantasies and "primal guilt" universally experienced by modern generations *toward parents*, he never speaks directly of the parental generation as having phylogenetically inherited, so to speak, tendencies to use, manipulate, control, and deceive the young. However, regardless of whether Freud directly recognized the implication that aspects of the psychological organization of both children and parents might have been shaped by ancestral, relational conflicts, he did find in the evolutionary idiom a way of concretizing and bolstering his essential vision of the human condition. It is hard to conceive of a more heightened view of the clash of individual interests, the intrinsic deception, ineluctable inner tension, and predestined repetition than is depicted in Freud's accounts of the ancestral relational world and its enduring contemporary legacy (Slavin, 1988). In effect, the simple "survival of the fittest"[3] metaphors of his day both fit and helped shape the central narrative account that Freud created—the account around which the classical model was constructed (Kriegman, 1988, 1990).

---

[3] A rather misleading phrase coined in 1864 by Herbert Spencer. We shall discuss the more accurate modern conception of "inclusive fitness" in Chapter Four.

As this book unfolds, we hope to persuade the reader that evolutionary theory has come far enough since the limited models available in Freud's day so that we can, in fact, now use the Darwinian perspective in a far more critical fashion than Freud (and, indeed, most subsequent analytic writers) have been able to do. We hope to create a platform, a vantage point, an evolutionary idiom that is sufficiently removed from the clash of psychoanalytic narratives to cast more light on them at both the basic philosophical level as well as at the clinical level where psychoanalytic constructs actually live.

But, again, we are skipping ahead. Freud was by no means the only major psychoanalytic theorist to have sought in the Darwinian perspective an opportunity to address the assumptions about human nature that underlie the classical and relational narratives. It is instructive to look at how Darwinian versions of psychoanalysis developed once they were out of Freud's hands.

Freud's Search Extended and Altered: Hartmann, Bowlby, and Erikson

The history of psychoanalysis and Darwinism after Freud is marked by two major mainstream attempts at placing analytic concepts in an explicitly evolutionary context. In addition to Freud, both the essentially loyal Freudians (Hartmann et al. in ego psychology) and certain more critical relational theorists—Bowlby (1969) and the tradition of attachment theory (Peterfreund, 1972; Yankelovitch & Barrett, 1971), and Erikson (1956, 1964, 1968)—invented illuminating versions of an "evolutionary" or "adaptive" perspective that bear directly on the central divergences within psychoanalysis.

Making use of a biology that was slightly more updated than Freud's (in the sense of his restraint regarding Lamarckian metaphors and invented prehistoric scenarios) Hartmann (1939) must be credited with compelling psychoanalysis to appreciate that certain of the ego's adaptive efforts could usefully be understood as built-in biological functions, and that the equation of what is "biological" with the id and what is "environmental" with the ego was untenable. In trying to define the psychoanalytic meaning of "adaptation" he asks:

> Have we the right to exclude the processes of adaptation from biology? Biological functions and environmental relationships are not in antithesis. This is not merely a terminological correction: the terms imply an underestimation of the very areas of biology [adaptive psychological functioning] with which we are concerned here. (1939, p. 33)

Hartmann was reacting to the tendency to see the id as the biological part of the human psyche as contrasted with the environmentally focused and responsive "nonbiological" ego. Hartmann argued that this is a basic misunderstanding of biology in which all life forms, physical structures, patterns of behavior, and psychological processes and structures must be seen as *biological* (Kriegman, 1990).[4]

Hartmann's struggle with what is and what is not biological led him to the conclusion that the ego must be considered a biological structure and therefore must have its own growth and development independent of the id. Thus, some ego functions are not the result of conflict; that is, they could possess their own growth and energic motivational resources without being energized through repressed, sublimated, or neutralized drives. However, the conflict-free sphere that Hartmann defined is almost entirely *cognitive or psychomotor*:

> I refer to the development *outside of conflict* of perception, intention, object comprehension, thinking, language, recall-phenomena, productivity, to the well-known phases of motor development, grasping, crawling, walking, and to the maturation and learning processes implicit in all these and many others. (1939, p. 8)

Hartmann did not examine the possibility of there being *relational needs* that have their own motivational or energic sources, independent of conflicts over sexual or aggressive drives (Kriegman, 1990). While Hartmann's *cognitive and psychomotor skills* can be understood as *theoretically* free of conflict, he makes it clear that rarely are these faculties unaffected by conflicts and characteristic ego resolutions of conflict. Just so, might there be adaptive emotional/relational needs that should be considered as developing without being *derivative* of conflict, even though we may rarely see them without conflictual entanglements? As we shall see, evolution theory predicts just such a phenomenon (Kriegman, 1990).

Further, while Hartmann states that "[h]uman action adapts the

---

[4]Curiously, earlier in his attempt to define adaptation, Hartmann dismissed the role of adaptation *as defined by evolution theory* and natural selection as not having relevance for a *psychoanalytic* conception of adaptation:

> At this point we encounter the controversies about the relation of phylogenesis to adaptation and the solutions proposed by Darwinism, Lamarckism, and other biological theories. These theories, however, have no direct bearing on our problem. (1939, p. 24)

Possibly, by avoiding the issue of "Darwin, Lamarck, and other biological theories," Hartmann could avoid trying to defend or reject Freud's embarrassing Lamarckism (see Appendix B, p. 290). This may have been the "controversy" to which he was referring (Kriegman, 1990).

environment to human functions, and then the human being adapts (secondarily) to the environment which he has helped to create" (1939, p. 26), note that Hartmann is referring to the *nonhuman* environment. He never fully understood the concept of adaptation in reference to the relational world and of motivation within a world of others. Thus, Hartmann never turned his inquiry around to try to understand the adaptive origins of the id itself—indeed of the whole realm of the pleasure principle and so-called conflict-based or conflict-born development and dynamics. His exclusive focus on the ego as essentially the sole organ of adaptation *completely avoided an examination of classical assumptions about instinctual drives (motivation) in evolutionary terms.* Indeed, even though he broke down the conceptual barrier between the biological id and the supposedly nonbiological ego, Hartmann (and the whole ego psychological tradition) essentially created an artificial division between the realm in which one dealt with questions of *adaptation* (i.e., ego functions) and the realm of conflictual drive–defense dynamics. Psychosexuality—the core system of motivation and development—was treated as though it were ipso facto biologically based and needed no further analysis in adaptive, functional terms (Slavin & Kriegman, 1992).

Hartmann (1939) seems to have implicitly recognized the value of an evolutionary analysis of the origins and adaptive design of the drive–defense system. He made passing reference to natural selection as "the reality principle in the broader sense" (p. 44); and he went on to acknowledge that such a "reality principle in the broader sense would historically precede and hierarchically outrank the pleasure principle" (p. 44). However, he stopped far short of even posing the question of how the instinctual drives as universal, innate forces, and the id as an evolved structure, might have arisen and been designed to function within a complex *relational* context that, through the selective pressures such a context exerted, actually comprised the "reality principle in the broader sense."

Hartmann was operating within the canon of the classical psychoanalytic narrative. He thus took for granted what has become (and remains in certain quarters today) an almost doctrinal position articulated by Anna Freud (1936): that there is an "innate hostility between the ego and the instincts which is primary and primitive" (p. 158). He accepted the corollary assumption of an inherent antagonism between our deepest human motivations and the demands of a realistic, socially appropriate adaptation to the environment. In this view, as Mitchell (1988) notes, the drives and the id are seen as, somehow, atavistic relics of an earlier, more elemental animal level of existence on which adaptive ego processes—though biologically grounded in "primary autonomy" and equipped with the "neutralized energy" of Hartmann's model—are still overlaid, like a cloak of civilized realism upon the "beast." Hartmann's exclusive focus on the ego as essentially the sole organ that is discussed in terms of its evolved, adaptive functions, and his tendency to see the environment as a highly impersonal (non-social) "reality" essentially short-circuited a continued evolutionary examination of classical assumptions about instinctual drives as adaptations to a relational world composed of other individuals.

Bowlby (1969, 1973, 1980) and Erikson (1956, 1964, 1968) both made substantial efforts to introduce an evolutionary perspective on psychoanalytic theory with the aim of revising basic psychoanalytic theoretical assumptions. Both of these theorists were heavily influenced by the "ethological" (Lorenz, 1955; Tinbergen, 1951) and "ecological" work that became widely intellectually influential in the '50s. Both developed new psychoanalytic paradigms in which we can see a major shift into the relational narrative. In the work of both theorists, there is a heavy emphasis on the interdependent, mutualistic features of the developmental process. Individual development is seen as part of a larger, remarkably integrated reciprocal adaptive system.

Bowlby and Attachment Theory

Bowlby drew on aspects of ethology to show that a biologically appropriate adaptive explanation (and reformulation) of important universal aspects of attachment and its regulation could not be constructed from the classical drive–defense model—even with the elaborations of ego psychology. He chose to completely discard the theory of instinctual drives in favor of an exclusive focus on attachment and its vicissitudes, substituting the language of information processing for energic, drive-based concepts.

Bowlby's contribution to the larger, complicated process of revising drive theory was of great value; his approach has ushered in a major area of productive research on early attachment and loss. Yet, Bowlby's ethological model shifts us thoroughly into the relational narrative. His model tends to portray the dynamics of attachment as if they operated in a relational world wherein the interests of parents and children predominantly overlap. His view of the relational world is philosophically rooted in a perspective within which conflict and deception are not intrinsic features of the human condition in general, and certainly not an essential part of the most basic familial attachments.

Thus, what is termed "biological" in attachment theory and considered under the rubric of an adaptive analysis serves to bolster a view of life that is, in many ways, directly opposed to Freud's Darwinian vision of power, deceit, and competition. Bowlby emphasized and used only those aspects of evolutionary theory that supported his particular psychoanalytic theoretical position with its underlying narrative of development and relational reality. We shall return repeatedly to versions of "attachment theory" as a presumably biological approach at several points in later chapters.

Erikson and the Psychosocial Model

Though Erikson's "psychosocial" model is portrayed not as an alternative to drive theory but as a complement to it, he supported his new views with a far-reaching critique (1964) of what he called the "pseudobiology" of the classical and ego psychological traditions—the "ideological and . . . unbiological [assumptions about] . . . the innate antagonism between the organism and the environment" (p. 221). He then went on to replace Freud's and Hartmann's evolutionary views of the relational world with a revised set of biological assumptions based on the notion that, "newer concepts of the environment (such as the Umwelt of the ethologists) imply an optimum relation of inborn potentialities to the structure of the environment" (p. 223). These newer ecological notions supported Erikson's shift into the relational narrative—his emphasis on viewing the individual as part of "a mutually supportive psychosocial equilibrium" (p. 223). In the process of creating his new psychoanalytic model, Erikson frequently recounted what are, in our terms, key elements of the relational narrative. The relational world, for Erikson, "can only be viewed as the joint endeavor of adult egos to develop and maintain, through joint organization, a maximum of conflict free energy in a mutually supportive psychosocial equilibrium" (p. 223).[5]

In its sweeping characterization of development as an evolved, innate, "epigenetic" process of intergenerational exchange and negotiation, Erikson's "psychosocial" perspective comes closer than any other psychoanalytic model to a genuinely "adaptive point of view" (Rappaport, 1960). As we hope to show, in some respects, Erikson's views are

---

[5]The emphasis on ecological systems as stable, interdependent, integrated wholes (prevalent in biological theories from the 1950s until fairly recently) found expression in biologically inspired metaphors that exaggerated how individuals behave in highly mutualistic ways that favor others, perpetuating and renewing the larger, longer-term interests of the system. Newer ecological findings (Stevens, 1990) stress the highly limited stability of biological systems, while the evolutionary biology of social behavior (Trivers, 1985) has recognized that structures of self-interest and related conflicts of interest play a vital role in the motives and strategies pursued by even highly interdependent individuals. We address this issue in Part II, Chapters Four through Six of this volume.

quite compatible with contemporary evolutionary theory. However, like Bowlby, and, in a sense, equal but opposite in direction to Freud, Erikson uses evolutionary thought to support his shift into the relational narrative. Because Erikson's and Bowlby's vision is essentially collectivist and mutualistic at its core, the model of the "joint endeavor of adult egos" toward a "mutually supportive psychosocial equilibrium" fully encompasses only one part of the human condition. It does not equip us to grapple with what they may too easily dismiss as the "unbiological" and "ideological" assumptions about an "innate antagonism" between the organism and the environment—assumptions in which, despite the superficial references to drive theory metapsychology, we can hear the deeper resonances of the classical narrative: the clash of individual interests in the good-enough, average expectable environment; the need for "endogenous" motives including the innate tendencies, aims, and expectations that may mediate this larger clash of interests; and the subjective distortions (deceptions and self-deceptions) that may be part and parcel of the design of our psyche.

Briefer Appeals to Evolutionary Biological Authority in Support of Classical and Relational Views

From Freud, at one extreme, to Kohut on the other, almost all psychoanalysts have made an evolutionary biological pitch in support of their theories. The psychoanalytic literature is replete with examples of attempts to find evolutionary rationales for specific psychoanalytic perspectives. Given Freud's explicit commitment to making psychoanalysis consistent with Darwinism and Hartmann's (1939) grounding of ego psychology in his conception of biological "adaptation," it is not surprising that a kind of taken-for-granted biological tone of authority became an expectable part of the broad classical analytic tradition. Some analysts, such as Schafer (1976), Klein (1976), and Kohut (1982), have argued for the separation of psychoanalysis and biology. While this latter group rejects traditional forms of "biologizing" psychoanalytic constructs, they may be less aware of their own implicit biological assumptions.

Anna Freud followed closely in her father's tradition of using evolutionary references to bolster the vision of the classical model. For example, several major biological assumptions are brought in to support her attempt to explain adolescent turmoil in terms of drive theory: "There is," she says, "in human nature a disposition to repudiate certain instincts, indiscriminately and independently of individual experience ... a phylogenetic inheritance ... continued, not initiated by individuals ... " (1936, p. 157). There is, she added, "an innate hostility between the ego and the instincts, which is indiscriminate, primary and primitive" (Sandler & Freud, 1985, p. 477).

More recently, in Blos's (1979) ego psychological attempt to broaden our conception of the adolescent potential for change (while, like Hartmann, retaining the basic classical/instinctual narrative of development), we find the characteristic appeal to biological authority. In searching for a reason to explain why the adolescent process is presumably able to create a highly autonomous, intrapsychically regulated adult, Blos claims that "we can view character formation *in an evolutionary perspective* and contemplate it as [creating] a closed system ... [a] ... transcendent feature—virtually taking humans out of the realm of environmental interdependence that lower forms of life must endure" (p. 191). The "evolutionary perspective," in this case, is an evocative metaphor that is used to support a view of the transformation of the adolescent psyche into an adult form that is tied to the highly individualistic (closed, one-person system) vision of classical psychoanalysis.[6]

Mitchell (1988) provides an excellent example of the explicit use of those aspects of the evolutionary adaptive perspective that selectively support a particular viewpoint. He unequivocally embraces biological conceptions that support his "relational–conflict" model but then vigorously rejects any notion of preexisting structures or motives other than the motive of relating itself; he rejects any concept of innateness that would necessitate modification of his specific relational view. For example he quotes Lichtenberg, who, according to Mitchell (1988),

> concludes that, "Study after study documents the neonate's preadapted potential for direct interaction—human to human—with the mother" (1983, p. 6). The phrase "preadapted potential" is crucial here. The evidence seems overwhelming that the human infant does not *become* social through learning or conditioning, or through an adaptation to reality, but that the infant is programmed to be social. ... the very nature of the infant draws him into relationship. (p. 24)

In the tradition of Bowlby, Mitchell acknowledges that "positing ... attachment or relatedness as primary suggests that it has motivational properties *within* the organism and might meaningfully be considered a "drive" (Mitchell, 1988, p. 24). Yet he insists that this is

---

[6]We discuss the views of adolescence held by both Anna Freud and Blos in greater detail in Chapter Nine.

very different from Freud's "drive." The latter presupposes motives and meanings in the individual a priori, in the tensions in bodily tissues themselves, which are brought to the interaction and which shape the interaction. (p. 24)

Mitchell's adaptive analysis stops short once he has posited the motive of attachment or relating. Once we are in a relationship with another, what do we want to do with the other? In rejecting Freud's specific notion of using others to obtain pleasure, tension reduction, and drive gratification—and in not substituting any other motive than attachment or the need to relate—Mitchell limits his notion of the adaptive utility of attachment:

> The other is not simply a vehicle for managing internal pressures and states; interactive exchanges with and ties to the other become the fundamental psychological reality itself.... Consistent patterns develop, but the patterning is not reflective of something "inside."

Mitchell is clear that he is trying to do away with the concept of the "a priori"—with the one exception of the *a priori* relational propensity:

> If we discard *Freud's* drive theory, is *some* concept of drive still necessary to account for that which is constitutional?
> In a broad sense, trying to locate the innate in the relational model is impossible, because it takes a term which is central from one paradigm and tries to locate it within another ... (p. 61)

Mitchell is at pains to limit the scope of what is innate to a central capacity—attachment and relating—that is compatible with relational narrative assumptions: (1) that the individual is an incomplete entity who seeks others and groups to become whole rather than to profit from the exchange that can enhance his or her *inner* state; and (2) that conflicts between individuals are not a function of inherent clashes between their intrinsic (inner) aims but, rather, of the failed coordination of their inherently compatible basic aim to relate to one another. Thus, all meanings and motives, besides the motive to relate, must derive from the relational context. He rejects Freud's notion of sexuality because he sees it as something that forces the individual to bring motives that are not derived from the relational context to interactions with others in order to employ "interpersonal experiences to express a priori themes and fantasies" (p. 72)—to *use* relationships for aims derived from some internal a priori motives.

Yet, Mitchell's objection to the a priori runs deeper than his critique of drive theory. He uses the same principle in his critique of self psychology. For example, he quotes Goldberg (1986) who states that:

> Since self psychology is so preeminently a developmental psychology, it is based on the idea of a developmental program (one that may be innate or pre-wired if you wish) that will reconstitute itself under certain conditions. (p. 387)

To which Mitchell (1988) responds:

> Here again is the assumption that there is an "inside" to the patient which can manifest itself more or less independently of the interactive situation in which it appears (p. 298)

For Mitchell, whose version of the relational narrative emphasizes the derivation of all psychological structure and human meaning from the vicissitudes of interpersonal experience, the use of evolutionary logic must be carefully limited in order to exclude those aspects that would challenge such a view by introducing notions of any other evolved, a priori structures and motives. Other than the a priori need/motive/drive/propensity to relate, no other a priori motives can be seen as being brought to the interpersonal context from within the individual.

Mitchell's significant but highly qualified acceptance of a certain limited degree of evolutionary adaptive logic is a far less extreme position than that taken by many other analytic theorists. Many analytic theorists have been particularly critical of the explicit use of any form of biology. There are calls to preserve psychoanalysis as a pure psychology (Kohut, 1982), by which is typically meant not only a psychology freed of the anachronistic, physicalistic, biological assumptions of the classical metapsychology, but, in the extreme, for a psychology grounded solely in the data and interpretive historical process of the clinical encounter (e.g., Klein, 1976; Schafer, 1983; Stolorow, 1985).

This major, contemporary "antibiological" theoretical strategy is particularly appealing to relational theorists (e.g., Guntrip, 1971; Kohut, 1982) who, for historical reasons, tended to equate the presence of biological reasoning in psychoanalysis with Freud's early physicalistic orientation and the problems of the classical metapsychology. Yet, more generally, many theorists of essentially quite classical sensibility (e.g., Klein, 1976; Holt, 1976; Schafer, 1974) also equate biological perspectives with a positivistic, mechanistic view of mind and reality; their fear is that explicit biological reasoning is incompatible with a proper clinical focus on human subjectivity, personal agency, and narrative (vs. objective and absolute) truth. Yet, some of these same theorists—across the classical–relational spectrum—are equally ready to make broad biological generalizations in order to support their particular views.

For example, in Kohut's (1982) attempt to understand parental devotion to offspring, he implicitly argued that, as in all species whose young must receive parental care, the human psyche must contain an inherent tendency to act altruistically toward one's offspring. Kohut described this congruent fit between the needs of the child and the innate emotional response of the parent, as "man's deepest and most central joy, that of being a link in the chain of generations" (p. 403). In this case, despite Kohut's biological disclaimers, he was actually following an evolutionary biological line of reasoning when he applied King's (1945) notion of normal healthy functioning to his conception of intergenerational relations: namely, that "man's deepest and most central joy" was seen as adaptive and healthy in that it represents an organism functioning "*in accordance with its design.*"

Thus, while he characterized classical theory as containing "insipid biology"—and thus took an explicitly antibiological stance—Kohut utilized aspects of biological logic that lie at the heart of evolutionary thought. His adoption of the notion of "functioning in accordance with design" was used to support a crucial element of the relational narrative—the intrinsically mutualistic motives (the overlapping interests) inherent in intergenerational relations. But, Kohut's reasoning was actually far more consistent than he might have thought with Freud's insistence that we must not exclude ourselves from the animal kingdom. It happened that, in this case, a valid, functional argument for the universally observed, innate satisfactions in parenting coincided with Kohut's need to radically alter the classical conception of what is most "primary" in human motivation: drive discharge versus naturally caring, relational ties to one's offspring.

Even the most thoroughly antidrive, antibiological theorists such as the intersubjectivists (Stolorow & Lachmann, 1980) make a pitch to the evolutionary court in support of their particular case. In arguing for the primacy of psychic organization, consolidation, and structuralization of the self as underlying all human motives, they suggest that psychosexual experiences, fantasies, and enactments can best be understood as psychic organizers. They support this notion by claiming that "Nature, in her evolutionary wisdom, has harnessed the exquisiteness of sensual pleasure to serve the ontogenesis of subjectivity" (p. 148).

Impressionistic Assumptions about Evolution and Adaptation versus Modern Evolutionary Theory

The appeal to "biological truth," within both the classical and the relational traditions, is thus a pervasive, perhaps inevitable aspect of psychoanalytic theorizing. Moreover, it is an appropriate aspect. The question is whether this appeal is done in an implicit, impressionistic fashion (and thereby is less subject to examination) or an explicit one in which the inevitable assumptions about our nature as a species are laid out for examination.

It should be clear that we assume that virtually every psychoanalytic position contains a host of implicit, and often explicit, assumptions about human nature that are, in turn, based on assumptions about the nature of the relational world. Thus, if a contemporary evolutionary vantage point can help us to explicate these underlying assumptions and to sort through them in order to identify those that may be consistent with features of the relational world as viewed from a relatively independent perspective, we would then be in a much stronger position to carry out a program of "comparative psychoanalysis," the fruitful approach to contemporary theory pioneered by Schafer (1983) and Greenberg and Mitchell (1983). While evolutionary biology certainly can not directly provide us with a specific picture of the psyche, we hope to show how it may help to clarify differing views and assumptions about the relational world in which our "complex mental architecture," as Tooby and Cosmides (1989) put it, must function and fit. Such clarifications can help us to perceive more accurately our own theoretical biases and their impact on our observations.

Our move now to explicitly consider the contemporary evolutionary perspective certainly guarantees no objective, value-free, scientific map of psychological structure—as the plurality of conclusions in the history of psychoanalytic biological approaches amply attests. Yet, such an approach is, in our view, a critical adjunct, a supplement, perhaps a corrective, to what we see as the inevitable *implicit* biologizing—the pervasive assumptions about mind and human nature in psychoanalytic theoretical discourse.[7] In addition, while seeking a vantage point outside the usual realm of psychoanalytic discourse on some of the

---

[7]To our knowledge, there are several theorists who have applied evolutionary theory to issues in psychoanalytic theory (Peterfreund, 1972; Badcock, 1986, 1988; Leak & Christopher, 1982; Lloyd, 1990; Nesse, 1990; Wenegrat, 1984). Though citations of some of their views shall appear throughout this volume, we have not treated their work extensively because it does not focus, as we do, on the implications of evolutionary theory for clinical and theoretical debates within contemporary psychoanalysis.

SELF PSYCHOLOGY, BIOLOGY, AND HUMAN NATURE

As we note (p. 51), Kohut explicitly rejected the relevance of biological concepts and then implicitly utilized a basic biological analysis in developing his conceptions of normal human functioning. As a group, the self psychologists have been united in their opposition to "experience-distant" biological drive theories. Yet, interestingly—possibly because they tend to see the individual striving toward growth and health, not toward resistance and immature instinctual expression (Kriegman & Slavin, 1989)—they have also been unanimous in taking an adaptive biological viewpoint that incorporates the notion of innate, pre-wired design and an underlying biological human nature. Consider the following:

Basch (1988) sees the central human tendency as a "search for competence [that] is ubiquitous and universal [because] it reflects the basic function of the brain which is . . . to create order out of the myriad stimuli impinging on it at any moment" (p. 48).

Goldberg (1986) states that "Since self psychology is so preeminently a developmental psychology, it is based on the idea of a developmental program (one that may be innate or pre-wired if you wish) that will reconstitute itself under certain conditions (p. 387). Later, he adds, "The many theories of psychoanalysis rest on some plan or program of normal development . . . Children are born with a whole set of abilities that enable them to comprehend and thus internalize the world" (1990, p. 93).

Lichtenberg (1983) states that "Study after study documents the neonate's preadapted potential for direct interaction—human to human—with the mother" (p. 6). Lichtenberg's (1989) motivational theory includes, "*innate programs* for the categorical affects of enjoyment, interest, surprise, distress, anger, fear, shame, and disgust [that] exist at birth" (p. 260, emphasis added). For Lichtenberg "each affect is an entity (a brain program)" (p. 269).

Ornstein (1991) acknowledges that "[T]here is always a . . . strongly influential aspect to our guiding theories that we rarely acknowledge and rarely, if ever, subject to close scrutiny. This is the image of Man and Woman our theories depict; the conception of *human nature* that is mostly implicit (rather than spelled out) in any particular psychological theory including psychoanalysis."

Wolf (1988) states "Human beings are born *pre-adapted* to actively participate in both physical and psychological interactions with the environment, which provide, respectively, for the individual's physical and psychological needs. Both are necessary for survival" (p. 10). [Also see our discussion of Stolorow and Lachmann (1980) on page 51.]

Thus, in a sector of analytic theorizing that is traditionally seen as uniformly opposed to the biological theorizing of the classical perspective, we consistently find what we would consider *appropriate* attempts to use biological reasoning that includes conceptions of innate design and pre-wired structure.

questions that are central to a comparative psychoanalysis, our goal is to remain close to clinically relevant theoretical issues and to preserve the language of human aims and experience in order to evaluate the fruits of this altered perspective.

Freud's flawed yet appropriate efforts to deal with the overall structure of the psyche within an adaptive context is the task that requires further exploration in light of contemporary evolutionary theory. As we shall see, in order to undertake this task, we must have a biologically grounded perspective on the place and role of conflict and mutuality in the relational world to which our psyche is adapted. Somehow, within this perspective, we must be able to account for some of the fundamental tensions and oppositions that are deeply a part of our inner psychic life. As we shall see, the world of modern evolutionary theory has not only moved dramatically beyond Freud's limited picture of it, but has equally superseded the understanding available to Hartmann, Bowlby, and Erikson as well as virtually all other attempts to keep an overt, explicitly evolutionary biological tradition alive within psychoanalysis.

Thus, the challenge: Can we develop a general psychodynamic model of the *overall* organization of the psyche, a model that is consistent with an evolutionary understanding of the mind and human condition? Can we use contemporary evolutionary theory to address some of the basic problems involved in a reconsideration of the human condition as depicted in the drive-based, classical narrative and the relational, social vision of human nature? Can we create an evolutionary psychoanalytic perspective that we can consult for help in the all-important process of sorting our way through the complex set of contrasting clinical theories, stories, and prescriptions? These are the questions that we address throughout the rest of this book.

CHAPTER THREE

The Modern Evolutionary Perspective: Some Basic Considerations

> Hold a baby to your ear as you would a shell:
> Sounds of centuries you hear
> New centuries foretell.
> Who can break the baby's code?
> And which is the older—
> The listener or his small load?
> The held or the holder?
> —E. B. WHITE

> The genotype (genetic program) is the product of a history that goes back to the origin of life, and thus incorporates the "experiences" of all ancestors . . . It is this which makes organisms historical phenomena.
> —ERNST MAYR, *Toward a New Philosophy of Biology*

Adaptation and Natural Selection

The notion of adaptation is obviously crucial to our whole discussion. In what sense do we use the term, and what does it mean? The term "adaptation" has a long, complicated history both within and outside psychoanalysis. In the interests of clarity and consistency, we limit ourselves to its specific evolutionary biological meaning. Whenever any structural (or behavioral) feature of an organism is called "adaptive" we mean that it has been shaped by natural selection in a way that was, by definition, advantageous to the unique configuration of genes comprising the genotype of the individual (Mayr, 1983b; Williams, 1966). The genotype is the novel, individual constellation of encoded ("historically acquired") information that "controls the production of the phenotype—that is, the visible organism which we encounter and study" (Mayr, 1988, p. 16).

Adaptive structures or systems, such as the complex wiring of our human linguistic capacity (or early, prototypal versions of it), represented, at one moment in our evolutionary history, an inner structural variation possessed by some individuals more than by others. Through its link to the complex interactions involved in human social adaptation, greater linguistic capacity represented a selective advantage to those individuals who possessed it. What this means is that, ultimately, these individuals survived and reproduced better than their compatriots whose inner structure took a different, less effective form—perhaps one that was less capable of manipulating word symbols that stood for realities far removed from current observation. By definition, then, the design of those with the greater linguistic capacity was more "fit," more able to promote their genetic, reproductive interests; in this sense, more adaptive. The variations common to their more successful genotypes came to be the basic genotype shared by our species as a whole (Williams, 1966). This is the process of natural selection, defined with elegant simplicity by Darwin (1858) as "the preservation of favorable variations and the rejection of injurious variations" (p. 81).

The overriding principle that characterizes contemporary evolutionary thought is the understanding that all life forms—structures and, to some extent, behavior patterns—can be understood in terms of the built-in push toward reproductive success of the genetic material contained within the organism (Trivers, 1985; Dawkins, 1976).[1] Considering that in most species, most individuals are not successful at reproducing viable copies of their genes, it is astonishing that every existing, living organism is the endpoint of an unbroken, 4–5-billion-year-old chain of successful reproducers. Every direct ancestor of every living organism was one of the minority of living creatures that successfully passed on copies of its genes to viable offspring. And this unbroken chain of

---

[1]Dawkins (1976) put this in a starker way when he said that evolutionary theory can "be read as though it were science fiction... but it is not science fiction: It is science. Cliché or not, stranger than fiction expresses how I feel about the truth. We are survival machines—robot vehicles blindly programmed to preserve the selfish molecules known as genes." (p. ix)

Such is the essence of evolution! All life forms are simply structures that enhance the survival and replication of copies of their DNA codes. Those that succeed become more common, while those that fail disappear. All life forms, structures (physical and mental), and certain behavior patterns can be understood in terms of the benefits these patterns provide to the genetic material underlying them (see box, "The Limits of the Adaptationist Perspective," on p. 58, where we discuss the current controversy surrounding this view).

success—represented by every creature alive today, goes back to the time that life first began!

Throughout this volume, we take the Darwinian, or neo-Darwinian, view that crucial, adaptive challenges faced by countless ancestral generations, through differential survival and reproduction (natural selection), have powerfully shaped aspects of the deep structure of our psyche before the individual child encounters the world for the first time. Using this perspective, we approach the problem of understanding the underlying structure of the psyche. From this "adaptationist perspective"[2] (Gould & Lewinton, 1979; Mayr, 1974) we listen to the "sounds of centuries," as the writer, E. B. White, put it. More literally, we shall consider the hundreds of thousands of generations of our "natural history"—our human, evolutionary heritage—in order to develop an understanding of how the forces of natural selection may have, in effect, "designed" us to have the capacity to deal with certain universal, relational dilemmas and paradoxes inherent in human development (Trivers, 1985; Williams, 1966).[3]

We assume that the relational environment, into which the child is born, powerfully influences and shapes its development. The child will psychologically, socially, and linguistically "speak" the "language" of its family and culture. Yet, the environment will operate on a highly prepared, active, individual organism, with a preexisting inner design consisting of "codes" or "schemas" that help organize and give meaning (conscious and unconscious) to experience (Mayr, 1974, 1983b, 1988).

The Evolutionary Approach, Psychoanalytic Experience, and Human Meanings: The Issues of Reductionism and Determinism

Psychoanalysts are often concerned that "biological" approaches, including evolutionary theory, are harbingers of the mechanistic, somatic, "physiologizing" from which psychoanalysis is still struggling to free itself. In its attempt to become a science of complex psychological *experience* (Kohut, 1982)—to focus on human *meanings* (Klein, 1976; Schafer, 1976) rather than mechanisms such as hydraulic, tension-reducing

---

[2]See box, "The Limits of the Adaptationist Perspective," p. 58, for a discussion of criticisms of the adaptationist perspective.
[3]See also Appendix A, "Has There Been Sufficient Time for the Evolutionary Design of Complex Psychological Structures through Random Variation and Natural Selection?" p. 287.

THE LIMITS OF THE ADAPTATIONIST PERSPECTIVE

Gould (1977), Mayr (1983a), Kitcher (1985), and others have repeatedly emphasized that it is scientifically unsound to take every organismic feature and assume that a specific adaptive evolutionary explanation can be derived to explain its current role and functions. Any given feature may be a part of another feature or a larger adaptive organismic system; thus, it may not have been selected because it, in itself, was adaptive. Or, the feature may be an accidental creation (i.e., one that is fortuitously connected to some other adaptive feature) that persists in the genotype because it does not seriously impair the organism's functioning. These are important caveats. However as Mayr points out, "the adaptationist question, 'What is the function of a given structure or organ?' has been for centuries the basis for every advance in physiology" (1983a, p. 328).

In other words, the adaptationist hypothesis is the source of our most fruitful hypotheses. It has also, of course, been the source of many incorrect hypotheses. What is crucial is that the logical structure of adaptationist hypotheses combined with a dedication to empirical investigation has enabled us to sort out the difference. For psychoanalysis, by using the adaptationist's logic we may achieve new insights into our clinical observations—always remembering that, in the final analysis, it is the clinical data to which we must return to assess the validity of these insights.

Because the adaptationist hypothesis coupled with empirical research has been the source of virtually every major advance in biology, Mayr (1983a) concludes that, contrary to what is claimed by its more radical critics, when it is carried out in a fashion that correctly identifies the *level of organization* on which selection has operated, "it would seem obvious that little is wrong with the adaptationist program as such, contrary to what is claimed by Gould . . . " (p. 332). Mayr goes on to argue for the attempt to identify as "adaptations" those structures that truly play a central role in the organization of the individual, not incidental or secondary aspects of the "whole." And, even in the case of phenomena that are generally labeled maladaptive or pathological (see, for example, our discussion of the "distortions" in transferences and resistance in Chapter 10) the *search for hidden functions*, as R. D. Alexander (1990) notes, has been "among the most fruitful procedures of evolutionary science ever since Darwin" (p. 20).

We are in agreement with Mayr, and believe that it is *psychodynamic structure* that represents, precisely, this *correct level of organization*. Thus, we believe that the concerns of well-intentioned critics such as Gould are probably far less applicable to a *psychoanalytic* use of the concept of adaptation than to what is essentially the naive behaviorism of certain proponents of "sociobiology." Indeed, even an outspoken critic of psychobiology such as Kitcher (1985) acknowledges that "the main criticism I level against . . . sociobiology is that it introduces evolutionary considerations in the wrong way by focusing on behavior and not on the underlying mechanisms . . . " (Kitcher, 1987, p. 91). All too commonly, sociobiologists isolate specific, discrete behaviors out of the context of the whole organism and its psyche, as the focus of adaptive hypotheses in which specific gene=behavioral links are assumed to exist. We attempt to look at the fundamental organization of the psyche as a "*deep structure*," so to speak, that underlies myriad specific individual, social, and cultural variations and values.

This essential human inner organization and propensity to follow certain developmental paths is not a minor feature that could potentially be accidental. We are examining the basic organization of the psyche and its major psychological components. Certainly such universal features (e.g., the self, major motivational themes, and the mix of relational needs with other motivations to pursue narrower individual interests) that seem to underlie human behavior—and often can mean the difference between a successful life and abject failure (even death)—are unlikely to be accidental features of the organism that lie outside the ken of the adaptationist hypothesis. Only an antianalytic approach would dismiss these dynamic features as being nonessential aspects of our psychological equipment. Whether or not one wishes to pursue any specific biological analysis of these features, we clearly cannot rule out the effort to understand them from an adaptive perspective.

drives—the very use of the word "biology" can be threatening. As we noted in Chapter Two, this aversion to biology has, in fact, quelled the use of biological assumptions by analytic theorists. Thus, it is quite relevant to ask if evolutionary biology is truly applicable to a psychoanalytic model of the human psyche? We must question whether a theory that suggests a relationship between human psychology and genes is too reductionistic for psychoanalysis? Are we not in danger of "rebiologizing" psychoanalysis in some pejorative sense with this evolutionary approach? Haven't we come to understand the need for psychoanalysis to stand as a psychology, through and through (Klein, 1976; Kohut, 1959, 1982), grounded in clinical data—the empathic apperception of complex mental states (Schwaber, 1979; Ornstein, 1979)—and not as forces and motivations rooted in some somatic substrate?

The Problem of Reductionism

Such concerns lead many analysts to react extremely cautiously toward any "biological" approach to psychoanalysis. In part, these objections are often associated with a narrow usage of the term "biology." Biology in this narrow sense conveys the somatic, the physiological, the biochemical, that is, the study of the physical substrate of life phenomena. In this sense, we believe critics like Klein (1976), Kohut (1982), and Schafer (1976) are correct in saying that psychoanalysis ought to be a psychology, true to the data (clinical and other forms) of observation of the human psychological "phenotype." Psychoanalytic constructs should not be derived from psychophysiological concepts. That is, as the science of complex mental states, psychoanalysis ought to be clearly differentiated from biology in the narrow sense.[4] This important point is clarified in Figure 1, the need for which was suggested by Paul Ornstein (personal communication, 1985) who noted that psychoanalysts' resistance—especially on the part of self psychologists—to the evolutionary perspective is largely grounded in this fear of a reductionist biologizing of human experience.

An evolutionary biological analysis is the study of how life was shaped by natural selection. It is an attempt to understand the *distal causes* (selective pressures) in the evolutionary history of species. These historical, functional pressures (shown on the left in Figure 1) shaped the phenotypes (the observable types) we see today. In understanding the relationship between these selective pressures and the *proximal mechanisms*—the existing mechanisms (like appetites, affects, cognitive systems) that control and shape current phenotypic attributes—we are able to understand the adaptive value of features of the phenotype. In so doing, we can perceive these features in a manner that sheds new light on their nature. An evolutionary biological analysis, then, is the attempt to find the *distal causes* for *proximal mechanisms*. It is not an attempt to *reduce* the psychological into the somatic.

Using Figure 1, we can see where the confusion develops. Arguments have been made (Kohut, 1959, 1982) for what one could term a psychoanalytic *reductionist barrier,* the call for a psychoanalysis to be a "psychology, through and through." Yet, no modern analyst would claim that there is no relation between psychological, relatively experience-near constructs, and extremely experience-distant psychophysiological constructs. However, despite significant recent strides (Winson, 1985; Reiser, 1984), the crossings of the reductionist barrier continue to be limited in helping to understand *the psychoanalytic field.* Mixing neuroscientific observation and theory with psychoanalysis can obscure and distort the data and constructs derived from analytic observation. The danger, as we view it, is that the clinical data of experience and meanings can lose its primary status as the arbiter between competing theories, becoming a secondary source of authority behind the attempt to develop interesting neurophysiological constructions.

---

[4]The point is that we cannot even begin to capture the complex mental states that we consider clinically relevant in the terms of modern, physiological biology. Complex mental experiences can only be known through equally complex observational mental processes. At this point in time, physiological descriptions of subjective experience come as close to capturing the essence of the experience being observed as a chemical analysis of pigments comes to capturing the experience that one may have when viewing a great painting (see Kohut, 1977).

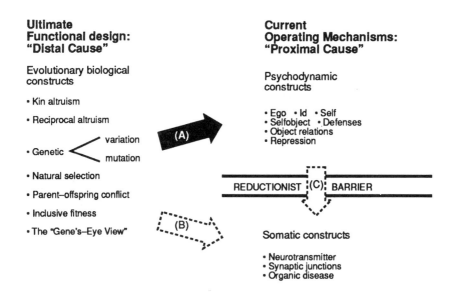

FIGURE 1. The problem of "reductionism" in evolutionary explanations. Adapted from Kriegman (1988).

The solid black arrow (A) in Figure 1 represents the nonreductionist, nonsomatic relationship between evolutionary biology and "pure" psychological constructs.

In an evolutionary analysis of psychological phenomena, the "reductionist barrier" not only does not need to be crossed, it doesn't even need to be approached; that is, an evolutionary approach can be used to guide either a psychophysiological investigation and/or what is called a "pure psychology" of experiential meanings that utilizes no somatic conceptions. Both the psychological and the somatic must "fit" into the evolutionary biological perspective (arrows A and B). Freud's earliest biological thinking (culminating in *The Project for a Scientific Psychology*, 1895) and the whole classical metapsychology was geared to a somatic, reductionist biology of "*proximal mechanisms*" (instincts, cathexes, and defensive forces) divorced from any "*ultimate*" or evolutionary functional rationale. Freud's later attempts to find an evolutionary, functional rationale shifted markedly away from the reductionistic attempt to create a proximal neuroscience of the mind to a search for ultimate (distal) historical and functional explanations. Though *Totem and Taboo* (1912–1913) and *A Phylogenetic Fantasy* (1915d) (and the other briefer references discussed in Chapter Two) were based on an inaccurate and incomplete understanding of evolutionary theory, they represent a very clear shift from Helmholzian, physicalistic neuroscience toward a Darwinian (historical) approach to the origins of the mind with emphasis on social selection pressures.

In this book, we seek evolutionary (ultimate, historical) explanations for psychological constructs (arrow A only); we do not attempt to reduce the psychological into the physiological (arrow C). Psychoanalytic theory has fostered major advances in our understanding of human behavior by providing a language and theoretical framework that enables us to envisage some of the patterns and structures involved in the internal workings of the human mind. While the patterns and structures referred to in any form of psychoanalytic metapsychology are hypothetical constructs (Turner, 1967), for the purpose of an evolutionary analysis they can be considered as real as the physical structures that we readily analyze in evolutionary terms. Mental structures and behavior patterns, as well as physical structures, can be profitably conceptualized in terms of their adaptive functions (i.e., their effect on inclusive fitness). For example, intelligence is a hypothetical construct. Given sufficient respect for the complexity of the construct and the range of the forms that intelligence may assume (Gardner, 1983; Bruner et al., 1976), we can speak about the adaptive value of intelligence just as we try to understand the evolutionary development of upright posture or the opposable thumb. Though intelligence is a psychological construct, we do not need to refer to its physiology in order to apply an evolutionary analysis to the history and function of intelligence. What is biological to the evolutionist includes but is not equated with or restricted to what is physical or material (Mayr, 1983b).

The dotted arrows (B and C) in Figure 1 refer to other types of evolutionary and nonevolutionary biological explanations, which we shall not pursue in this book. Arrow B represents evolutionary biological explanations of molecular mechanisms that represent the biochemical substrate of higher level adaptive structures, for example, the evolution of the neurotransmitters. Arrow C represents neurobiological (reductionist) descriptions of the mechanisms that comprise the somatic substrate for higher level psychological phenomena, for example, attempts to explain the collapse of self-esteem found in depression in terms of biogenic amine activity (Kaplan & Sadock, 1988). Thus, following the solid arrow (A) we can consider the evolutionary development (selective pressures, adaptive advantages) of ubiquitous patterns and structures in the human psyche, for example, repression (Slavin, 1985, 1990) or empathy (Kriegman, 1988, 1990) without endangering the empathic stance, the "purity" of the clinical data, or any danger of reversion to 19th century physiological biologizing of the psyche.

The Problem of Genetic Determinism

We assume that human psychology has some basis in our evolutionary history. What we pursue is not a simplistic "evolutionism" in which human traits are seen as unfolding from a prerecorded set of genetic instructions. Such overly deterministic attempts to explain the psyche in terms of genes—such as the early "sociobiology" of Wilson (1975), which is often erroneously equated with the field of sociobiology as a whole—end up bypassing much of the complexity of the human mind and eliminate psychodynamics (see also box, "The Limits of the Adaptationist Perspective," p. 58). As one of the most forceful and influential critics of sociobiology noted,

> The main criticism that I level against [a certain type of] sociobiology is that it introduces evolutionary considerations in the wrong way by focusing on the behavior and not on the underlying mechanisms. (Kitcher, 1987, p. 91)

We believe that like much of the tradition of academic psychology (and for many of the same reasons), the focus on *behavior* (bypassing psychodynamics) characterizes many sociobiological views which end up adding little of direct interest or utility to a psychology of human inner life, and especially little of clinical interest. We briefly note several such studies throughout this volume because they raise important questions and sometimes provide extraordinarily interesting data that, nevertheless, need to be translated into a dynamic model of human subjectivity. Yet, any evolutionary biological analysis—whether dynamic or behavioral—is at such a level of theoretical remoteness from the inner workings of the psyche that it cannot be used to derive a psychology. As we view it, the development of a model of the psyche that effectively deals with the organization of human subjectivity remains the proper province of dynamic psychology or psychoanalysis.

Looking at Figure 2, you can see psychodynamic "deep structure" as the crucial set of proximal mechanisms that mediate between genes and psychological experience. We are referring to those innate, universal psychodynamic structural features (e.g., the inherent design needed to fully develop a self or an ego, repression, primary object attachments) that must exist in our psyche in order for it to be capable of processing interactive, developmental experience and to use this experience to build a functioning, adult identity. *A crucial feature of our model lies in its focus on inner, dynamic structures and processes as the targets of selection pressure, rather than on the direct selection for specific overt behaviors.* Though in some senses it may be useful shorthand to understand whole patterns of behavior (e.g., toddler night awakenings, see

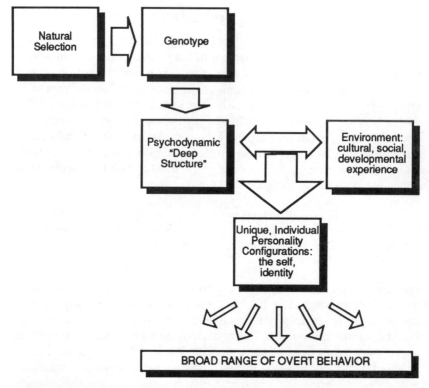

FIGURE 2. The path of determinism in evolutionary explanations. Adapted from Kriegman & Slavin (1989).

box on page 65) as potentially having, somehow, been directly shaped by natural selection,[5] in our view it is vastly more useful to conceptualize intrapsychic deep structure as a system for organizing experience (perceptions and affects) that, depending on the current environmental context, will generate a vast range of overt behaviors.

Evolutionary biology can provide a perspective on the nature of the world in which the basic functional design of the psyche (its deep structure) was shaped over evolutionary time. Evolutionary theory can help us understand the relationship of the deep structure of the psyche

---

[5]See Appendix E, "Specific, Functional, Affective, Motivational Systems from an Evolutionary Perspective," for a discussion of a need for an understanding of subsidiary "programs" or strategies that operate under an overarching psychic structure. An understanding of such subsidiary programs may make use of some of the suggestive findings of a more behavioral evolutionism.

A NONDYNAMIC EVOLUTIONARY EXPLANATION FOR A COMMON CHILDHOOD "PROBLEM"

An article by Blurton Jones and da Costa (1987) that recently appeared in the evolutionary biological literature made the case that the common, widely observed "night waking" behavior of 1- to 3-year-old children may be understood as a particular kind of adaptive, behavioral strategy. They assume that, given a chance, children who commonly wake and cry would normally suckle to soothe themselves. Yet, citing a good deal of both developmental and cross-cultural empirical data, they note that suckling under these circumstances is very *unlikely* to be directly *nutritionally* important. They conclude that it is more likely that the function of the suckling is, in fact, to stimulate the maternal breast and maintain lactation. By demanding closeness to the mother who is kept lactating and in a relatively exhausted state (in which her energy is focused on the demanding, needy infant) the behavioral strategy may, on average, succeed in preventing maternal ovulation, perhaps impeding intercourse, and, ultimately, delaying the mother's next pregnancy.

While by no means a surefire way of preventing ovulation or intercourse and thus impeding the conception of a sibling during the toddler's early years, there is well-documented cross-cultural evidence that, in general, continued suckling does have this effect on the maternal reproductive system. More crucial, perhaps, for us to keep in mind is that in the condition of primitive hunter gatherers (the condition during which we lived for something like 99% of our history) it has been demonstrated that "lengthening the interbirth interval from two years to four years reduces mortality (for the older infant) from over 70% to around 10%" (Blurton Jones & da Costa, 1987, p. 137)! Such an increment in survival is an enormously powerful "selection pressure." Since by delaying the next birth the present child's chances of surviving are much better, the authors speculate that natural selection favored the development of an innate disposition in children between the ages of 1 and 3 to frequently awaken at night demanding to be suckled and comforted.

Clearly there is much in this kind of evolutionary reasoning that may confuse or disturb the contemporary psychoanalytic observer. The intricacies of the process of attachment and separation-individuation are being worked out in their most heightened form at precisely this 1- to 3-year-old period. Do not the longing for proximity to the mother, for "holding," the desire (in drive terms) for the innately pleasurable affects associated with suckling provide a sufficient range of analytic explanation of the motivations and intentions of the child under these circumstances? We believe that, in fact, those analytic formulations are, ultimately, not enough (Blurton Jones and da Costa try to consider them and dismiss them).

We do not cite this study to discuss the complexities of innate strategies of parents and children at this point, since we do so in Chapters Four and Five and then throughout the rest of this book. We cite Blurton Jones and da Costa's hypothesis about the toddler's innate strategic wisdom, as it were, to point out (1) how despite an intriguing evolutionary hypothesis and some interesting anthropological data, some evolutionary explanations make huge causal leaps from genes to large complex, overt patterns of behavior; (2) these leaps amount to a kind of phylogenetic behaviorism that will often alienate an analytic audience; and (3) inferential leaps

> over the complexities of human subjectivity and the related tendency to fragment the psyche into discrete bits of behavior must (and we believe can) be avoided by an approach that focuses on how natural selection may have targeted the child's inner dynamic structure. In short, the evolutionary perspective may greatly enhance our appreciation of how phenomena that are problematic (or pathological) from the parent's point of view may, for the child, represent complex, evolved strategies to garner and preserve maximal parental investment (see Chapter Five for a full discussion of this issue).

to the ancient environment in which this deep structure evolved and those aspects of the current environment that this design must enable us to "read," evaluate, and respond to in a fashion that relates to the pursuit of our interests. An exploration of the functional design of the human psyche—the relationship between our psychological deep structure and the environment of evolutionary adaptedness—can allow us to evaluate and compare psychoanalytic paradigms from their foundations up: where each contains many specific (if implicit, or even disavowed) assumptions about the nature of the object world to which the human psyche is adapted.

When we return to the concept of "deep structure" as depicted in Figure 2, we shall see that it provides a useful way of conceptualizing an "evolutionary psychoanalysis," minimizing some of the dangers of somatic reductionism, genetic determinism, and the loss of the central importance of clinical data. But, first let us look briefly at a related caveat.

Genetic Influences versus Culture

A frequent objection to applying evolutionary explanations to humans is the assumption that new, nonevolutionary forces began to operate once our large brained intelligence evolved. Surely, we are told, since we are the only species that uses language and can be dramatically influenced by culture, adaptive functional explanations—applicable to other species in which behavior is more reflexive and less influenced by experience—may have little to do with many of the actual patterns of behavior (and the related inner psychological experiences) exhibited by humans (Gould, 1983).[6] Further, such critics note that much human behavior appears to be patently maladaptive. Over and above issues of psychopathology, if we consider nationalism, political ideology, certain

---
[6]See also the related discussion in the box on p. 58.

religious beliefs, other customary rituals and beliefs—all of which can appear in quite healthy individuals—they seem, at times, to offer no conceivable adaptive advantage. Indeed, such beliefs are often destructive to those that hold and act on such views (Kitcher, 1985).

Although part of this claim is certainly true—our species is more profoundly influenced by the environmental vicissitudes of developmental experience than is any other species—we must not lose sight of the fact that (1) such influences are unlikely to eliminate the influence of our evolutionary past, and (2) the very way in which experience can shape our psyches has, itself, been shaped by natural selection (Mayr, 1988).[7] Once freed from reflexive, tightly programmed reactions to the environment, developmental and cultural experiences can certainly shape behavior that is maladaptive. Yet, as Badcock (1986) and Alexander (1979) point out, such observations in no way contravene the view that (1) the tendency to be influenced by cultural experience must rest on a complex set of innate capacities, and (2) these innate capacities may well include universal expectations and predispositions that, on average, are quite adaptively advantageous yet can be overridden and exploited by other individuals, groups, and institutions in ways that undercut their inherently adaptive potential. We can examine both constructive (adaptive) tendencies, such as the healthy need for idealized others, values, and beliefs, as well as examine what happens when such tendencies go awry, as can be seen in regard to such needs in fanatic religious cults (Kriegman, 1980; Kriegman & Solomon, 1985a).

The essence of this critique of evolutionary applications to human psychology is based on the notion that because of the linguistic, cultural, and experiential influences that shape the human psyche, what we glean from ethological studies of animals with more fixed "closed programs" (Mayr, 1974) should not be applied to the human situation. Indeed, as do many discussions of evolutionary concepts, we at times draw on ethological data. Of course, data from other species must be applied in careful, qualified ways to our own species. However, observation of animal behavior across a great range of species can serve to illustrate and test hypotheses. It is not the *data* from other species that can be applied to humans, but rather the logical structure of the theory of evolution—a logical structure that can be known through the comparative study of many species.

When we refer to "reciprocal altruism" among fish (e.g., in Chapter Four), we are not suggesting that people behave as fish. Rather, we are suggesting that evolutionary principles (e.g., the adaptive value of a behavior pattern in regard to the "fitness" of the organism), which may

---

[7]See also the box on p. 70.

be illustrated by examples from other species, also apply to humans. Unquestionably, evolutionary theory is often misused; behavior patterns in another species are taken to indicate that such patterns must exist in humans. This misuse justifies a degree of caution in the application of evolutionary data and theory to human psychology. However, we must also be cautious and scrutinize our readiness to dismiss or deride observations of our close connection to the world of nature.

Evolutionary theory, as Freud (1917) well knew, represented a massive "blow to human narcissism" (p. 140). As we well know, narcissistic injuries will be vigorously defended against. Challenges to the boundaries of one's own, or group, identity will typically be met with vigorous elaborations of one's essential distinctness and, often enough, one's inherent superiority. As Alexander (1979) and Ruse (1983) note, there is a long, venerable history to precisely such attempts at distancing ourselves from the rest of nature—before Darwin and only slightly diminished since Darwin's theory. We suggest that the emphasis on culture (or language, morality, religion, tool use, curiosity, etc.) as essentially separating humans from the rest of the animal world—thus making evolutionary explanations less applicable to people—may, to some extent, be motivated by this type of defensiveness. While retaining a critical stance toward evolutionary explorations, the psychoanalytic reader should be especially attuned to the variety of conscious and unconscious ways in which vital, highly charged issues of our "human identity" are at stake in the kinds of exploration on which we have embarked. Ultimately, the defensive distortions that may be entailed in the protection of our "human narcissism" can only be revealed and tempered by a readiness to listen to and (echoing Kohut) vicariously participate in other perspectives on the larger world of nature.

The Concept of Evolved Deep Structure

Returning now to the concept of deep structure, let us elucidate a crucial bridging concept that may help to span the gap between human subjective experience and other forms of "objective" insight into our nature. Let us draw a clearer picture of what we are calling the "adaptive design," or "the evolved deep structure" of the human psyche (Slavin, 1985; 1990; Kriegman & Slavin, 1989, 1990; Slavin & Kriegman, 1990, 1992). Originally inspired by the work of Chomsky (1972) and expanded with a major set of new data by Bickenton (1990), our notion of "psychological deep structure" parallels, yet radically extends, the idea that innate structures underlie linguistic ability; there is a complex set of innate, *universal* processes, essentially built into our mind as a species that render us capable of learning and using *specific*, culturally based languages (see also Ogden, 1990; Yankelovitch & Barrett, 1971). The underlying human capacity that enables us to learn to encode and process meanings using a specific language is a built-in, adaptive design—an aspect of the innate deep structure of our minds. Similar to the way in which Chomsky (1972) has viewed our capacity to learn language as necessitating an *innate* linguistic ability to apprehend complex rules, subtle meanings, and grammatical structures—without specifying the actual language itself—the capacity to be shaped by a social environment is clearly hard-wired into the human psyche, even though the specific content of this socially shaped psyche is, to a significant degree, variable and open ended.

In recent decades, numerous developments in psychoanalysis (Stern, 1985; Emde, 1980; Bowlby, 1969; Ogden, 1990; Bollas, 1989), perceptual, linguistic, and cognitive theory (Minsky, 1985; Fodor, 1983; Bickenton, 1990), neurobiology (Winson, 1985; Edelman, 1987), and, especially, evolutionary biology (Tooby & Cosmides, 1990; Mayr, 1983b; Dawkins, 1976; Hamilton, 1964; Trivers, 1985) combine to make a convincing case for the fact that the human, innate linguistic capacity is but a part of a much broader, innate, inner "ground plan," or complex, preexisting psychic architecture that regulates many of our key interactions with the world and guides the process of organizing experience, including such processes as the creation of a subjective sense of self. As we noted previously, "The world is itself complex in ways that are not . . . analyzable or deducible without an enormous amount of a priori knowledge: in order to solve a task, you must already know a great deal about the nature of the circumstances in which the task is embedded" (Tooby & Cosmides, 1990, p. 12).[8]

Fundamentally alien to British and American environmental-de-

---

[8]Plato attempted to develop a logical argument similar to this evolutionary conception of necessary innate knowledge. In the Dialogue, "Meno," Plato sets the stage for Socrates to discuss this issue by having Meno ask:

> How will you try to find out something, Socrates, when you have no notion at all of what it is? If you meet it by chance, how will you know *this* is that which you did not know? (Plato, Dialogue 9: Meno)

This need for some form of a priori knowledge in order to "know that we know," and to "know what we know" appears to have been a cornerstone on which Plato built his argument for the existence of innate knowledge itself. Though he, of course, lacked any conception of the historical (phylogenetic) and organic (genetic) process by which information can become encoded into the mind, Plato's logic is in some ways consistent with what we now know about infant development and how the human mind creates structure from sensory perception.

DOES "HARD-WIRING" REDUCE ADAPTIVE FLEXIBILITY?

In order to examine the relationship between "hard wiring" and psychological flexibility, let us use a metaphor from the world of artificial intelligence. For simple "closed programs," relatively simple computers with simple information processing instructions built into the hardware can provide standard reflexive responses to differential input. However, for complex "open programs" that try to simulate human intelligence with its interactive flexibility and capacity to learn from experience, a much more powerful computer, with a great deal of additional hardware and much more complex sets of a priori instructions for processing information, is necessary. Thus, somewhat paradoxically, we see that greater flexibility and the capacity to learn requires, along with a greater "capacity," a greater amount of "hard-wired" pre-set basic rules for information processing. The same principle can be gleaned from the phylogenetic tree: as one moves from reflexive responses to complex flexibility that is highly dependent on situational nuances and individual history, one moves not away from "hard-wiring" per se. Rather, we move from simple nervous systems toward greater, more complex and different types of hard-wired rules. It is a mistake to equate "hard-wiring" (a relatively fixed, predictably reliable "deep structure") with reflexive inflexibility (Mayr, 1974). Humans are "hard-wired" for a relatively invariant deep structure that makes possible adaptive variability and flexibility (Tooby and Cosmides, 1989).

terminist thought—in which the mind has been conceived of as, if not a blank slate, then an initially undifferentiated entity that is patterned by individual experience—the notion of a priori knowledge (arising prior to experience) is very difficult to grasp (Mayr, 1988). Epitomized in Locke's belief that "No man's knowledge can go beyond his experience," the intellectual tradition in which most of us were educated is especially hard to integrate with this notion of preexisting, innate knowledge. An understanding of those aspects of our mind that are patently and powerfully shaped by experience, or socially constructed in a given relational context, is usually fitted awkwardly, at best, with attempts to delineate our legacy of universal, a priori knowledge. Let us turn now to see if we can develop a clearer understanding of the concept of "evolved deep structure" that can help to bridge this conceptual abyss.

Observing Deep Structure: Infant Research

Approached from the direction of infant research, we now recognize that a child must be born with a great deal of preexisting knowledge about the world in order to key into certain kinds of information while ignoring the vast amount of less relevant stimuli that impinge upon us at every moment. As Stern (1977) points out, the fascination with the human face for the newborn infant far exceeds its interest in other, equally complex visual patterns. Also, as Ogden (1990) notes, the existence of inborn capacities to recognize the "constancy" of entities and make inferences or deductions about their behavior in the world can be seen as underlying later, complex developmental steps, or even, perhaps, representing another, primordial, intuitive kind of knowledge that must precede what is learned, or inferred, on the basis of experience. For example, the fact that an 8-week-old infant anticipates the continuing movement of an object that, for a moment, has temporarily disappeared behind a screen indicates a certain "intuitive" sense of the continuity of objects from birth. Such preexisting conceptions of objects may be necessary in order to carry out the complex interactions and inferences about the nature of the world that are integral to the development of more complex experiences of object constancy at 8 months, 18 months, adolescence, and probably on into adulthood.

Implied Deep Structure: Psychoanalytic Theoretical Models

Quite distinct from the quest for innate deep structure that characterizes current, empirical infant research is the more speculative tradition of psychoanalytic assumptions about the innate tendencies, structure, and design of the psyche. This tradition of speculation about deep structure is vastly more ambitious and it has important consequences because, as we noted, assumptions about deep structure are directly built into the basic narratives of the mind that underlie the major psychoanalytic schools or models. Within the systematic elaborations of, as well as the briefer appeals to, evolutionary theory (see Chapter Two), we can find specific assumptions about the innate substrate that humans bring to (and through which they structure) experience. Let us briefly look at three such psychoanalytic sets of assumptions about the deep structure of the human psyche.

Freud

If we recast our discussion (Chapter Two) of Freud's (1905, 1912–1913, 1915d, 1917, 1923, 1926, 1939, etc.) evolutionary thinking using the term "deep structure," we can see that, though he did not use the term, he was, in fact, engaged in speculation about the nature of the deep structure of the psyche. The psyche—especially in the later drive/structure model—was essentially built on a deep structure rooted in what he conceived to be the highly conflictual dramas of our evolutionary history (Sulloway, 1979; Ritvo, 1990; Slavin & Kriegman, 1988; Slavin,

1988). Many aspects of our psychodynamic structure represented, for Freud, an "archaic heritage" that was seen as the "precipitate" of historical, adaptive challenges, including invented, relational scenarios like the dramatic conflict in *Totem and Taboo* (1912–1913). According to this account of the origins of our superego, Freud envisioned a group of brothers killing and eating their father—the tyrannical "primal father." Their guilt that, somehow, was then inherited by future generations was viewed as the result of this traumatically formative phylogenetic event.[9]

But, beyond the specific drama of *Totem and Taboo*, Freud strongly believed that, over and above the particular experiences of the individual, certain predispositions and expectations about relational reality (deriving from ancestral experience) were built into the psyche. These universal ways of organizing experience might be represented by fantasies [e.g., parricide or castration (1913)] or affective signals [e.g., guilt (1912–1913, 1915d, 1923) anxiety (1915d, 1926)] or particular instinctual pleasures [e.g., the infant's innate desire to suckle at the breast (1905)]. It was not primarily the *specific* contents of the *individual* superego or its intensity that Freud viewed as the expression of evolved deep structure. Rather, it was the *universal* predisposition in our nature for such intrapsychic entities to arise—and the capacity to use them to process experience—that Freud viewed as part of our "archaic heritage." In Freud's view of mind, "something inherent, wired in, prestructured, is pushing from within" (Mitchell, 1988, p. 3) and is brought to all subsequent experience.

Inferences about reality that could not be attributed to the child's actual experience, such as the "Wolfman's" (1918) apparent knowledge of parental intercourse without having observed it, indicated to Freud more than the existence of innate, intuitive, "a priori" knowledge. In addition, such inferences suggested the function of such knowledge in penetrating the covert or deceptive appearances of the relational world to illuminate "phylogenetic truths"—basic, universal truths of which the child may

> catch hold ... [when] his own experience fails him. He fills in the gaps in individual truth with prehistoric truth; he replaces occurrences in his own life by occurrences in the life of his ancestors. (Freud, 1918, p. 97)

Freud did not seem to view this primal knowledge, so to speak, this "archaic heritage" as simply a vestigial phenomenon in the sense of an atavistic remnant of a prior adaptation that no longer retains a real,

---

[9]For a discussion of Freud's Lamarckism, see Appendix B, p. 290.

adaptive value in the contemporary relational world. Although he was never able to develop a clear or accurate understanding of the way such innate modes of organizing experience might actually work, he appears to have believed that they were tied in some fashion, like the generational conflict in *Totem and Taboo* (1912–1913), to vital truths about the relational world that, indeed, continue to be relevant to human development (Slavin, 1988). Freud's "archaic heritage" is, in short, an adaptive heritage given the limited knowledge about crucial issues that one could hope to obtain from an average expectable environment in which, as Freud deeply believed, secrecy, deception, and illusion were the rule.

Klein

At the heart of Kleinian theory it is also possible to discern assumptions regarding what is innately known about the world. These assumptions include certain relational scenarios concerning the *inherent conflicts or tensions* in object relations that will be experienced by the individual child quite apart from inferences made from its own direct experience in the relational world. Grotstein (1985) and Ogden (1990) have clarified the role played by such "innate preconceptions" (Bion, 1962) in the Kleinian perspective, that is, how the child's mind is essentially "designed" for "the creation of persecutory and dangerous objects which are re-introjected" (Ogden, 1990, p. 17) to form an inner map of the relational world.

It is crucial to recognize that although the Kleinian infant is innately involved in the paranoid (projective) creation of the dangerousness of its inner object world, the actual (external) relational world—the world in which the near delusional infant, in fact, lives—is, after all, in the Kleinian narrative, composed of objects who, however normal and civilized they may be, have psyches that are built on a concealed core of primal envy, rage, and destructiveness. This Kleinian vision of the nature of the *external* world is essentially what the infant, in its own fashion, anticipates quite accurately with its paranoid, phantasy-based, *internal* constructions. Thus, as Grotstein (1985) suggests, Klein described the rudiments of an adaptive system in which inborn schemata serve to orient the individual to realities (especially dangers) that he or she must know about despite the fact that, for whatever reasons (not the least of which are the deceptive defenses employed by the object), these dangers might be too subtle, covert, or hidden to be inferred from the child's own direct experience.

Thus, in the Kleinian view, phantasied, projective–introjective scenarios, characteristic of the child's mind may serve, in part, as an innate way of penetrating to crucial, covert, conflictual truths about the relational world. The resulting a priori grasp of certain aspects of relational reality is thus independent of what the child might conclude from inferences based solely on its actual experience of interactions within its own family. The Lockean child, so to speak, equipped only with knowledge that cannot "go beyond his experience," would be at an extraordinary disadvantage in understanding and meeting the challenges of life in the Kleinian relational world. In this sense, we can think of Klein's version of innate deep structure not as consisting of "inherited thoughts" (a vague, mystical notion) but of a "powerful predisposition to organize and make sense of experience along specific lines" (Ogden, 1990, p. 15)—lines, we must add, that actually alert the child's mind to an often covert, yet crucial core of motives, tensions, and conflicts operative in the relational world. In effect, certain constructs in the Kleinian model can be viewed as providing the needed mechanisms (or proximal causes) by which an "archaic heritage" and "phylogenetic memories" could be seen to operate in the modern psyche. Or, as the philosopher, Alan Watts (1964), put it "you are unbelievably more wise in your nature than you ever will be in your conscious thoughts."

Relational Models

As we discussed at the end of Chapter Two, relational theorists such as Kohut (1971, 1972, 1977, 1982), Erikson (1968), Fromm (1941), Sullivan (1953), Guntrip (1971), Winnicott (1965), and Bowlby (1969) tried to alter our expectations about the nature (not the existence) of innate knowledge and expectation. No less than Freud and Klein, these versions of the modern relational paradigm assume—both implicitly and explicitly—the existence of a fund of essentially *a priori* knowledge and expectation contained within the deep structure of the psyche.

To be sure, as creators of an alternate psychoanalytic vision (a far more social vision) of human nature, the relational theorists shift our attention from the dominant role of imperative, instinctual forces in the shaping of the mind to the prominence of perceptions, expectations, and inferences *based on actual, interactive experience* with objects, selfobjects, family patterns, and culture (Greenberg & Mitchell, 1983). Yet, the capacity of the psyche to form structures using the information derived from actual experience within the relational world must itself be built into the human mind from the outset (see the box on p. 70). In addition to the highly complex, preexisting capacity necessary to make even simple inferences from experience, individuals must be equipped with an immense amount of expectation and anticipation about the relational world (and a predisposition to interact with it in certain ways). Such innate expectations, combined with inferences from direct experience, serve to guide the extremely complex process of creating a representational world that includes an inner working model of relational reality.

If we "deconstruct" the assumptions implicit in much of the relational tradition, we find that the relational model has redefined the nature and content of the a priori—the intuitive knowledge and capacities that were assumed to underlie anticipation and expectation about relational experience. Because, as we discussed in Chapter Two, they viewed the normal, relational world as characterized by a fundamental harmony and mutuality among individuals and between the generations—a mutuality that at least for "sunnier" relational theorists (such as Kohut, Erikson, and Guntrip) did not invariably conceal a hidden core of clashing, inherently conflictual realities—they viewed the deep structure of the child as designed simply to anticipate and elicit this "mutuality," "attachment," and "attunement." The relational child, so to speak, does not need to function in the same way as the Freudian or Kleinian child for whom conflict and covert danger are always a critical dimension of developmental reality to be regularly reckoned with. While the relational child may have to protect itself, self-protective measures can safely be *responses* to a dangerous or failing environment. Only for the Freudian or Kleinian child do the dangerous often hidden truths in their world have to be *anticipated* by means of primal phantasies, or drawn somehow from a phylogenetic archaic heritage.

Thus, while the notion that we possess complex forms of preexisting knowledge about the world has found considerable support in the observations of psychoanalytic infant researchers, specific theoretical inferences about the nature of that knowledge—the way it is designed to anticipate and fit in with the fundamental nature of relational reality—raise far broader questions about which different theorists come to quite contrasting conclusions. These contrasting conclusions are rooted in the narrative structures underlying the two major psychoanalytic traditions. The most prominent elements in these narratives consist of assumptions about how individual interests inherently converge or compete (the origin and nature of conflict) and assumptions about the prominence of a covert dimension of deception and self-deception in relational reality.

What is critical is that we realize that there is no way to separate our assumptions about the nature of a priori *inner* knowledge that exists in deep structural form from what we assume to be the features of the *external* relational world that our psyche is designed to fit—that is, the nature of the relational problems or tasks that can only be perceived and solved through the use of preexisting structures. In short, the Freudian child is, by and large, designed "to fit" and function in the same inherently conflictual, deceptive relational world for which the Kleinian child was designed. Neither of these children would function well in the essentially mutualistic world with its valid perceptions and communications conceived of by most relational theorists. And, of course, the relational child would be at a significant disadvantage in the Kleinian or Freudian world. In psychoanalytic discussions of innate deep structure, we must be able to spell out our underlying assumptions about the complexities of the "average, expectable" relational world for which a priori knowledge is designed to serve as our inner guide.

Two Components of Deep Structure: Individual Variation within a Universal Architecture

There appear to be two components of the preexisting inner design that each individual brings to his or her life. The first is the universal, invariant deep structure that all humans share; the second are those *unique* elements of *individual* psychic structure, disposition, and attributes that vary from individual to individual. We believe that decades of research have now shown rather convincingly that, just as there is clearly genetic variation in eye color, leg length, physical strength, susceptibility to certain diseases as well as resistances to such, etc., there are genetic variations in personality characteristics.[10] In studying and reviewing the many studies of identical twins reared apart, Neubauer and Neubauer (1990) concluded:

> ... that in the earliest observations of the first few weeks of life the similarities between separated twins reared apart were greater than we expected, and the differences in parenting, though always considerable and sometimes powerful, did not have the effect we imagined we'd find.... It appeared that an infant's unique way of engaging the environment and making it respond to his needs—plus the stability of these features throughout life—was influenced by genes and not only created by mother, father, or school.... Of course, the significance of nurturing isn't minimized by considering the role of genes.... [T]he importance of providing appropriate care, the

---

[10]The definition of what such heritable characteristics consist of is an extremely complex issue. We in no way assume that personality "types" (or types of pathology) represent meaningful heritable units, that is, can be meaningfully understood as inherited in any global way. Indeed, our whole approach in this volume is based on the view that what we inherit (as unique individuals and in our shared, universal inheritance as a species) are largely internal structural and dynamic features (ways of organizing and disguising experience, sensitivities, vulnerabilities, etc.) that through interaction with a given social environment will be manifested overtly (behaviorally) in a whole range of different ways (see Figure 2, p. 64). This complicated topic will return as a leitmotif throughout this volume.

power of human interaction, and the positive effects of therapeutic intervention are all alive and visible.... [Yet,] we must... understand that we respond to the environment in ways biased by our innate makeup, and furthermore, that this makeup helps select the environment we respond to. (pp. 6, 8)

Again, using the analogy to physical attributes, we can mix genes influencing different eye colors, leg lengths, physical strengths, etc., and have a unique, viable individual with unique strengths and weaknesses. However, there are *universal, invariant* deep structural aspects to physical features. For example, we expect a failure to thrive when there is variation in the number of eyes, where the longer or shorter legs are attached, how cells will differentiate into different organ systems, etc. There is a universal structure to physical characteristics within which there is genetic variability.

> Of the six *billion* base pairs of DNA in human chromosomes, unrelated strangers differ in only six *million* of them. The vast majority of our genes, then, are shared, identical in each person and representative of our species... Even brothers and sisters—conceived by the egg and sperm from the same sources—differ in two million base pairs of DNA, and depending on the combination of their genes may appear strikingly similar to each other or astonishingly different. (Neubauer & Neubauer, 1990, pp. 18–19, emphasis added)

Just so, we expect a universal architecture of the deep structure of the human psyche within which there will be individual variations. In fact, we would expect that a major feature of the universal deep structure must consist of mechanisms or strategies for dealing with inevitable individual variation. As the evolutionary theorists Tooby and Cosmides (1990) note:

> Every generation, each developmental or psychological mechanism finds itself born into a genetically and biochemically unique individual, whose precise properties could not be predicted in advance. The mechanisms that comprise an *individual's* psyche, must discover "who" and "what" he or she is, which impulses and perceptions to trust, and which to disregard. Because [universal] psychological mechanisms are inherited from individual to individual, from generation to generation, they must be designed to promote adaptive outcomes in spite of the fact that they are sequentially embedded in an unpredictable series of genetic backgrounds, environmental conditions, and situations defined by genetic-environmental interactions. This variation is a persistent property of the evolutionary landscape, and, as an enduring selection pressure, should call forth

adaptations that allow the psyche to function properly in spite of this unpredictable variation.

This selection pressure creates a systematic difference in how machines and humans are engineered, even at the level of motor control, for example. A precisely engineered robot can have a hard-wired guidance system because its guidance system can rely on all of the parts fitting an exact design specification. A human cannot have such inflexible guidance programs, because the length of one's arms, the strength of one's muscles, the acuity of one's eyesight, the speed of one's reflexes, and so on, will all shift with the life span, and especially from generation to generation . . .

[A]n evolved psychological mechanism . . . can rely on its own components and on other components in the psyche to emerge in only an approximately adaptive arrangement, rather than in absolutely reliable fixed relationships. . . . For this reason, psychological mechanisms should have the ability to tune themselves, by adjusting their properties, weightings, levels of activation, and so on. (pp. 56–57)

As we show in Chapters Four through Seven, evolutionary theory suggests that this recognition of unique individual elements and the "tuning" of the psyche to maximize the adaptation of each unique individual to a specific, unique environment occurs in a complex, ambiguous relational context, one that includes both overlapping interests (genuine love and nurturant care) as well as inherent conflict. Thus it is not a matter of indifference to others how this tuning process occurs. The inner tuning becomes, in part, a social act. The vitally needed relational surround, itself, contains others whose interests both overlap and conflict with those of the individual. Individual differences occur and develop within a larger context of *universal, internal* deep structure—which in turn, functions within a specific, *external*, relational context which, itself, contains both *unique and universal* features. We shall attempt to show how the complexity of the adaptive challenge this entails necessitates some conception of evolved, universal deep structural elements that serve to maximize the individual's ability to utilize his or her unique characteristics in order to protect and promote his or her unique interests.

In this view of the development and maintenance of intrapsychic structure, the basic (universal) organization of the self and its motives will be seen as an individual's elaboration of a *universal*, innate, human strategy for negotiating *universal relational* dilemmas in *specific environmental* contexts drawing on that *individual's unique* attributes. Most directly, the relevance of this issue for psychoanalysis lies in the highly evocative yet vaguely defined concept of the "true self" (Winnicott,

1960a), "nuclear self" (Kohut, 1984), or "authentic self" (Fromm, 1941). In our view, the dynamics surrounding these notions of individuality need to be understood as a crucial part of our evolved deep structure. Though we shall not directly address the question of the "true self" until Chapter Eight, the whole of Part II (Chapters Four through Seven) is devoted to an exploration of the average, expectable, good-enough environment as seen and defined in contemporary evolutionary theory. It is in the context of this environment—this relational world—that the dynamics of a "true self" would be designed to operate. Specific families, cultures, etc., represent variations on this universal context.

An evolutionary view of the psyche thus clearly revolves around the complicated issue of how we define the nature of these presumably "universal, relational dilemmas" that our mind is designed to be able to know and manage. How does the long-range, historical vantage point of the evolutionist alter our view of the nature of the recurrent, ancestral situations that we are designed to be able to know? *In short, what is it that is adaptively crucial for us to anticipate about relational reality?* Approaching these questions in a way that relates directly to psychoanalytic concerns, we ask the reader to join with us as we ponder the answers to questions that we believe are central to the challenge of "comparative psychoanalysis." Ultimately, our response to these questions shall guide us toward the rapprochement of classical and relational narratives.

The Evolutionary View of the Human Psyche in the Relational World

We believe that evolutionary theory can provide a perspective on the nature of the world in which the basic functional design—the functional architecture—of the psyche was shaped over evolutionary time. An understanding of this functional design can thus allow us to evaluate and compare different psychoanalytic paradigms, each of which, as we have argued, contains specific assumptions about the basic nature of the world in which the psyche must function. Looking at the psychoanalytic comparative landscape from this evolutionary platform can help us become more aware of and clarify our biases and their impact on our observations.

In Chapters One and Two we suggested that there are biases present in every psychoanalytic position; that each entails implicit assumptions about human nature that are, in turn, based on assumptions about the nature of the relational world in which we function. The key contribution of the evolutionary perspective lies in the degree to which it can help to better clarify the *central aspects of the relational world (in psychoanalytic terms, the average, expectable, good-enough environment) to which the design of our psyche may well represent a critical adaptation.* Thus, as we noted earlier, evolutionary theory does not, in itself, provide us with a specific model of the psyche. This is the province of a psychology such as psychoanalysis. Evolutionary theory clarifies our assumptions about the nature of the relational world, and provides a set of criteria for generating and evaluating the adaptive validity of different models of the mind—the degree to which the varying models fit with our only scientific theory of creation. There are certain features of the social world that are universal for all life forms. Assumptions about relational reality that do not conform to these universal features need, at least, to be questioned.

In Chapters Four and Five, we take a close look at two important, ancient, universal aspects of all relational environments: the inherent clash of interests between individuals and their convergent or shared, mutual interests. We explore the implications of these ancient aspects of the relational world for the evolved design of our psyche—our psychic structure, motives, and expectable modes of communication within ourselves and in our interactions with others.

PART II

Contemporary Evolutionary Theory and the Relational World: The Average, Expectable (Good-Enough) Environment Reconsidered

CHAPTER FOUR

Conflict and Mutuality in the Relational World: Kin and Reciprocal Altruism

> Man, who can subsist only in society, was fitted by nature for that situation for which he was made. All the members of human society stand in need of each others assistance, and are likewise exposed to mutual injuries.
> —ADAM SMITH, *The Wealth of Nations*

> The condition of man . . . is a condition of war of everyone against everyone.
> —THOMAS HOBBES, *Leviathan*

Individualistic Competition versus Collectivist Altruism in Nature

As the noted evolutionary theorist Robert Trivers (1985) points out, the classical Darwinian conception of fitness posed relatively few significant problems for understanding the evolution of physical features of organisms. When it came to patterns of behavior, or psychological mechanisms, however, certain major conceptual problems appeared. Both psychoanalytic and evolutionary theories, each in its own way, have had to struggle with the existence of prosocial behavior that appears to be motivated in a primary way by care and concern for the interests of others. Both theories have had difficulty explaining behavior that appears to be primarily geared toward others' interests, often in ways that create a cost to the self (Darwin, 1871; Hamilton, 1964, 1969; Trivers, 1971, 1985; Kriegman, 1988, 1990). In a word, the problem seemed to revolve around an apparently irresolvable clash between "individualist" and "collectivist" perspectives on social behavior. Are the dynamics of relating to others organized around and based on the satisfaction of built-in individual need states (as in drive theory and the classical narrative)? Or, are such dynamics geared toward seeking ways of relating to others in terms that (like object relations and self psychological theories) inherently go beyond individual needs and encompass shared mutual or group goals? A close parallel exists between the evolutionary and psychoanalytic problems—and, we believe, their solution. Through an evolutionary biological lens we try to reexamine the basic psychoanalytic theoretical clash between classical and relational narratives.

Ever since Darwin's initial observations in this area (Darwin, 1858), evolutionists had become increasingly aware of a huge number of reports of highly mutualistic, helpful, or even altruistic, behavior in a vast range of species (Trivers, 1985). For example, consider the extremely cooperative, self sacrificing social insects (Trivers, 1976b). Individual bees will commonly sacrifice their lives to protect a hive (Dawkins, 1976). Even non-social aphids, when diseased, will altruistically "commit suicide" rather than risk infecting their relatives (McAlister & Roitberg, 1987). Dolphins will make great efforts to help sick or injured compatriots remain near the ocean's surface (Trivers, 1985). Many birds will incur a grave risk to their own life by issuing a predator warning call to neighbors, thus attracting increased attention to those sounding the warning (Trivers, 1971). In addition, there are countless human examples of extreme self-sacrifice (Durkheim, 1897).

Explanations for behavior that often did not seem to promote the fitness of "altruistic" individuals (that is, those individuals whose acts appeared to benefit others while incurring a cost to themselves) seemed to require the introduction of the idea that evolution favored larger goals. Presumably, group goals or group success took precedence over individual success or individual fitness. This collectivist evolutionary view explained behavior that was costly to the individual, yet benefited others, as having evolved for the good of the "group" or, often, for the good of the species as a whole (Wynne-Edwards, 1962). Although systematically codified in the 1960s, such thinking had long infused evolutionary discourse. Parallel to collectivist social theories, group selectionist explanations—in which natural selection operates on groups, not individuals—viewed the individual as ultimately subordinate to the group. Thus, penguins, for example, might be seen as controlling their birthrate because having fewer offspring under overcrowded conditions benefits all penguins. Groups of penguins that do not control their birthrate are less successful than those that do.

Yet, how could the inner design of the "altruist" (the innate cooperator, the self-sacrificer) be more successful at replicating itself (surviving and leaving more viable altruistic offspring) than other, more selfish competitors? Would not the competitors, with their essentially "selfish" inner design, regularly have the evolutionary advantage? For example, a selfish penguin could take advantage of altruistic penguins' voluntary "birth control" and easily outreproduce them. Indeed, in the group selectionist explanation, individuals were expected to have been *designed to sacrifice themselves, as it were, for the larger, collective good*. The problem with this view was that except in very special, limited circumstances, the inner design (along with the underlying genotype) for a group-oriented, self-sacrificing type of individual could not ultimately survive successfully in competition with more "selfish" individuals who would quite readily take advantage of the former's self-sacrificing orientation. Thus, group selectionist explanations of social behavior ran into a major conceptual difficulty.

A leading evolutionary theorist, G. C. Williams (1966), concluded that a true understanding of such social behavior could not be based on such a collectivist model. The inner design of social organisms could not primarily revolve around such an individually costly orientation toward the benefit of others. Many biologists who recognized both the validity of this argument as well as the need for an explanation of observed "altruistic" behavior, saw a need to conceptualize an evolved design for social creatures that would work simultaneously to reliably enhance individual fitness (individual interests) and serve to promote the interests of others even when it was costly (inimical to one's direct interests) to do so. Yet, lacking such a solution, many clung to collectivist, group selectionist beliefs. Others, like Williams, insisted that Darwinian evolution was an individualistic process that simply could not regularly work on any other level. Like the dialectic between the two psychoanalytic narratives we have been discussing, evolutionary thought flipped for decades between seemingly irreconcilable individualist and collectivist paradigms (Mayr, 1988).

A solution required the clear recognition that a *phenotypic* view of this conflict between individual and group aims was inadequate. In an extraordinarily productive decade of theoretical breakthroughs, the evolutionists W. D. Hamilton (1964) and R. L. Trivers (1971, 1976b) began to recognize how, from a *genetic* point of view, the selfish pursuit of self-interest and mutualistic cooperation with others, though genuinely clashing aims, were not, in the long run, necessarily antithetical. When the vast range of data was examined, it turned out that extreme versions of individually costly behavior that were helpful to others were virtually always found to exist in the relational context of *kinship*. Such altruism, it turned out, actually benefited one's *genetic* self-interest because of the high degree of overlap between one's own genes and those of closely related others. This meant that what appeared (phenotypically) to represent a foregoing of self-interest by some individuals actually represented a net benefit to them (or, more accurately to their genes) through related others who carried copies of their genes. In other cases, among unrelated individuals, "altruism" could predictably bring benefits to the "altruist" if complex, internal mechanisms (e.g., friendship, the affective and cognitive elements of a moral system, etc.) existed to facilitate and regulate longer-term, sometimes indirect reciprocal "repayments" and exchange (Trivers, 1971). These evolutionary advances ("kin altruism" and "reciprocal altruism") enabled evolutionists to explain how natural selection, operating, as always, to favor the most reproductively favorable genotypes, could have produced inner designs that could work in both selfishly individualistic and other-directed, altruistic ways. Never satisfactory to begin with, the collectivist, "for the good of the species," group selectionist model was, by and large, discarded (Trivers, 1985; Mayr, 1983b, 1988). Let us take a closer, more careful look at the vital social evolutionary concepts that solved the thorny, longstanding conceptual problem: inclusive fitness, kin altruism, and reciprocal altruism.

Inclusive Fitness and Personal Fitness

The new clarifying conception of fitness developed by W. D. Hamilton, in 1964, resolved the problem of kin altruism by introducing the concept of "inclusive fitness." Inclusive fitness is based on the recognition that survival of copies of an organism's genes *in other individuals*, and in the resultant future gene pool for the species, is the only measure of evolutionary success or ultimate fitness. The success and survival of the *individual* is *not* the ultimate focus of selective pressures. Natural selection has shaped organisms that maximize their *inclusive* fitness, not their *personal* fitness. Although at times these dimensions overlap, the important distinction between them will become clearer as we proceed.

At this point, note that if we were to define fitness in the narrower, personal sense, then behavior that increases another's fitness while decreasing the fitness of the performer would always be self-destructive and unfit; it would create a selective pressure toward the removal from the gene pool of any associated genes. However, we can easily see that parental care, which benefits the child often at considerable cost to the parent, fits this definition and would therefore appear to be unfit—a conclusion that is obviously false. It is false because the "cost" to the parent in reduced *personal* fitness must be diminished by the degree of relatedness to the beneficiary of the parent's behavior (the child) in order to assess its net adaptive (genetic) success. Parental care may reduce a parent's ability to survive and thrive (i.e., may reduce the parent's personal fitness) while actually increasing the parent's *inclusive* fitness since the beneficiary carries copies of the parent's genes.

Since our genotype is, literally, shared with kin, the phenotypic (or overt, observable) behavior that is actually most advantageous to our genes will necessarily include the welfare of these other individuals, albeit always in somewhat "discounted" form relative to one's own interests. The fact that inclusive fitness does not refer to the *individual's* fitness per se, but rather to the survival and successful reproduction of the *genetic material* carried by the individual, simply means that inclusive fitness ultimately is a measure of one's net contribution to the success of one's relatives, especially offspring.

Hamilton's breakthrough was a mathematical formulation of an idea first presented by J. B. S. Haldane, the famous population geneticist (Trivers, 1985; Hamilton, 1964). According to an already legendary story, Haldane was sitting in a British pub when someone asked him if he would ever be willing to sacrifice his life for another on evolutionary grounds.

> Haldane is supposed to have grabbed a beer mat and a pencil and, after a few quick calculations, to have declared that he would willingly lay down his life if he could save more than two brothers, four half-brothers, or eight cousins. (Kitcher, 1985, p. 79)

Haldane (1955) published some further thoughts on this matter in an attempt to understand the way in which genes leading to kin altruism may have been favored. He considered

> a gene that led an individual to try, at the risk of drowning, to save a drowning individual. He argued that you would have to save a cousin more than eight times as often as you drown yourself in the attempt in order for this gene to be positively favored. He then did a very curious thing: having stated the general principle, he at once severely restricted its scope. He said that the two times he himself had jumped into the river to save a drowning individual, he had acted at once without considering the possible degrees of relatedness. (Trivers, 1985, p. 46)

Trivers goes on to consider the reasons for Haldane's "curious restriction." While egotism—his focus on his own heroic acts—may have been a factor, Trivers believes that the real reason may have been the political consequences of arguing:

MY FAMILY/MYSELF: DEGREES OF RELATEDNESS IN THE FAMILY

Using our understanding of genetics we can even specify the average degree of genetic overlap between individuals. Consider the following "gene's-eye" views of relative relatedness within a family.

In Figure 3, we visually represent the fact that each parent shares 50% of his or her genes with each child. But since, on average, half of the genes that each sibling receives from a specific parent will be identical to the genes a sib received from that same parent, we can also represent relatedness between sibs.

In Figure 4, we see that each sibling shares 50% of his or her genes with every full sibling. If later, one sibling, let's say the sister, marries and has children, we can also specify the degree of relatedness between her brother and her children.

In Figure 5, we see that since the sister passes one half of her genes to each of her children, on average she will also pass one half of the genes that she shares with her brother. So an uncle shares 25% of his genes with each niece or nephew. Just so, we can look at the degree of relatedness between a grandparent and a grandchild.

In Figure 6, we see that since each grandparent shares 50% of his or her genes with their daughter, and the daughter passes on 50% of her genes to her children, each grandparent will share 25% of his or her genes with each grandchild. In Chapters 4–6, we shall examine some of the implications of this gene's-eye view of relatedness.

FIGURE 3. My parent/myself.

FIGURE 4. My sibling/myself.

FIGURE 5. My nephew/myself.

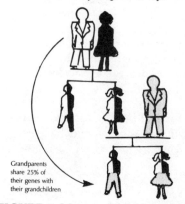

FIGURE 6. My grandchild/myself.

that selection will rapidly favor the ability to *discriminate* between intended recipients on the basis of degree of relatedness.... If this latter possibility is the reason, then it was certainly not the [first or] last time someone blinded himself for fear of the consequences of an idea. If Copernicus dethroned us from the center of the universe and Darwin from the center of organic creation, then work on the evolution of altruism has dethroned us once again, making altruism more general [among many species] and more deeply self-serving. This has been a painful realization for some, generating minor spasms of resistance to this way of thinking.

In any case, it fell to William Hamilton, a lowly British graduate student ... to appreciate the importance of this reasoning. Curiously enough his [work] ... was barely appreciated at the time. Indeed, he was told in 1963 that his work was not up to the standards of the University of London and he could not receive his Ph.D. for it. In order to gain employment, he rushed into print in 1964 a paper entitled "The Genetical Evolution of Social Behavior." This turned out to be the most important advance in evolutionary theory since the work of Charles Darwin and Gregor Mendel. (Trivers, 1985, pp. 46–47)

Hamilton's work spurred on many new investigations into kinship effects on behavior. There is now ample evidence that, throughout nature, organisms are quite adept at not only detecting kin, but detecting degrees of kinship. Even bees and tadpoles in experimentally manipulated circumstances—circumstances that made it unusually difficult to detect degrees of kinship and simultaneously ensured that only kinship detection could account for the results—could make distinctions based on degrees of kinship (Trivers, 1985).

Yet, similar to Haldane's reservation regarding the effect of kinship on altruism, we frequently face skeptical challenges to the notion that kinship is such a crucial dimension in human relations. In one clinical setting, following a presentation of the evolutionary view of kinship, a senior clinician said, "I think you are making too much of this kin/non-kin dimension of relationships. At this point in my life, friends and colleagues are a much larger part of my life than biological kin." Pointing around the room, she continued, "These people constitute my family now." To this, one of us responded by asking, "How many of these people—as opposed to biological kin—are mentioned in your will?" In getting at a similar sense of the limits of non-kin ties, Mark Twain once said:

> The holy passion of friendship is of so sweet and steady and loyal and enduring a nature that it will last through a whole lifetime. If not asked to lend money.

It is not hard to gather substantial evidence from our own lives and clinical practices to support the notion that kinship is a powerful guide for altruism. Many patients who have truly awful relationships with their family of origin will find themselves going "home" in times of material crisis (financial, health, etc.). Though there may be powerful, painful psychological strings attached, most people can count on family for substantial aid (if the family has the resources) if the situation is serious, even if the quality of the relationship is poor. It is not hard to find numerous examples of a wide range of investments directed toward related others (brothers, sisters, cousins, nieces, nephews, aunts, uncles, etc.) that we simply cannot account for based on what we know about the quality of the relationships. In some cases, the individuals may not even know each other, for example, a cousin or an uncle or aunt who helps kin whom they have never met settle into a new city, find a place to live, and find employment.

From the evolutionary perspective, important universal psychological design features were likely to have been favored by natural selection because they represented functional designs that, over evolutionary time, were advantageous—that is, they increased the inclusive fitness of individuals who possessed such types of psychological organization. The distinction between the older notion of fitness in the narrow sense (personal fitness) and the concept of inclusive fitness has broad and important implications for psychoanalytic theory.

Inclusive Fitness: The Gene's-Eye View of the Self in the Relational Matrix

In Figure 7 we present a simplified, schematic drawing of an individual in relationship to others. The left side of the figure shows the socially observed, or apparent level of relatedness—the everyday phenotypic perspective. The overlapping and nonoverlapping circles on the right show a more accurate and complete picture of relatedness from an evolutionary point of view—the "gene's-eye view."

The gene's-eye view illustrates the way in which parts of the individual genotype are literally shared with others who are kin. At a level that may be quite relevant to matters of the adaptive design of the psyche, to its evolved deep structure, the self quite literally transcends the individual. Self-boundaries include parts of other individuals. Although this may sound like a psychotic version of reality, or, if you will, the penetrating truth recognized by many in psychotic perceptions, it is perhaps most simply the biological reality that underlies Winnicott's (1952) telling aphorism, "There is no such thing as a baby, only a nursing couple." Winnicott was attempting to force us out of our cus-

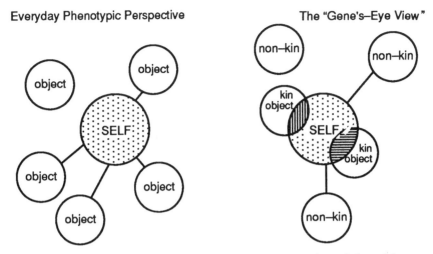

FIGURE 7. The "self": Phenotypic and genotypic perspectives. Adapted from Kriegman and Slavin (1989).

tomary (adult, Western) way of looking at individuality; in an evolutionary sense, this customary stance falls prey to the *phenotypic illusion of greater uniqueness and boundedness than in fact exists*. We argue that aspects of the "evolved self," so to speak, do appear to exist a priori, in a psychologically meaningful sense as a distinct entity, as the infant researchers (Stern, 1985; Emde, 1980; Beebe & Lachmann, 1988) have contended for some time. Simultaneously, the evolved self also appears to be inextricably intertwined within a "relational field" very much as it has been viewed in the relational narrative. As we shall see more clearly as we proceed, at its foundation, the entity we call the self cannot be a strictly individual or a thoroughly social entity. The self can, in effect, be seen as a "semisocial" entity. The implications of this view are enormous. We spell them out throughout the rest of this volume.

Kin Altruism and Inclusive Fitness: The Innate Concern for Related Others

The conception of kin altruism—that is, genetic success through the pursuit of personal attainments that enable one to *invest* in others who carry one's genes—was added to (and thus revised) the existing evolutionary framework that traditionally emphasized the central concern with simple (more "selfish") reproductive success—that is, genetic success through personal attainments that enable one to replicate the self through the *creation* of others who carry one's genes. The structure of the self is thus seen by evolutionists as innately bounded by the individual (with personal goals unrelated to the well-being of others) *and* intrinsically contained within the selves of related others. If such a self has been the essential target of selection pressures—the form that was selectively favored by the evolutionary environment—what kind of inner design (motives, dynamics) is likely to have been constructed in this evolutionary shaping of such a semisocial self? What are the evolutionary, environmental realities to which the semisocial self is likely to be adapted?

Because of the tremendous theoretical importance of the replacement of the older notion of personal fitness with the new conception of inclusive fitness, let us examine some of the implications of this transition for our view of the overall organization of the psyche. Evolutionary theory not only tells us that all life forms are essentially designed to maximize their own self-interest, it specifically defines the ultimate strivings of living organisms as, in effect, *the maximization of one's "inclusive self-interest."* This overarching principle precedes and underlies any other goals such as the pleasure principle, the search for relatedness or for self-cohesion, etc. Remember that in Chapter Three, we referred to the impact of natural selection as Hartmann's "reality principle in the broader sense." In effect, the notion of inclusive fitness enables us to define with greater specificity this overarching "reality principle"—the contingencies of the real world to which all life must inevitably conform. In light of this principle (of inclusive fitness), let us look at the issue of altruistic action (or altruistic modes of relating) as Hamilton understood the issue. Hamilton's realization that individuals must be understood as literally transcending their own skin altered the fundamental metaphors—the central narrative structure—of evolutionary thought. The Hobbesian view of "nature, red in tooth and claw," the social Darwinist dog-eat-dog, survival of the fittest perspective on the evolutionary process was retired not by some new, non-Darwinian, or anti-Darwinian theory but by this new conception of fitness that led to a fundamental deepening and broadening of the Darwinian perspective itself (Mayr, 1983b).

In other words, the alternative to the narrative of individualistic competition was not a new story line built around collectivist assumptions about the existence of a larger, normative harmony of interests in the relational world. On the contrary, the individual self, as we shall see more clearly in Chapter Five, was still conceived as having significant interests that diverged from others—interests that would fundamentally shape the perceptual and motivational structures of the self. What

THE COMPLEXITY OF KINSHIP: A CLOSEUP OF THE GENE'S-EYE VIEW

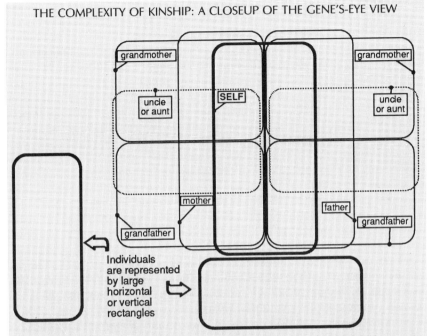

FIGURE 8. A closeup of a gene's-eye view showing the actual genetic overlap within a part of the extended nuclear family.

While the above presents a more accurate map of the basic biological/genetic overlap in the nuclear family, note that due to the extraordinary complexity, there is no way to add a sibling to this two-dimensional map. The complexity of kinship—that our psyches must be able to, to some degree, intuitively sense—is quite marked.

was radically and ineradicably changed, however, was the basic conception of that evolved self—its basic definition and boundaries. Whereas we may see distinct, skin-bounded individuals,[1] natural selection sees, so to speak, the genotypes that both underlie our skin-bounded individuality and, to some degree, are shared with related others. In short, individual selves are inherently designed to compete with others; promote and protect themselves, whenever their interests

---

[1] Any anthropologist could tell us, of course, that our Western perceptions of individuality are extraordinarily exaggerated relative to most of the world throughout human history.

(inevitably) diverge. But there are many complex distinctions to be made about who, in fact, is an "other," and especially to what degree this "otherness" exists and is relevant to the self.

Classical psychoanalytic narratives, like the older, social Darwinian narrative, have no way of accounting for expressions of human mutuality without reducing them to inherently selfish proximal aims. Freud (1915a, e) repeatedly emphasized the classical position on this issue. The capacity for concern for others based on an empathic sharing of the other's experience (e.g., in the form of compassion) was typically seen as a false facade over our true nature, that is, as a reaction formation against sadism:

> Reaction-formations against certain instincts take the deceptive form of a change in their content, as though egoism had changed into altruism, or cruelty into compassion. (1915e, p. 281)[2]

Altruism, empathy, and aim-inhibited love (Freud, 1921) were seen as achievements in opposition to our primary human nature—pleasure seeking, instinctual drive discharge; they certainly were not seen as having their own motivational source.

The classical position on inherent mutuality is particularly starkly highlighted in Freud's discussion of parental motivation:

> [Parents] are under a compulsion to ascribe every perfection to the child . . . They are inclined to suspend in the child's favor the operation of all the cultural acquisitions which their own narcissism has been forced to respect, and to renew on his behalf the claims to privileges which were long ago given up by themselves. The child shall have a better time than his parents; he shall not be subject to the necessities which they have recognized as paramount in life. Illness, death, renunciation of enjoyment, restrictions on his own will, shall not touch him; the laws of nature and of society shall be abrogated in his favor; he shall once more really be the center and core of creation—"His Majesty the Baby," *as we once fancied ourselves.* . . . At the most touchy point in the narcissistic system, the immortality of ego, which is so hard pressed by reality, security is achieved by taking refuge in the child. Parental love, which is so moving and at bottom so childish, is nothing but the parents' narcissism born again. (1914a, p. 91)

---

[2]Strachey (1957) points out that the word "pity" or "compassion" could be used here as a translation for the German *Mitleid*, which literally would be translated as "suffering with"—an excellent choice of words for a form of empathic union with another (p. 129).

From an evolutionary perspective, Freud was accurate in his observations: the parental attitude *is* a narcissistic one; the child is, in the most literal sense, a selfobject to the parents—an external object that is part of their genetic "flesh and blood," whose well-being and success unquestionably enhances a parent's self (inclusive fitness). However, the evolutionary view demands that we see humans as *inherently* social (or, as we have been using the term, inherently "semisocial") as opposed to *reluctantly* social. The classical view of narcissistic and selfish/instinctual motivations finds an echo in the colloquial expression, "looking out for number one." This clearly captures a paramount evolutionary truth: personal survival and well-being are necessary for any effective action, purpose, or goal. Yet, the goal of all evolved life forms, all motivations, must be part of a system that *in the end* enhances the fitness of *others*! Only others can extend oneself in time and space carrying forth copies of one's genes—the evolutionary conception of adaptive success. Any life form in which personal fitness served as the paramount goal would have been a very brief, aberrant mutation indeed. In this sense, the evolutionary view leads us away from Freud and the classical narrative toward the relational perspective.

Kohut's views on parental motivation provide a useful illustrative contrast, not only to Freud's particular views but to the essential sensibility regarding these issues within the classical perspective. In Kohut's view of what he characterized as Tragic Man (as opposed to Freud's Guilty Man), he saw the parent as:

> striving, resourceful man, attempting to unfold his innermost self... and warmly committed to the next generation, to the son in whose unfolding and growth he joyfully participates—thus experiencing man's deepest and most central joy, that of being a link in the chain of generations... Healthy man experiences... with deepest joy, the next generation as an extension of his own self. It is the primacy of the support for the succeeding generation, therefore, which is normal and human, and not intergenerational strife and mutual wishes to kill and to destroy—however frequently and perhaps even ubiquitously, we may be able to find traces of those pathological disintegration products... which traditional analysis has made us think [of] as a normal developmental phase... 1982, p. 403–404)[3]

---

[3]In fact, the evolutionary view suggests another major selfobject line that needs to be more fully developed: the selfobject function of the child, the student, the next generation to the parental generation. Kohut describes in detail the misuse of the next generation in the

(continued)

There is no need to restrict our consideration of the issues Kohut raises to the question of parental motivation. The challenge Kohut poses for the classical narrative in the realm of parental motivation is emblematic of broader issues concerning human motivation and its relational structure. Freud and the classical narrative penetrated as close to the heart of the biological evolutionary reality as one can get and still missed a very significant point: those activities that are biologically necessary for the minimally necessary inclusive fitness of almost all living things (e.g., parental care as the most central, dramatic example) are likely to spring from the most profound, biologically ancient motivational sources. From an evolutionary perspective, the model of the ego as adversary and antagonist fighting to suppress and repress infantile narcissism only to allow its expression when one is a parent, as the motivational source of parental love, is highly questionable. For one thing, such a convoluted motivational system would only be possible in a species with an extraordinarily highly developed neocortex, overlaid on the hungering, instinctual, tension-reducing brain. It is not a model that is consistent with the universal parental investment made by thousands of other species at virtually every level of evolutionary complexity.

It is unlikely that the basic motivational forces that lead to the direct expression of parental care in other animals have completely disappeared from the evolved design of our psyche and have no role in human parenting. It is thus quite unlikely that our motivational system is designed to generate mutualistic concerns for others entirely or even primarily through a regressive reversion to infantile motives or by way of reaction formations against antagonistic impulses that are, in some sense, more primal. On the contrary, one would expect to see at least the "residue" of such widespread, evolved designs for parental investment in humans. It is unlikely that homo sapiens would completely drop an adaptive mechanism that was used in our relatively recent evolutionary

---

service of the parent's pathological narcissism, but only briefly mentions, and does not elaborate, the self-enhancing joy of sharing in the next generation's growth and development. That is, he does not focus on the parent's healthy narcissistic use of the child. (Of course, this is probably due to the fact that self psychology, like classical theory, developed as a clinical theory.) A related phenomenon occurs in a properly conducted analysis wherein the analyst's self is enhanced when the patient is enabled to successfully remobilize resources in the pursuit of further self-development; even though the patient is not asked primarily to mirror the analyst's self-experience, or to imitate or idealize the analyst (Kohut, 1971; Kriegman & Solomon, 1985a, b).

history (de Waal, 1982, 1989; DeVore, 1971). We must not only face the evolutionary truth when it is unpleasant, that is, when it argues for basic selfish, animalistic aspects of the human psyche. We must also face the hard evolutionary truth when it is pleasant, that is, when it argues that an aspect of our human self—like our animal relatives—is innately loving and nurturing and allied with the interests of others (Kriegman & Knight, 1988). Thus, we should expect that in the adaptive design of our psyche there is a natural, adult disposition to experience a sense of inherent mutuality with one's offspring that will lead to altruistic action. To a lesser degree, because of a smaller shared amount of genetic material this disposition should extend to all of our kin.

At root, kin altruism cannot be understood as a "higher development" of cortical processes in opposition to more primary instinctual motivations. Rather, such altruism is understood as deriving from its own independent sources that are, in all senses, a primitive biologically based wellspring. The fact that "kin altruism" is only altruistic in the special sense of indirect benefit to the self (through overlapping genotypes) should not lead us to conclude that the motives (affects, wishes, expectations) that provide it are less real, less an innate evolved design feature, than more selfish impulses. Certainly we would not claim that all, or even most, of our evolved motivational design is likely to be geared toward promoting the interests of others no matter how much indirect gain we will intrinsically obtain through genetic relatedness. Indeed, as the classical model stresses, we have a substantial component of narrowly self-interested motives many of which we subjectively equate with what is infantile in our character. The point is that neither the presence of such self-interested motives nor the ultimately self-enhancing function of mutualistic motives should obscure the fact that, as proximal mechanisms, primary concern for the interests of others can be built into the evolved deep structure of our psyche.[4]

---

[4]In Appendix D, "The Confusion of Proximal and Distal Causes," we illustrate a common error in evolutionary applications to human psychology using our understanding of mutualistic human motivations. In this error, a failure to distinguish between the evolved functional system (the *proximal* product) and its adaptive function (*ultimate* explanation of the selective pressures that created the proximal feature) can lead to a mistaken notion of what is currently occurring in the organism. In the clinical situation, essentially accurate explanations of what the individual is striving to accomplish in an ultimate sense—the ultimate/evolutionary function of the motivation and action—may fail to capture the genuine meaning of the act on a personal level, the meaning of the individual's actual motivations (both conscious and unconscious) in the here and now.

Reciprocal Altruism and Inclusive Fitness: The Innate Basis of Concern for Unrelated Others

Kin altruism explained the vast majority of unselfish acts found throughout nature (Trivers, 1971). Yet, there was still a surprisingly large reservoir of observed altruism that could not be explained on the basis of kinship. At first glance, the idea that individuals might be designed with a disposition to act in the interests of unrelated others at a cost to themselves appears to violate the very process by which natural selection favors those designs that best propagate themselves. Having largely dispensed with "group selectionist" narratives to explain such altruistic ways of relating in regard to kin, was it now necessary to invoke collectivist, nonindividualistic notions to explain how individuals might somehow benefit from mutualistic relating in ways that are potentially quite costly to the self? Again, the solution turned out to lie not in a reversion to unworkable collectivist narratives, but rather, in this case, to a model that focused on the complex process of negotiated exchange that is found in social species. Robert Trivers (1971), in what is already considered a classic paper (White, 1981), developed the concept of "reciprocal altruism" explicating the way in which individuals could have been selected for a disposition toward relating altruistically with unrelated others. He was even able to do this in regard to altruistic exchanges between members of different species which, of course, is an extreme example of a lack of genetic relatedness.

The concept of reciprocal altruism is based on the notion that an altruistic act can at some point be "paid back" to the altruist. For example, Trivers describes the relationship between certain host fish and unrelated, relatively tiny cleaner fish. The cleaner's diet consists of parasites removed from the host which can often involve entering the host's mouth. Trivers describes how a host fish will go through extra movements and delay fleeing when being attacked by a predator in order to allow a cleaner extra time to leave its mouth! One would assume that it would be adaptively advantageous for the host, at such moments, to simply swallow the cleaner. Instead, the host delays its departure in order to signal to the cleaner that it is time to get out of its mouth. This type of altruistic behavior seems to reduce the fitness of the host in two ways: (1) it increases the chance that it will be eaten by a predator, and (2) it forgoes a meal of the cleaner. Yet, Trivers was able to show that there is an adaptive advantage to this altruistic act: the increased chance of having severely debilitating parasites removed in the future. One certainly would not posit that an affect such as guilt serves as the motivation for such behavior in the host fish. One assumes that this kind of altruistic behavior is directly embedded in the evolved

"TIT FOR TAT": THE COOPERATIVE WINNING STRATEGY

Based on Trivers's model, Axelrod and Hamilton (1981) used game theory to prove that a reciprocal strategy can outcompete more selfish strategies. They used a modified version of a game that has been extensively studied by social scientists, The Prisoner's Dilemma.

This game presented players with opportunities (1) to attempt to cooperate, (2) to selfishly attempt to take advantage of the other player's willingness to cooperate (to cheat), or (3) to protect oneself or retaliate by refusing to cooperate with a cheater. Successfully "cheating" a cooperator "paid off" the best (+5). Successful mutual cooperation was next (+3). Punishing or self-protecting refusals to cooperate enabled the player to protect his or her own "investment" (+1). Finally, being cheated when one attempted to cooperate resulted in the worst outcome (+0).

	I/we cooperate		I/we defect	
They cooperate	I/we cooperate They cooperate (cooperation)	+3 +3	I/we defect They cooperate (we cheated)	+5 +0
They defect	I/we cooperate They defect (they cheated)	+0 +5	I/we defect They defect (we both cheated)	+1 +1

FIGURE 9. Tit for Tat.

In Figure 9, the top number in each box indicates my/our payoff and the bottom number indicates theirs. There were a number of rounds of play in which each player made a decision prior to knowing what the other had decided to do on that particular round. After each round the results were made known to the players.

Axelrod conducted a computer tournament of strategies submitted by game theorists in economics, sociology, political science, and mathematics. Though some of the strategies were quite intricate, the winning strategy (the highest score averaged against all challengers) was one of the simplest—a basically cooperative strategy called Tit for Tat (TFT). TFT's strategy was simply to cooperate on the first move and thereafter to do whatever the other player did on the preceding move. Thus, the strategy is to initially "announce" an intention to cooperate and then to let the other player know that cheating will not lead to a gain. TFT is quick to forgive no matter how many times the other player has cheated—just one indication of a willingness to cooperate from the other player leads TFT to try cooperation again.

> TFT is a strategy of cooperation based on reciprocity . . . it was never the first to defect, it was provocable into retaliation by a defection of the other, and it was forgiving after just one act of retaliation. (Axelrod & Hamilton, p. 1393)

> After circulating the results of the first round, sixty-two entries were received from six countries to compete in the second round. The winner of the second round, TFT! Axelrod and Hamilton went on to show how a "population" of strategies could be invaded by TFT. If high scores were indicative of differential success in survival and replication, after a number of rounds of play (a number of "generations"), TFT would replace the original strategies.
>
> They essentially used game theory to present a mathematical model of how, under certain conditions, cooperation between unrelated individuals could evolve. They were able to show that TFT could hold its own and win in a population of mixed strategies. In addition, they were able to show that once TFT became a significant part of the existing population of strategies, it would become an even more successful strategy as the reciprocal altruists would benefit from their mutual cooperation while being able to protect themselves from cheaters who would be excluded from reciprocal exchanges. Indeed, they were able to show that selfish strategies would be unable to successfully reinvade!
>
> Recent simulations suggest that a more generous strategy may actually outcompete TFT (Browne, 1992). In this variation of Tit for Tat, there is no fixed response to cheaters. "Generous TFT" responds in kind to cheaters only 1 out of 3 times. If a population consists mostly of betrayers, Generous TFT does poorly. But once TFT is established in the population, Generous TFT out-competes it.

structure, the innate design of the perceptual and motivational system of the host fish.

Another paradigmatic demonstration of reciprocal altruism was provided by Wilkinson's (1984) study of vampire bats. Starvation occurs in these bats within days of failure to feed, and failure to feed is not an uncommon occurrence. He was able to demonstrate reciprocal food sharing (via regurgitation) between unrelated bats when the "altruist" was in good condition and the beneficiary was in serious need. Wilkinson showed that such a system is a great advantage for the altruist because the altruistic act is frequently reciprocated when the current altruist is the one in need. In the box on page 99, we present a report of a computer simulation of competition between reciprocal and other strategies in which a reciprocal strategy showed surprising robustness and competitive advantage.

Of course, human interactions and the design of the psychological system that underlies them are not as straightforward as simple computer simulations or even the more complicated arrangements between fish or bats. In terms of the relational system, as it were, the evolved design of the human psychological system, its innate deep structure, must be attuned to a range of complex meanings and ambiguities. For one thing, there are frequently kin ties present simultaneously with non-kin reciprocal ties—kin ties that are not dependent on reciprocity though they too are apparently influenced by it. Also, for reasons we shall discuss at greater length in ensuing chapters, with humans it is not always possible to detect "cheaters"; and it is much harder to evaluate the degree of cooperation one is getting. Opportunities for subtle cheating on the reciprocal system are widespread; and there are many other motivations, values, and meanings at play.

Yet, it is easy to think of numerous situations in which reciprocal exchanges benefit individuals disposed to relate mutualistically and cooperatively by providing advantages simply unavailable to individuals acting on their own. Reciprocal altruism can be markedly adaptive when there are frequent situations in which the cost to the altruist is significantly less than the beneficiary's gain. If this sounds like a perpetual motion machine where the output magically exceeds the input, consider just a few human examples such as a traditional "barnraising," the act of helping an unrelated child find its way back to its parents, providing sustenance and child-care to temporarily impaired neighbors (e.g., after a new birth, during a sickness, after an injury, following a tornado, flood, or fire, after the death of a close relative, etc.), and most forms of human philanthropy and charity (Kriegman, 1988, 1990). In such situations the cost to the altruist may be significant, but it is frequently far less than the recipient's benefit. It has been shown that those individuals who are disposed to trade such acts have a significant advantage over nonaltruists or those excluded from reciprocal arrangements (Axelrod & Hamilton, 1981).

Even more remarkable is the somewhat counterintuitive fact that, like vampire bats which, without reciprocal exchanges, would be in constant danger of starvation, people living on the verge of starvation often will share food with unrelated others. Lapierre (1985) documents numerous examples of regular, reliable reciprocal exchanges among non-kin residents in a Calcutta slum. In such circumstances, powerful friendship/reciprocal exchange systems can develop in and around competing systems of cheaters (see also Levi, 1986, and others who have written about the Holocaust concentration camps). During our phylogenetic history when conditions were extremely harsh there was no "cushion." There was no Medicaid, welfare, life insurance, or disability insurance. Given the physical dangers, it was unlikely that one could go through a normal lifetime without needing help from genetically unrelated neighbors. It is only in times of surplus and physical security that one can "afford to" live independently of one's neighbors: thus, the demise of the "community" as affluence allows people to risk not even knowing their neighbors. Somewhat paradoxically this suggests that the less one has, the more likely one is to be willing to share a greater portion of what one does have (Kriegman, 1990).

Another powerful demonstration of reciprocal exchanges in humans occurred during the early years of World War I in the trenches. Much to the consternation of the generals, the "disease of cooperation" between the opposing soldiers broke out all along the line. Stationed week after week and month after month in the same spot facing the same "enemy," the ideal conditions were present for the generation of reciprocal altruism. Soldiers shot to miss. Troops left the trenches and worked at repairing them in full view and within range of enemy soldiers who passively looked on. Christmas was celebrated together! One striking incident occurred when in one area an artillery burst exploded sending both sides diving for cover. After several moments, a brave German soldier called out to the other side and apologized saying that his side had nothing to do with it, "It was those damn Prussian artillerymen." Note that artillery is farther from the line and the mutual association with trading of beneficial acts (shooting to miss) was not available to them. Finally, the generals solved this thorny problem by ordering random raids (and shooting those that refused) that broke down the mutual trust and cooperation that had evolved.

The skeptic could argue that the human examples are anomalies, and of course we do not expect humans to behave as fish, vampire bats, reciprocally altruistic monkeys (de Waal, 1982), or relatively simple computer programs. However, what is demonstrated by the evolutionary analysis are the necessary prerequisites for the evolution of reciprocal altruism: (1) high frequency of association, (2) reliability of association over time, (3) the ability of two organisms to behave in ways that benefit the other, and (4) the ability to recognize and remember cheaters and altruists. The prerequisites for the evolution of reciprocal altruism, which have been shown in other species to be capable of shaping extremely cooperative behaviors, are present in our species and are possibly present *to a greater degree than in any other species.*

Trivers was able to show that altruistic behavior between human beings can confer a powerful adaptive advantage to the altruist and may have been selected by the same selective pressures that were forcing increased cranial development.

> A long memory and a capacity for individual recognition are well developed in man. We might therefore expect reciprocal altruism to have played an important part in human evolution. Trivers goes so far as to suggest that many of our psychological characteristics— envy, guilt, gratitude, sympathy, etc.—have been shaped by natural selection for improved ability to cheat, to detect cheats, and to avoid

being thought to be a cheat. Of particular interest are "subtle cheats" who appear to be reciprocating, but who consistently pay back slightly less than they receive. It is even possible that man's swollen brain, and his predisposition to reason mathematically, evolved as a mechanism of ever more devious cheating, and ever more penetrating detection of cheating in others. (Dawkins, 1976, p. 202)

Beyond simple direct exchanges, adults in all cultures will engage in a great many "reciprocally altruistic" exchanges with unrelated individuals who are bound into an elaborate, culturally mediated system of reciprocation. Such altruistic acts can be advantageous even though they are directly costly to the actor's own immediate, personal fitness because reciprocation is either (1) delayed, and/or (2) "paid back" to other individuals to whom the original altruist is genetically related (kin) or who form part of an extended, reciprocal network. Thus, temporarily costly altruistic actions that reduce personal fitness can have long-term inclusive fitness benefits, though in some cases the return may not occur for many years and ultimately may not even be directed toward the original altruist.

A moment's reflection will demonstrate that much of human object relations exists within a complex web of kin and reciprocal altruism. In this analysis, rather than being an outcome of conflict brought into being by recent cortical evolution—the Freudian view in which altruism is a culturally induced reaction formation (Kriegman, 1988, 1990)—the tendency to act altruistically is seen as a historically primitive feature. In fact, recent primate studies (de Waal, 1982) suggest that reciprocal altruism existed at an early stage in primate evolution and that in humans the selective pressures exerted by the system of reciprocal exchange led to the rapid development of intelligence (Kriegman, 1988). (In Appendix C, "The Evolution of the Human Cortex and Its Relation to Civilization and Guilt," p. 298, we take a closer, more technical look at the evolution of altruism, cooperation, civilization, and intelligence.)

Figure 10 indicates just a small part of some of the extraordinary complexity that enters the "evolutionary relational matrix" when we add multiple kin and non kin (or reciprocally altruistic) ties to the simple "gene's-eye view" we presented in Figure 7.[5]

---

[5]Human cortical development becomes even more clearly a product of relational complexity (see Appendix C, on page 298) when one tries to combine a further elaborated version of the relational matrix (Figure 8, page 93) including more siblings and other relatives to generate a more fully elaborated version of Figure 10. As a rational process, the complexity clearly goes beyond our conscious, linear thought capacity—beyond our capacity to consciously label and verbalize all the factors being weighted simultaneously; clearly, humans have been designed to intuitively feel their way through such intricate relational webs.

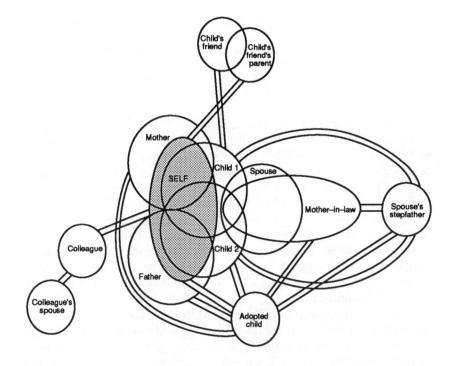

FIGURE 10. The self: some of the complexity of kin and nonkin (reciprocal) ties. *Note:* each circle (individual) has an equally complex set of kin (overlapping) and reciprocal (lines) ties surrounding itself. In addition, each kin tie is simultaneously a reciprocal tie. Only a small part of the quagmire (or rich tapestry) of complexity surrounding the self can be illustrated.

Implications for the Deep Structure of the Individual Psyche within the Relational Matrix

Our very general characterization of the relational matrix, for which the psyche has evolved its adaptive design, deep structure, and strategic capacities, is only a first glimpse at the relational world from an evolutionary perspective. Our inferences about the complicated matter of adaptive design and evolved deep structure must be quite preliminary and tentative. Nevertheless, as we go on to give more psychological meaning to these issues and look further at the relational world through an evolutionary lens, let us keep in mind the possibility that as a strategic way of organizing the relational world our psychological deep structure may well include (1) an evolved, inner attunement to the

DARWINIAN ALGORITHMS: TO WHAT EXTENT IS OUR LOGICAL REASONING AN ADAPTATION TO THE COMPLEX CHANGES IN THE RELATIONAL WORLD?

An interesting experimental investigation of the human capacity to reason logically about social interactions was conducted by Leda Cosmides, the evolutionary psychologist, whose theoretical work on the evolution of inner adaptive design distinguishes her work from the emphasis on overt behavior in most of the sociobiological literature. Using the Watson Selection Test, Cosmides observed that people are far more able to make highly complicated logical inferences when the logical problems posed to them were embedded in hypothetical social situations rather than posed as logical abstractions (i.e., if A implies B and all B's imply C, does A imply C). Neither the simple fact of concreteness nor the degree of familiarity with the context could explain the striking differences between our ability to correctly assess complicated sets of "rules" when those rules were cast as rules of social behavior (including the accurate detection of who was "cheating") than as rules embedded in other, nonsocial contexts. In different versions of these experiments, what appeared to be crucial was the subjective relevance of reasoning about problems involving reciprocity—whether particular characters were following the rules or cheating.

Cosmides interprets these results as indicating crucial aspects of our reasoning ability evolved as adaptations (Darwinian algorithms) to recurrent challenges in the ancestral world. (Cosmides & Tooby, 1987). As such, it provides another significant link in the picture of the psyche as a deeply, intrinsically relational entity—one that is, however, designed to monitor all relational transactions as they relate to the differing and overlapping interests of the interacting individual.

Tooby & Cosmides (1989, 1990) go on to construct a picture of the mind as composed of sets of "domain specific mechanisms," (like reasoning about social reciprocity). Although in some of their work (1990) they seem to recognize the need for overarching inner principles that organize such discrete systems into a relatively cohesive subjective whole, their work still tends to yield a far more fragmented theoretical model of the inner workings of the psyche than fits with the experiential and clinical data of the analytic observer.

The analytic observer, is especially attuned to the relationship of the whole psyche to its parts. The subjective sense of wholeness, or self-cohesion, is an exceptionally critical subjective sign that the intricate interrelationship between, what Cosmides and Tooby call "co-evolved domain-specific, mechanisms" are working purposefully as an overall, functional entity. We might note though, that their observational methods are limited to the laboratory and to the study of domain specific cognitive mechanisms outside the powerful emotional contexts that more closely replicate the conditions of our actual development. Perhaps laboratory observations in the context of extremely short-term, highly superficial relationships vastly underappreciate the central importance of phenotypic structures and processes designed to create the overarching organization of the self.

existence of certain inherent connections (overlapping interests) between self and others; (2) the simultaneous, innate sense of others as entities with whom, ineluctably, one has certain competing aims; (3) an adaptive capacity to assign different meanings to different categories of relatedness, for example, between kin and non kin, true reciprocators and varieties of less authentic, less trustworthy characters (nonreciprocators); as well as (4) an intuitive conviction of the need to get at the hidden dimensions of relatedness—to distinguish between appearances and more reliable perceptions of others within the relational world; between apparent (narrowly defined, individualistic) self and object boundaries and those that fully and accurately embrace the wider, unseen, long-term world of (inclusive) overlapping interests in which we live. The fuller meaning and relevance of these subjective distinctions for an adaptive psychological design will become clearer after we take a look at the inherent conflict and deception in the evolved relational world of the family in the next three chapters.

CHAPTER FIVE

Conflict and Mutuality in Development: Parent–Offspring Conflict Theory

> Conflict during socialization need not be viewed solely as conflict between the culture of the parent and the biology of the child; it can also be viewed as conflict between the biology of the parent and the biology of the child.
> —R. TRIVERS, *Parent-Offspring Conflict Theory*

> Every time you say it's time to stop I feel like my mother just had another baby.
> —Analytic patient at the end of an hour

Conflict and Mutuality in the Family

As we discussed in Chapter Four, because of the very close genetic relationship between parents and their offspring, the evolutionist knows that throughout the millions of years of our development there has been, in each generation, a large degree of overlap between the interests of parents and their children. Parents and children share a large percentage of their genes. As we have seen, what has been good for the "fitness" of children has also, to a high degree, been good for their parents. And vice versa. The child's essential biological being has always been deeply connected—literally shared—with its parents and, indeed, by extension, with all its close kin. In this respect, the generations have been evolutionary allies, so to speak, not competitors (Trivers, 1974).

The reality of this alliance is particularly obvious from the parents' point of view: a parent's own reproductive success, his or her "fitness," is intimately tied to the future reproductive success of his or her offspring. What this means at a very general level is in some ways obvious, but needs to be stated because it has been lost from many analytic discussions of human desire. Namely, that this literal "identity of interests," so to speak, of parent and child is what has made parental love—the affection, protection, and devotion that parents experience toward their offspring—an essentially universal part of the design of the adult human psyche (Trivers, 1974, 1985; Kriegman, 1988, 1990). These affects, and the huge investment in offspring to which they lead, would simply never occur with any regularity were it not that we are all literally descendants of hundreds of thousands of generations of the most effectively devoted parents. Effectiveness is, of course, defined primarily in terms of the relatively greater survival and reproductive success of their offspring. As Kohut (1982) put it (without knowing that he was making a biological argument), we are indeed functioning "in accordance with our design" when we experience as our "deepest and most central joy, that of being a link in the chain of generations" (p. 403).[1]

Yet, the acknowledgment of the adaptive significance of any inherent pleasures and meanings in the costly investment by parents in their children (and other types of kin altruism) always exists in the context of divergent interests or conflict. For no sooner do we begin to think in terms of the inherent overlap of interests and the innately mutualistic motives that are an intrinsic feature of kin relations, than we encounter a second, equally basic feature of the biology of relatedness—one that, in this case, implicates us in the essential dilemma at the very core of human object relations. Despite the overlap—the literal identification with each other's interests—parents and their offspring have throughout the eons of our evolutionary history, from the very moment of conception onward, been genetically distinct, unique, separate individuals. What this means from an evolutionary standpoint is

---

[1]Does the genetic relationship between parents and children really account for "parental altruism"? The prevalence of widespread child abuse would seem to contradict the notion that we have an innate predisposition to invest altruistically in offspring. We must remember, however, that any evolved inner dynamic and affect that promotes altruism is invariably intertwined with motives designed to promote the self-interest of the individual in narrowly personal (vs. mutual) ways. Evolutionary accounts invariably include *both* primary motivational elements and the tension between them. However, the fact that the existence of such balancing tensions (not a pure altruism by any means) is far more likely in kin relationships is strongly supported by the important research of the evolutionary, social psychologists Martin Daly and Margo Wilson. Their large epidemiological studies of child abuse and other forms of violence tend to show that the presence or absence of genetic relatedness is the variable that most powerfully accounts for the relative likelihood of child abuse. Daly and Wilson (1988) found that the incidence of fatal child abuse is 100 to 150 times higher when, instead of biological parents, unrelated adults (such as step-parents) are in the parental role. This pattern holds across cultures (Daly, 1989).

that at some fundamental level the interests of normal parents and their children necessarily diverge as well as potentially compete and conflict.

Parent–Offspring Conflict Theory

The preeminent evolutionary theorist Robert Trivers was the first to grasp the powerful implications of the biological realities of familial conflict. In his initial description of the relational matrix in the family, Trivers, argued as follows:

> ... conflict is ... an expected feature of ... (parent–offspring) relations ... in particular, over how long the period of parental investment should last, over the amount of parental investment ... and over the altruistic and egoistic tendencies of the offspring. ... One expects the offspring to be programmed to resist some parental teaching while being open to other forms. (1974, p. 249)

Trivers cited an enormous range of data from ethological studies on a wide variety of species as he developed this now broadly accepted evolutionary model. Referred to commonly by biologists as "parent–offspring conflict theory," this perspective on the universal features of intrinsic intergenerational conflict has now been extensively applied to numerous phenomena in nature (Trivers, 1985). Trivers initially used the example of a caribou calf and its mother to illustrate parent–offspring conflict theory, though the theory is broadly applicable to all sexually reproducing species. Note that in this example, as in all evolutionary analyses, the costs and benefits refer to the individual's "inclusive fitness."

> Consider a newborn (male) caribou calf nursing from his mother. The benefit to him of nursing (measured in terms of his chance of surviving) is large, the cost to his mother (measured in terms of her ability to produce additional offspring) presumably small. As time goes on and the calf becomes increasingly capable of feeding on its own, the benefit to him of nursing decreases while the cost to his mother may increase (as a function, for example, of the calf's size). ... At some point the cost to the mother will exceed the benefit to her young and the net reproductive success of the mother decreases if she continues to nurse. ... *The calf is not expected, so to speak, to view this situation as does his mother, for the calf is completely related to himself but only partially related to his future siblings.* (1974, p. 251, emphasis added)

The mother is equally related to all of her offspring while the calf clearly is not. Thus, the cost to the mother in terms of a decrease in her ability to bear and rear additional young, and the benefit to the current suckling, should have very different "subjective" meanings to the mother and the calf. The resultant "weaning conflict" has been well documented in many species. Trivers points out that weaning conflict is simply a specific paradigmatic example of the much larger category of conflict over parental investment. This argument holds for all forms of parental investment, for example, feeding the young, guarding the young, cleaning/grooming the young, carrying the young, teaching the young.

In a similar fashion, Trivers spelled out predictions of parent–offspring conflict over the social behavior of offspring:

> Parents and offspring are expected to disagree over the behavioral tendencies of the offspring insofar as these tendencies effect related individuals. (p. 257) ... [For example, in interactions among siblings] ... an individual is only expected to perform an altruistic act towards its full sibling whenever the benefit to the sibling is greater than twice the cost to the altruist. (1974, p. 259)

We expect this probability of altruistic relating because each sibling carries one half of the genetic material of the actor. Thus, if an altruistic act benefits a sib more than twice the cost to the actor, the actor actually receives a net benefit to his inclusive fitness.[2]

> Likewise ... [the actor] is only expected to forego selfish acts when [the cost to their sibling] is greater than [twice the benefit to the self] (where [a] selfish act is defined as one that gives the actor a benefit, while inflicting a cost on some other individual, in this case, on a full sibling). (Trivers, 1974, p. 2)

We illustrate parent–offspring conflict in Figure 11 where we compare the parental view of children with the child's view of self and sibling.

In this "gene's-eye" view (top of Figure 11) we clearly see the overlap between the inclusive fitness of individuals. In our diagram the skin-bounded (everyday, phenotypic) selves are represented by the circles that are the child, parent, and sibling, respectively. Note that while they do overlap, each skin-bounded individual is unique. This uniqueness is the source of significant conflict between parent and child, an example of which is depicted on the bottom of Figure 11.

Viewing the situation from the subjective perspective of the child's self, we see (on the lower left) that the child values itself twice as much as it values its full sib. Yet, the parent has a subjective view (depicted on

---

[2]Remember the "gene's-eye" views of the self in the relational world (presented in Figures 7, 8 and 10) and the concept of kin altruism from Chapter Four.

the lower right) in which, on average, each child is valued relatively equally (half as much as the parent's self). Keeping in mind these differing, genetically based "gene's-eye" views depicted in Figure 11, let us reconsider some of the complexities of mutuality and conflict during the developmental process within the family.

Parents, who are equally related to all of their offspring, are likely to value and promote altruistic relating between their offspring whenever the benefit is greater than the cost; and children only want to act altruistically whenever the benefit to their sibling is twice the cost to them. Likewise, parents are likely to value and encourage their children to forgo selfishness whenever the cost exceeds the benefit; children, on the other hand, ought only to be willing to readily forego selfishness when the cost exceeds twice the benefit. In Figure 12 we visually represent this inherent intrafamilial conflict in regard to a child's selfish and altruistic tendencies. Similar graphic schemes can be presented for the expectable patterns of individually self-interested versus mutualistic forms of relating among other relatives as well as for our expectable disposition toward the interests of others who are linked in reciprocal ways with our kin.

The vertical axis of Figure 12 represents the cost to a sibling of an act, while the horizontal axis represents the benefit to the actor. Starting

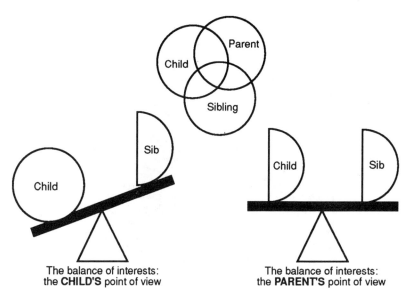

FIGURE 11. Primary conflict in the relational world: the "gene's-eye" view of the balance of interests (subjective realities). Adapted from Slavin and Kriegman (1990).

on the horizontal line (at the point labeled I), note that the cost to a sibling is zero, while the benefit to the child of such a hypothetical act is great. As we follow the arrows that point counterclockwise up to the shaded area (to II) we pass through a realm in which, because the cost to the sib is less than half the benefit to the actor, the self-interests of the parent, child, and sibling all coincide: they agree that the child should act in a self-interested manner.

However, when we enter the shaded region, the situation changes dramatically. Here, in this zone, the cost to the sibling has risen and is no longer less than half the benefit to the actor. Thus, the sibling, who is wholly related to him- or herself and only 50% related to the child, no longer finds it advantageous for the child to continue to act selfishly. The parent still sides with the child because the parent is equally related to both offspring and the benefit to the child still exceeds the cost to the sibling. As we move through this region to III, the sibling becomes more and more distressed by its disadvantage. At III, the cost to the sibling starts to exceed the benefit to the child. Thus, the parent, who is equally related to both children, ceases to find it advantageous for the child to act self-interestedly. The child is now alone with the parent allying with the sibling. But the child does not see it their way until we leave the shaded region (at IV), entering, once again, a region of overall agreement. Here, because the cost to the sibling exceeds twice the benefit to

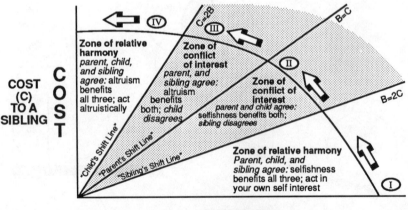

FIGURE 12. Parent–offspring conflict theory: Intrinsic conflicts of interest between parent and child (in regard to sibling's fate). Adapted from Kriegman and Slavin (1990).

the child, it is in the self-interest of the parent, child, and sibling for the child to forgo selfishness and act altruistically. The resultant prediction matches the common observation of intense and ubiquitous sibling rivalry with parents engaged in major struggles with their children over their egoistic impulses.

In sharp contrast to the common, relatively unquestioned assumptions about the socialization process found in earlier evolutionary views as well as certain views in the social sciences and psychoanalysis, parent–offspring conflict theory goes on to depict how, in Trivers's words:

> ... it is a mistake to view socialization in humans as only or even primarily a process of "enculturation," a process by which parents teach their offspring their culture ... [Instead, Trivers suggests, developmental socialization is] ... a process by which parents attempt to mold each offspring in order to increase their own inclusive fitness, while each offspring is selected to resist some of the molding and to attempt to mold the behavior of its parents (and siblings) in order to increase its inclusive fitness. Conflict during socialization need not be viewed solely as conflict between the culture of the parent and the biology of the child; it can also be viewed as conflict between the biology of the parent and the biology of the child. Since teaching (as opposed to molding) is expected to be recognized by offspring as being in their own self-interest, parents would be expected to overemphasize their role as teachers in order to minimize resistance in their young. (1974, p. 260)

We believe the implications of parent–offspring conflict theory are immensely important for psychoanalytic views of the relational matrix to which our psyches are preadapted from birth—the matrix within which normal development takes place. The evolutionist, Trivers, expects that the recurrent, ancestral (phylogenetic) features, as well as the salient contemporary (ontogenetic) features of the human relational matrix are such that:

1. The realities of conflict during development are intrinsically rooted in the evolved deep structure, the normal design of the motivational system of both parent and child (as distinct from presumed "distortions" or resistances to socialization introduced by the presumed immaturity and instinctuality of the child, or parental inadequacy and pathology). In certain respects this narrative of relational conflict, if you will, corresponds precisely to Freud's conviction that:

> A small child does not necessarily love his brothers and sisters; often he obviously does not. There is no doubt that he hates them as competitors, and it is a familiar fact that this attitude persists for long

years, till maturity is reached or even later, without interruption. Quite often, it is true, it is succeeded, or let us rather say overlaid, by a more affectionate attitude; but the hostile one seems very generally to be the earlier. (Freud, 1916, p. 304)

Freud's observation that hostile sibling rivalry persists until maturity, until children no longer need to turn to their parents for a major investment (for the largest part of their material/emotional needs), is quite consistent with one crucial facet of parent–offspring conflict theory. Competing individual interests are a fundamental part of the normal relational landscape in even the closest, most loving, nurturing relationships. However, this clash of interests (and the subjective aims, affects, and biases that ensue from it) are in no fundamental sense derived from or built on the operation of the instinctual drive system. Freud's appreciation of and his willingness to face and recount the narrative of conflict remained rooted, unfortunately, in the drive model in which only such narrowly self-interested motives could exist at the bedrock level—the level on which an "affectionate attitude" is "overlaid" or developmentally "succeeded." From the evolutionary perspective, *both* selfishly competitive and affectionate attitudes are inherent from the beginning. Kin altruism based on inherently overlapping interests is as deeply embedded in the evolved bedrock of the psyche as competitive rivalry; it is as conceptually primary as competitive aggression. When it is not in the self-interest of individuals to aggressively compete with their sibs the affectionate attitude comes to the fore. Thus, when these individuals become independent of their parents, the affectionate attitude *appears* to replace the other.[3]

2. The whole process of development will normally be colored by an evolved, pre-existing subjective bias in all parties' accounts of reality. There will exist, as well, continuing efforts at deception, or concealment of that bias by all family members. In nature, the representation and communication of "reality" are always strategically tailored, or biased, to some degree to fit with the interests of the different individuals involved (Goleman, 1985; Mitchell, 1985; Trivers, 1985, 1976a). Individuals in all social species will, as a matter of course, tend to conceal certain information and selectively accentuate other information de-

---

[3]Though sibling rivalry is often quite intense, as Freud described, it is commonly observed that sibling rivalry is most pronounced in the *vicinity* of the parents; it is most frequently engendered by competition for parental investment. In clinical work, we have treated families in which the physical well-being of children was seriously jeopardized by their sibs—extreme cases of sibling rivalry. Yet, even these children would be able to play together quite well in the absence of their parents. What is more, they would occasionally risk injury to themselves in attempting to aid one another in difficulties outside the home.

pending on what is in their best interest to have others perceive. Over evolutionary time, the complex capacity to encode information in a self-enhancing way and to detect that process emanating from others (and correct for it) will comprise a substantial part of our psychological deep structure. Deception is—must be—an intrinsic part of the story in any narrative of conflict.

What are the broader implications of the universality of this intrinsic juxtaposition of shared and distinct identities, mutual and conflicting interests, and deception within the family matrix? The evolutionarily successful parent, that is, the parental design that was likely to replicate its own genes more effectively than did its evolutionary rivals, had to be inclined to invest heavily in his or her child and the child's success, and simultaneously, thoroughly inclined to invest in the maximization of the parent's own fitness. This meant that, at the level of evolved deep structure, parental motives were designed to enhance their own interests in all of the ways that these interests *may differ from* and at times literally *compete with* those of their child. Investing or preparing to potentially invest in other offspring is a major expression of these competing interests. For there is no such investment in others (to whom the parent may be as highly related as to the child) that does not entail some costs to the child (who inevitably is *more highly* "related" to him- or herself than to the other relative involved).

On virtually every crucial psychological issue in the course of development (indeed, of the full life cycle), the evolutionarily successful way of organizing a parental "self" is likely to have entailed an essentially two-sided, or inherently divided, strategy: treat any child as a "flesh of my flesh" ally, and, when our interests diverge, as a competitor. Such a strategy would entail not only an enormous investment in the child and the child's own intrinsic welfare but, also, a subjectively biased attention toward the parent's own interests. The parent's self is constituted of a web of relations the weighting of which differs from that of all other individuals, including their own children. Likewise, the child's self is, in part, composed of its own uniquely weighted web of overlapping interests with others.

The design of the child's psyche has been fashioned, in part, by hundreds of thousands of generations of such alliance and conflict. The modern child must be one whose ancestors had a relatively more effective deep structural design for coping with a relational world that shares some of the child's interests while being biased toward its own (often competing) interests. He or she comes into the world with a unique genetic identity, and, ipso facto, a distinct, intrinsic self-interest: a novel creation of nature with a host of innate features, dispositions, capacities, etc. The evolutionarily successful child must have been prepared to organize its subjective experience of reality in a fashion that best represented and promoted its own unique, genetic interests. The universal features of the modern child's evolved, psychological deep structure "anticipates" such a relational world and is prepared for it.

This, of course, implies that we expect to find that the child's subjective world is organized in ways that will, often enough, conflict with parental aims and constructions of reality.[4] Both psychoanalytic narrative accounts of human nature and development—the classical conflict/deception narrative and the alternative relational/mutualistic vision—have tended to relate the universal story of development in ways that are biased toward one or another version of how and why conflict occurs between parent and child. We believe that the evolutionary account can help to illuminate these biases and provide a more "symmetrical" picture of developmental interactions. The resulting picture more fully accounts for the active mutual influence exercised by parents and children in the intricate, intersubjective negotiation process we call development. In Chapter Six, we take a closer look at the psychoanalytic developmental narratives using the evolutionary perspective on the relational world as we've outlined it thus far.

Oedipal Conflict versus Parent–Offspring Conflict

The widespread acceptance of Trivers's P–O Conflict Theory by social biologists and many nonpsychoanalytic "evolutionary psychologists" has generated several critiques of Freudian theory that partially dovetail with our own critique of the classical narrative (Bower, 1991). As we noted in Chapter Three and Chapter Ten (see boxes, pages 65 and 240), many such critiques overly dichotomize the opposition between the psychoanalytic and evolutionary narratives, and thus fail to see the

---

[4]It is likely that the parental psyche has been selected for the capacity to overcome attempts on the child's part to decode the biases and deceptions in the parent's view of reality. Parents who were better at influencing their children to act, believe in, and identify with the parent's self-interests were probably genetically more successful than those who were less well designed to exercise such influence. (This should not be confused with pathologically narcissistic parents who destructively demand that a child overidentify with the parent's interests and in the process cripple the child. In so doing, the parent has *not* acted in his or her interests.) In turn, we would expect there to have been a selective pressure for children whose psychic structure is designed to be more adept at countering such influences. This type of "arms race" is frequently found throughout nature. In humans, selection to influence, detect influence, deceive, detect deception, etc., is now considered to be a major component of the selective pressures that led to the rapid development of human intelligence. (Also see Appendix C, "The Evolution of the Human Cortex and Its Relationship to Civilization and Guilt," on p. 298.)

ways in which evolutionary and psychoanalytic perspectives and methods may complement each other.

Daly and Wilson's Challenge

One of the clearest, most well-reasoned expositions of evolutionary psychology is found in the critique of the classical Oedipal model by the Canadian psychologists, Martin Daly and Margo Wilson. Daly and Wilson (1990) argue that (1) parent–offspring conflict theory—viewed as a "gender blind disagreement about resource allocation" (p. 163)—explains the phenomena of rivalry, competition, and hostility within the family better than does Oedipal theory; (2) evolutionary theories of relational conflict based on fitness-related strivings better explain the motives for most overt manifestations of adult conflict (as well as fantasies and dreams about conflict) than do interpretations that view adult experience as an "acting out" of childhood Oedipal issues. Citing a range of epidemiological data on violence between parents and children as well as among unrelated adults, they argue that "psychoanalysts may mistake substantive conflicts between nonrelatives for symbolic manifestations of family conflicts" (p. 163).

Daly and Wilson focus on the classical analytic tendency to view the child's instinctual, incestuous desire for the parent of the opposite sex as the ultimate motivational explanation for conflict between the child and the same sex parent as well as to interpret "later conflicts between nonrelatives . . . as symbolic actings out of Oedipal conflicts" (p. 184). Equally important, Daly and Wilson claim that interpreting the Oedipal themes of adulthood (particulary male–male competitive rivalry and aggression) as motivated by repressed childhood Oedipal impulses overlooks the overwhelming occurrence of such behavior throughout the world of mammals:

> Male mammals from bull seals to billy goats behave as if they "wished to possess females attached to other males" while being "jealously possessive of the females to which they themselves are attached." The ontogeny of this ubiquitous mind set has nothing to do with frustration of "the son's wish for an exclusive relationship with his mother" by his father: Father has bade mother farewell long before their son was even born . . . [M]ale sexual rivalry is endemic to the 4,000 or so species of the class Mammalia, whereas father-son interactions of any sort are rare and secondarily involved. It is therefore absurd to maintain that such rivalry constitutes evidence for a particular developmental theory in which paternal influence is crucial. (p. 175)

In other words, Daly and Wilson claim that the evidence of pervasive, conflictual themes in adult sexuality throughout nature contradicts the notion that an infantile complex is the underlying motivational source of the vicissitudes of sex and agression in adult humans. Moreover, the vast majority of conflicts with nonrelatives are, in an evolutionary sense, analyzable in terms of the individual's subjective assessment of competing and overlapping interests in the adult relational world; such conflict is, as they put it,

> genuine and not merely symbolic of conflicts with parents. . . . Lacking this insight, psychoanalytic writers regularly misinterpret genuine conflicts of interest as nonadaptive symbolic manifestations of "primal" conflicts. (1990, p. 180)

Daly and Wilson treat Oedipal theory as if it were still the core expression of conflict theory in psychoanalysis. In a sense Oedipal theory did hold such conceptual status within the classical Freudian tradition until recent decades when, in even relatively loyal Freudian circles, it has given way to the far broader conception of the narrative of conflict that we have been discussing (Simon, 1991). However, Daly and Wilson's argument is perhaps best understood, not simply as a critique of Oedipal theory but, rather, as a critique of (1) drive theory (as opposed to parent–offspring conflict theory) as the basis for understanding the origin and the role of psychic conflict; and (2) a pervasive tendency in psychoanalytic explanations (both classical and relational) to reduce certain universal relational dynamics in adult behavior to their supposed individual infantile sources, thereby overlooking the realities of competing interests in the adult relational world and the internal adaptations (deep structures) that have evolved to manage them.

A Less Dichotomized View

Classical Oedipal theory does, indeed, represent the most extreme version of the classical narrative, the one that is most obviously biased toward the parental point of view and towards viewing selfish, asocial motives (especially in the child) as primary with altruistic, mutualistic motives seen as defensive reactions to castration anxiety. Though some analytic writers (see Badcock, 1990; Lloyd, 1990) have claimed that evolutionary theory can support the Freudian Oedipal model, we believe that parent–offspring conflict theory represents a significantly different and better way of conceptualizing the evolved, relational origins of the child's conflicts with parents and siblings than either classical psychoanalytic drive theory—Oedipal or pre-Oedipal—or the alternative psychoanalytic relational theories of environmental failure.

Yet as psychoanalytic observers, it is inconceivable to speak of conflicts of interest within the family that are "gender blind" in origin without recognizing, simultaneously, that the relational matrix of the family is absolutely permeated with gender and sexual issues. And, though we agree with Daly and Wilson that much genuine conflict of interest in adulthood (both within and outside the family) is mistakenly reduced to what are presumed to be the "real" childhood determinants, it is equally inconceivable to overlook our daily clinical observations that the *particular* form that *individual* gender identities and sexual identities take is profoundly influenced by the prior negotiation of conflicts within the family. Indeed, the human child develops in a highly gendered environment in which parental and social gender roles and expectations infuse the very way in which parents experience and define their own and their children's interests. Relationships within the human family inherently entail one or another version of the sensuality, lust, competition, rivalry and jealousy that, as Daly and Wilson aptly note, characterize most of mammalian social life.

However, the recognition that human identities are powerfully constructed by sex and gender does *not* imply that sexual meanings and agendas within the family primarily originate in the *child's* instinctual drives as in the classical Freudian model. Ultimately, "Oedipal" sexuality (in the sense of the multiple enacted and symbolic manifestations of childhood sexuality) might well be seen to function (within the parent–offspring paradigm) as part of an evolved, adaptive program that is highly sensitive to parental sexual and gender strategies. Affects and behavior that are initiated by the child may be responsive to parental needs, serving, in turn, to (1) increase parental investment (through the child's capacity to use his or her own sensuality, attraction and gendered behavior to elicit reciprocal parental affect); (2) make use of the relative safety (inherently overlapping interests) in the normal family setting to experiment with the forms and limits of heterosexual attachment and within-gender rivalries, including the development of an inner sense (prototype) of a desirable sexual object [for which Daly and Wilson actually cite supportive evolutionary data on the similarities between parental phenotypes and mate choice (pp. 172, 173)].

We do not see such an evolved human childhood strategy for the negotiation of a gendered self as the "cause" of adult sexuality—that is, as a complex that is universally "acted out" in adulthood with the implication that the conflictual themes of adult sexuality are merely symbolic derivatives of childhood experience. Rather, the child's evolved strategy is a preparation for and fine tuning of the pursuit of a more "inclusively fit" path through the intrinsically complex (often deceptive and conflictual), arena of adult sexuality and object choice.

There is abundant data in every analytic practice that childhood/familial sexuality and rivalries profoundly color adult behavior in related areas.

We shall go on to present a view of the developmental process that gives far greater emphasis to the ongoing assessment and adaptive negotiation of the universal conflicts of interest in adult life than is found in any of the existing psychoanalytic paradigms. In Chapter Six we spell out the ways in which the evolutionary perspective on conflict, mutuality, and deception in the family requires us to consider a radical revision of the way in which both the classical and relational psychoanalytic narratives have envisioned the "average, expectable" or "good-enough" developmental environment.

CHAPTER SIX

A Revised View of the Average, Expectable Environment

> The first lesson that innocent Childhood affords me is—that it is an instinct of my Nature to pass out of myself, and to exist in the form of others.
> The second is—not to suffer any one form to pass into ME and become a usurping Self in the disguise of what the German Pathologists call a FIXED IDEA.
> —SAMUEL TAYLOR COLERIDGE

> The problem is one of individual integrity versus individual relatedness ... There is no escape from this theoretical prison, because the opposition of subject and object dictates it. The dilemma seems to call for an alternative, a new description of the person in the world of others.
> —A. GOLDBERG, *The Prisonhouse of Psychoanalysis*

Evolutionary theory suggests that even responsive, attuned, facilitative familial environments will inevitably be characterized by a highly ambiguous mixture of overlapping, mutual interests, intrinsic conflict, and ongoing deception. Much deception concerns the very existence of conflict, its meaning and sources. Simultaneously, there is an equally ancient, universal overlapping of interests and aims within the family; and this aspect of the ancestral environment (still encoded within the structure of object relations in the modern relational world) represents a natural, intergenerational alliance based on overlapping identities and shared interests—the shared interests that make mutualistic action a central feature of human life.

Embodied within the "evolutionary narrative" are elements of the classical psychoanalytic narrative of intrinsic conflict, deception, and self-deception. Conflict and deception are inherent in the clashing interests and covert strategies that are part and parcel of all motivated interactions in the relational world. However, the conflict and deception in the evolutionary narrative is a far more two-sided process; in effect, a more mutual process than it can ever be conceived to be within classical accounts of development and motivation. It is *not* conflict that emerges from somatic states of tension or a blind cauldron of innate impulses, the satisfaction of which, according to Freud, "expresses the true purpose of an individual organism's life" (1940, p. 148).

At the same time, embodied within the evolutionary narrative are elements of the relational narrative of inherent relatedness, mutuality, and valid subjective truth. However, while there are substantial similarities, there are also essential differences between the evolutionary and the psychoanalytic relational narratives. In the evolutionary narrative, the innate mutuality and relatedness—the evolved, universal tendency to incorporate in one's self and to promote the interests of others—remains in constant, essentially unresolvable, tension with individuals' efforts to shape relationships toward their own biased, subjective perspective. Let us look more closely at how the evolutionary narrative may help us deconstruct several psychoanalytic models revealing what we believe are the partially valid yet incomplete and biased assumptions on which they are built.

The Classical Psychoanalytic Model and the Evolutionary Narrative

On countless occasions, Freud articulated a broad vision of the human condition at the center of which could be found the inherent conflict between children and parental, familial authority. This conflict was directly paralleled by the wider social conflict between individual interests and group or collective aims. The drama enacted in the family precisely parallels the social, historical, and phylogenetic narratives. In these accounts, as Mitchell (1988) notes:

> The innate needs of the individual organism necessarily and inevitably clash with other features of social and interpersonal relations.... it is not the objects or the discordant experiences which necessitate repression and self fragmentation; the child brings to his objects and experiences an array of rapacious desires which make social living and loving, by definition, a process of self-restraint and concealment. (pp. 72–73, 77)

Separated from the misleading connotations of its supposedly "somatic" sources in instinctual drives (a topic we shall discuss further in Chapter Eight), Freud's vision unquestionably captures a vital aspect of relational reality as construed in modern evolutionary theory. The evolutionary perspective suggests that conflict and its accompanying strategic deceptions and self-deceptions have been (over millions of years), and continue to be, intrinsic features of all object relational ties and interactions. This inherent, universal, conflictual dimension (relational theorists such as Kohut, Guntrip, Fairbairn, Fromm, and Sullivan notwithstanding) is understood to be a consequence of the absolutely normative clash of interests, and thus motives, between parents and children in even the fully "good-enough" environment.

Kleinian theory has, to a great extent, transmuted Freud's physicalistic drive theory (and the id as a structure) into relational metaphors of interactions between fantasied internal objects (Greenberg & Mitchell, 1983). As a result of this shift, Kleinian theory depicts—in perhaps the clearest, least encumbered form—the classical narrative of innate conflict and deception. In this sense (as we noted briefly in Chapter One), Kleinian narratives are quintessentially classical in terms of the continuum of psychoanalytic visions we have been discussing. And it may be particularly instructive to spend a moment considering the Kleinian view of parent–offspring conflict from an evolutionary perspective. We then move on to consider those "relational" theorists who, unlike Klein, broke away from the underlying classical narrative structure to envision a familial environment in which the interests of the parties largely overlap, inherent subjective biases are minimal, mutualistic intergenerational relations are the rule, high levels of inner unity are achievable, and deception and self-deception are signs of pathology and breakdown. First, though, to the Kleinian vision of innate conflict and complex strategies for dealing with the hidden, deceptive aspects of the average, expectable, good-enough family.

In what she called the "paranoid-schizoid position," Klein contended that, rather than accept the fact that the breast may have no more milk, the infant angrily demands that the "withheld milk" be brought forth. Although the use of the term "paranoid" connotes, of course, a pathological (irrational) process, it is now well accepted that the Kleinian "positions" (paranoid–schizoid and depressive) are best understood as referring to a universal developmental sequence in the organization of internal object relations (Grotstein, 1985; Segal, 1964). We believe that the Kleinian dynamic of paranoid suspicion and rage may be interpreted as a way of constructing and probing aspects of the relational world; a way that contains fundamental truths about the relational world's actual underlying nature: sometimes the milk (as well as other forms of parental investment) is, in fact, there (available) and the mother may not *want* to give it. That is, sometimes it may be in the best interest of the mother to withhold investment in order to pursue her other interests that do not coincide with those of the infant. For example, rather than devoting herself to the care of the infant to the degree that the infant wishes, it may be in her interest to direct her energies toward the care of the infant's sibling, to whom she is equally related and typically considers of essentially equal importance as the infant; however, from the infant's point of view the sibling is (as we note in Chapter Five), only half as important and her withheld investment may indeed conflict with the infant's own interests, aims, or needs.

The Kleinians cast their narratives in destructive oral and phallic imagery, which they often appear to take literally: the devouring infant projects sadistic impulses onto the mother and through introjection perceives her as dangerously devouring. While the degree of emphasis on sadistic, oral impulses as the source of such experiences certainly needs to be questioned (as we note in Chapter Eight), the Kleinian narrative of destructive, engulfing, and devouring infants and mothers may be an integral aspect of most mother/child experiences.

It is not unusual for children (and adult patients also frequently report this experience) to feel that their parents' agendas threaten to overtake their own, leaving them in states ranging from "confused" or "overwhelmed," to the terrifying feeling that their sense of self has been "lost" or is in the process of being "destroyed." From an adult perspective (that is, for the evolutionist from an adult *bias*) the chronic struggles in which the child engages to ensure that his or her will is not overpowered often seem far out of proportion to the issue at hand, for example, the adamant refusal to get dressed or go to bed, the demand or refusal of certain foods, the refusal to take a bath. While sometimes patient exploration with the child reveals specific fears, at other times it appears that the child is involved in a highly complex, interpersonal negotiation process. The child often experiences the negotiation as a crucial fight to protect and promote itself; as if the child's tiny, tenuous will and sense of self will be annihilated if the parent's agenda is not fought with every ounce of determination.

On the other hand, mothers (and, to a degree, fathers too if they happen to be the primary caretaker) *at some point* almost universally report feelings of a loss of a sense of self, or frightening exhaustion, or overwhelming depression—literal destruction of their identity—as the infant's and/or toddler's needs threaten to override and displace their own (Wolfson & DeLuca, 1981). This starts with the child's actual growth within the mother's body which, though often accompanied by a powerful sense of pleasure and well-being, frequently is fraught with significant anxiety regarding loss of control, body changes, weight gain, etc. This is followed by an actual experience of being eaten (sucked) which, though this too is frequently experienced as immensely pleasurable, can be extremely draining, so to speak, for it is accompanied, at times, by profoundly exhausting, sleep-depriving demands for attention (holding, stimulation, etc.) all of which can lead to feelings of being dangerously depleted (Wolfson & DeLuca, 1981). Conflicting agendas based on the divergence in the self-interests of parent and child are almost universally experienced at times as literal threats to the continued existence of both parties' selves. The notion of devouring and being devoured may not be too strong a way to express an essential aspect of the parent–child relationship, at least at certain points in time.

To the extent that the Kleinian narrative quite accurately depicts aspects of this relational scenario in metaphorical terms, it captures an aspect of the vital negotiation for investment in which all human infants must engage. If (over and above the adult bias) the Kleinian child's metaphors are ultimately somewhat overly concrete, exaggerated, and distorted, it is not unlikely that such extreme projective characterizations may be functionally necessary to counter the advantages of power and experience possessed by parents. Parents are far more able to remain smoothly and subtly unconscious of some of their own motives. They are able to organize and portray their own subjective biases in ways that, by and large, can be extremely convincing to the child. Thus, fantasied (projected and reintrojected) representations of dangerous objects may put the child in touch with what are essentially hidden realities: realities about parental identities and motives that are otherwise inaccessible to observation and inference based on direct interactive experience. A "Lockean child," so to speak, designed to function only as an empiricist, would be hopelessly lost in the ambiguities of the intricately organized and biased world of parental affect and interactions. In short, as Freud (1918) put it (trying to formulate an insight he had no adequate theoretical model to sustain):

> A child catches hold of phylogenetic experience where his own experience fails him. He fills in the gaps in individual truth with prehistoric truth; he replaces occurrences in his own life by occurrences in the life of his ancestors. (p. 97)

The child's management of his or her knowledge of highly ambiguous, vitally important (and thus highly emotionally charged), conflictual relational realities will be directly related to the child's later ability to cope effectively with the relational world—an ability that is eventually referred to as health or pathology. The environment plays a major role in fostering such health or pathology. Unlike Klein's (and, to a great extent, Freud's) emphasis on the playing out of inner fantasies *independently* from the responses and responsiveness of the environment, in the evolutionary narrative, the evolved, universal capacity to use primal fantasy, projection, and introjection would serve as a way of *interacting with* or probing the relational world. Inner, innate expectations (probabilities, potentials) are mixed together with relational realities to probe, provoke, test, and evaluate parental responses. This serves to ferret out truths about parental identities and to assess the extent and limits of overlapping interests).[1]

Ascribing what are sometimes terrifying, overwhelming, and at times traumatic parent–offspring conflicts to parental failure—the attempt to explain Kleinian scenarios as the pathological results of failed environments—may not be consistent with the evolutionary view. Freud and Klein certainly overplayed the deceptive, conflictual dimension of human relationships at the expense of the recognition of equally innate forms of mutuality and relatedness (Kriegman, 1988, 1990). Yet, we need to find a way to retain their vision of a psyche organized to deal with the enormously difficult and dangerous problems associated with the child's attempt to maximize his or her self-interest in the context of fundamentally necessary relationships with others whose agendas differ significantly from the child's and who are simultaneously engaged in the pursuit of their own interests.

The Normal Environment in the Relational Narrative

The basic narrative accounts of development given by most relational theorists—Winnicott, Fairbairn, Balint, Guntrip, Sullivan, Fromm, Kohut, Stolorow, et al.—are far more closely attuned than those of Freud or Klein to the realities of overlapping interests in the family. Drawn from a collectivist philosophical paradigm, their theories illustrate the inherent place of *others*, especially related others (the family), in the very structure of the individual self. In their emphasis on an "innate search for relatedness" (Mitchell, 1988), relational narratives do, indeed, penetrate beneath individuality. Human "sociality" (Sullivan, 1953), primary object love (Balint, 1965), and the embeddedness of human life within "a self-selfobject milieu" (Kohut, 1984) are concepts that, in our view, can be deconstructed as attempts to carry our understanding of the innate structure of the psyches of parent and child beyond what is, in effect, a social Darwinist (the struggle of all against all) sensibility portrayed in the child's psyche by Freud and Klein.[2]

---

[1]In Chapter Eight, we discuss the evolved relational functions of "endogenous" dynamics at greater length. And in Chapter Ten, we shall apply this perspective to the phenomenon of transference in the clinical situation.

[2]Though both Freud and Klein developed explanations for loving attachments (reaction formation, anaclitic love, reparation), in their systems these loving motivations are reactive or are secondary to more primary selfishness. For example, the reparative motivation to heal the mother is a result of the child's horror of his or her own destructive rage. It is not a primary motivation in its own right. Though it does modify the social Darwinist

Indeed, relational narratives are often told from the child's point of view, from the child's perspective. As a consequence, there is a tendency to deemphasize the child's quest to promote its own interests in favor of the (equally genuine but one-sided) child's search for relatedness. In Winnicott's system, for example, the role of the "transitional object" is to enable the child to deal with separation from the mother, to create a bridge between "internality" and "externality" that allows a true sense of the external (or otherness) to develop. The child's experience revolves largely around "missing mother"; the child substitutes a controlled, child-animated object as a precursor to a more fully internalized ability to experience separateness and manage separation anxiety. The Lacanians (P. Johannet, personal communication) question the accuracy of the exclusive emphasis on the effort to master separation and relatedness. They point out that it is not simply loss with which the child must struggle. Rather, the child comprehends that mother is pursuing *her own desires*. The child must do more than struggle to come to terms with a world in which one must deal with loneliness, separation, internality, and externality, fantasy, and reality. In addition, the child must deal with the fact that the needs and desires of those on whom he or she is dependent will frequently compete and conflict with those of the child.

The evolutionary narrative echoes this Lacanian emphasis on the ineluctable, universal presence of the "third force," the "third presence" (Lacan, 1977)—that is, an "other" who is external to the dyad and who represents the fact that the mother is not just invested in the infant/child—in all relational scenarios. From an evolutionary perspective, triadic or triangular (social) modes of construing relationships are present from the beginning of life. The psyche of the modern, homo sapiens child took shape over countless generations in which a significant degree of competition for investment with current siblings—or conflict with the parental capacity to invest in future siblings, and myriad other parental kin ties and reciprocal altruistic social ties—was (and continues to be) the rule. To the extent that natural selection has favored any form of innate predisposition or evolved strategy for coping with the expectable relational world, the intuitive sense of the existence of the "third force" or "third presence" is, in our view, apt to be of prime functional importance for an individual facing the challenges of development.

The problem of accounting for both the innate tendency to seek

---

sensibility, in the Kleinian narrative the depressive position and the reparative motif are a result of the child's awareness of the fact that his or her own motives (as well as those of others) are, in fact, initially, at the most "primal" level, consistent with this "all against all" world view.

mutualistic, reciprocal ties *and* to inherently compete and deceive oneself and intimate others is thus at the heart of constructing a viable psychoanalytic narrative. A look at Kohut's brilliant but one-sided efforts to revise the classical paradigm and develop a more accurate view of the profoundly relational human psyche will help to further illustrate the dilemma of the relational narrative.

Kohut (1982) posed an important question when he asked:

> What stands in the way of the acceptance of our outlook, why can we not convince more of those who have espoused the traditional psychoanalytic outlook that intergenerational strife ... refers not to the essence of man but ... [is a] deviation ... from the normal, however frequently ... [it] may occur? Why can't we convince our colleagues that the normal state, however rare in pure form, is a joyfully experienced developmental forward move in childhood ... ? (p. 402)

While agreeing with the estimate of the near ubiquity of the occurrence of the Oedipus complex, Kohut "disagrees completely with traditional psychoanalysis concerning the *significance* of human intergenerational strife" (1982, p. 402). He adds:

> the normal state ... [includes] the step into the oedipal stage, to which the parental generation responds with pride, with self-expanding empathy, with joyful mirroring, to the next generation, thus affirming the younger generation's right to unfold and to be different.... We believe ... that in the last analysis we are not dealing with an uninfluentiable conflict of basic opposing instincts ... but with, at least potentially remediable, interferences that impinge on normal development. (pp. 402–403)

Kohut was reacting to the one-sidedness of the classical narrative, or as he put it more specifically, to:

> Freud's ... view that man's essential nature is defined with reference to intergenerational strife, above all and in particular ... the paradigmatic intergenerational conflict between father and son.... (p. 403)

In contrast, Kohut gave us a narrative in which the father is

> warmly committed to the next generation, to the son in whose unfolding and growth he joyfully participates—thus experiencing man's deepest and most central joy, that of being a link in the chain of generations (p. 403).

It is crucial that in order to sustain this account (in which, from our point of view, only kin altruism is stressed) it is necessary to relegate many ubiquitous manifestations of competing aims, self-deceptions, and deceptions to pathology. For example, Kohut (1984) suggested that while intergenerational conflict is ubiquitous, this does not imply it is normal. He notes that dental caries are also ubiquitous yet are clearly a disease process. Kohut suggests that intergenerational strife is avoidable:

> It is only when the self of the parent is not a normal healthy self, cohesive, vigorous, and harmonious, that it will react with competitiveness and seductiveness rather than with pride and affection when the child, at the age of 5, is making an exhilarating move toward a heretofore not achieved degree of assertiveness, generosity, and affection. And it is in response to such a flawed parental self which cannot resonate with the child's experience in empathic identification that the newly constituted assertive-affectionate self of the child disintegrates and that the break-up products of hostility and lust of the Oedipus complex make their appearance. (1982, p. 404)

Kohut was at pains to insist that he did not *equate* conflict and pathology. Conflict is ubiquitous, he claimed, but only leads to pathology when the parental response is deficient:

> self psychology does not consider drives or conflicts as pathological.... Three cheers for drives! Three cheers for conflicts! They are the stuff of life, part and parcel of the experiential quintessence of the healthy self.... But such experiences are not tantamount to the drives, conflicts, guilts, and anxieties of the Oedipus complex.... They do not, in other words, bring about the *type* of conflict... that constitutes the nucleus of the classical transference neuroses of adult life.... (Kohut, 1983, pp. 388–389)

Kohut (1984) states that it is "flawed" parental responses and "parental pathology" that result in the pathological Oedipus complex (p. 24). Yet he goes further to argue that conflict is not pathological, that the human psyche is structured to resolve conflict caused by optimal frustrations, and that some of the resolutions of fairly severe conflicts have yielded some of the most productive and creative psyches (1984, pp. 44–45). Kohut repeatedly tries to differentiate conflict from pathogenic conflict (e.g., 1984, p. 53).

> if the self is healthy, the drives are experienced not in an isolated fashion but as an imminent modality of this healthy self; and that, under these circumstances—even when we set ourselves up against

our aggression and lust—pathogenic conflicts will not arise, however great our pain, however absorbing our struggles. (1984, p. 208)

Yet it is equally clear that Kohut ultimately linked *significant intergenerational* conflict between the parent's self and that of the child as well as conflict within the self (dividedness or lack of cohesion) to environmental deficiency and pathology. Any notion of *inherent* conflict remained secondary to the vicissitudes of the self-selfobject milieu with its basically mutualistic form of relating based on shared interests and aims.[3]

In the work of the intersubjectivists (Stolorow, 1985; Atwood & Stolorow, 1984; Stolorow et al., 1987), we can observe an equally fascinating and instructive example of the difficulties the relational narrative encounters when one recognizes the existence of conflicts of interest between selves (nonmutuality) and what are essentially conflicts of interest within the self, that is, conflicting selfish and mutualistic aims. Like Kohut, the intersubjectivists clearly appreciate the presence of conflict in at least some of the ways in which human subjectivity and intersubjectivity are experienced. Yet, they invariably seem to see significant parent–offspring conflict as emanating from the pathological narcissistic needs of the parent. For example, consider the suggestive tone of the following:

> The selfobject concept provides a depth-psychological framework for understanding the developmental significance ... of the exquisitely coordinated reciprocal regulatory patterns disclosed by infancy researchers. ...
>
> Wolf (1980) alludes to this point [when he notes that] ... The cumulative effect of "a history of finely attuned reciprocal responses ... gradually [transforms] baby and mother into custom-fitted parts of a unique mother–child unit" (p. 124). (Atwood & Stolorow, 1984, p. 68)

Infant researchers no doubt observe such events, but the fact that overlapping interests lead to interactions characterized by harmony and mutuality is only part of the story. The same infant researchers have plenty of access to descriptions of disharmony and significant conflict. How the conflictual scenes are interpreted, that is, the narrative structure that underlies the interpretation, is crucial. For Atwood and Sto-

---

[3]Other self psychologists run the gamut from those who place intergenerational conflict—as classically conceived—offstage in the wings, to those who follow Kohut's wording when referring to conflict, to those who would try to bring a reformulated concept of conflict back to stage center (Kriegman & Slavin, 1990).

lorow (1984) the harmonious scenes are a result of healthy parental functioning and the disharmony is a result of parental psychopathology or failure.

> When the psychological organization of the parent cannot sufficiently accommodate to the changing, phase-specific needs of the developing child, then the more malleable and vulnerable psychological structure of the child will accommodate to what is available. A number of pathological outcomes are possible. (p. 69)

After listing some of these pathological outcomes, they describe some of the factors that can lead to parental failure to "sufficiently accommodate to the ... needs of the developing child."

> [These] factors may result in a weakened and unsupported sense of self in the mother, leading to rigidity, distancing, and other interferences in her function, and contributing from the beginning to a constricted field in which psychological patternings are thrown out of kilter and the child's development impeded. (p. 70)

Note that not only is the tone one of mutuality, reciprocity, and a lack of inner tensions or divisions in healthy functioning, but only pathological parental functioning is credited with causing significant problems for the developing child. Indeed, conflicts "within the self" are rarely, if ever, described in terms of the competing aims embraced by the child's superordinate self. Rather:

> the formation of inner conflict, whether in early development or in the psychoanalytic situation, always takes place in specific intersubjective contexts of selfobject failure. (Stolorow et al., 1987, p. 26)

The intersubjectivists clearly attempt to avoid blaming the parent or analyst by recommending the phenomenal term "selfobject failure" over the seemingly more "objective" assessment implied by the term "empathic failure." In their terminology a selfobject failure is

> a subjectively experienced absence of requisite selfobject functions. ... Misunderstandings on the part of the analyst may or may not be experienced as selfobject failures, depending on their specific transference meanings for the patient. (Stolorow et al., 1987, p. 17)

Yet, the implication that environmental failure is the source of pathology is unavoidable:

As we have stressed, the need to disavow, dissociate, or otherwise defensively encapsulate affect arises originally in consequence of the failure of the early milieu to provide the requisite, phase-appropriate attunement and responsiveness to the child's emotional states. (1987, p. 74; see also pp. 98–99)

Further, it is clear that in this view environmental failure is the result of pathology in the parents. In discussing parental failure to help a child integrate discordant affect states, they speak of a parent who "*must* perceive the child as 'split'" (p. 71, emphasis added) into "good" and "bad." In discussing "the route by which inner conflict becomes structuralized" they refer to parents who "*cannot* adapt themselves to the changing needs of their developing child" (p. 90, emphasis added). Nowhere do they mention problematic conflicts that could result from selfobject failures (using their definition) that result from *inherent conflicts in interest* between the parent and child. It is hard to see how the intersubjectivists, like Kohut, Fairbairn, and Sullivan before them, can escape the "pull" of the relational narrative in which the relational world is conceived as a place in which the interests of individuals normally converge and the self is not inherently torn by its own internal representation of different (selfish and mutualistic) facets of itself.

The assumption that the healthy parent–child relationship is essentially free of significant conflict is implicit in the discourse of most (non-Kleinian) relational theorists from Fairbairn to Guntrip to Sullivan and (after larger societal change takes place) Fromm. Healthy parents will respond helpfully, and without substantial conflict, to the child's developmental needs. Minor failures (nontraumatic, nonpathogenic) will inevitably occur. But, failures that produce pathology, or produce an inherently conflicted inner experience (with significant repressions or disavowals), are a result of parental pathology, generally pathology in the narcissistic sphere. Healthy parents are in tune with their child. Unhealthy parents have significant conflicts with their offspring.[4]

A good example of how the *inherent* conflict in interests is lost in relational attachment theories can be found in Mitchell's (1988) discussion of these issues in the context of his attempt to create a "relational conflict theory." In trying to explain and elaborate some of the com-

---

[4]In response to the evolutionary perspective, Stolorow (personal communication) acknowledges that it may universally be necessary for children to repress versions of their experience in order to adapt to parental biases. However, in Stolorow's view, such universal repression does not engender psychopathology. Pathology emerges from the child's accommodation to more extreme versions of parental bias that presumably cross a threshold at which point the parent's subjective bias becomes classifiable as a failure to meet the child's needs.

plexity inherent in applying Fairbairn's notion that "libido is object seeking" to the parent–child relationship, Mitchell states:

> Parents are highly variable emotional presences, whose experience of the child is entangled with their own security needs and therefore embodies narcissistic meanings. The parents' capacity to be reached and moved by the child is necessarily obscured by and hostage to the vulnerabilities and conflicts of the character structure of the individual parent. (p. 104)

Note that in Mitchell's lucid account of the universal tensions inherent in the parent–child relationship, the only parental tendency that conflicts with the needs of the child are the parent's security needs and parental pathology (vulnerabilities and conflicts). There is no *inherent* conflict in interests between the parent and the child; no hidden agendas; no biases or inherent motivations to manipulate in a way that could potentially be harmful to the other. Problems appear to arise primarily from certain unspecified difficulties (difficulties that can never be fully explained because, we suggest, once the source of the problems has been eliminated any explanation will be incomplete) in pursuing the one universal agenda: maintaining attachment. Parents have no healthy agenda to use the child (and vice versa) in ways that may be injurious to the other unless they are simply trying to manage their anxiety (security needs). Mitchell goes on to describe how:

> Freud's concept of the superego is predicated . . . on the assumption that the child finds it necessary to bend himself out of shape to accommodate himself to the social environment; but for Freud, accommodation is necessary because of more fundamental asocial drives. Relational-model theorists like Winnicott, Sullivan, and Farber assume a powerful primary need for interpersonal connection, which makes it necessary for the child, to one degree or another, to shape himself to the parents' vision of him; to present himself in a way that is both visible and palatable; to become, in Sullivan's terms, the parent's definition of "good me." Thus, the child inevitably becomes captive to the parents' world of meanings and values. (p. 106)

In the movement from Freud's narrowly defined, mechanistic drive conceptions to Mitchell's relational conflict model, a crucial dimension of the *inherent* conflict of interests is lost. Yes, we are a social species for whom relational attachment is both problematic and of primary importance. Yet this is simply a *description* of the problem. It does not account for *why* parents regularly attempt to mold the child or *why* presenting a "palatable" image to the parents is fundamentally inimical to the child's interests. We are motivated to do more with one another than just attach, stay close, and provide mutual support and comfort. We are also designed to *use* one another, and to guard against being used in ways that are dangerous to ourselves, in order to maximize our inclusive fitness. And more than the need to simply maintain attachment, it is the inherent conflicts in interests that account for self-distorting repressions and accommodations on the part of the child and later the adolescent and adult. Libido may be, as Fairbairn put it, "object seeking" as distinct from purely and blindly "pleasure seeking"—as in the narrowest version of the classical model. Yet, "libido" seeks objects for larger purposes (the achievement of which may sometimes be significantly signaled by the experience of pleasure). Overall, the adaptive problem of the human child is far more complex than has ever been dealt with by attachment theories.

The relational narrative corrects the classical overemphasis on conflict in human relations and, especially, the overemphasis on conflict that originates in the child's (immature and driven) ways of pursuing its interests. However, from an evolutionary perspective we cannot accept the view that healthy parents will not engage their children in significant conflicts that are at times deleterious to their offspring. Nor can we accept the view that problematic conflicts arise *solely* from parental *pathology*, usually in the narcissistic sphere. There is no basis for the assumption that the child's needs do not inherently conflict with other parental needs, including *healthy* narcissistic parental needs.

Revising the Classical Narrative: An Attempt to Remove the Bias

Some theorists who basically operate within the psychoanalytic classical model have attempted to retell that narrative from a point of view that focuses more clearly and directly on parental (particularly paternal) aims (see Munder-Ross, 1986). While Freud and Klein took what is basically the parental point of view on parent–offspring conflict—locating subjective distortions and destructive, selfish motivations in the instinctuality of the child—these later theorists emphasize envy and destructiveness in the father. Munder-Ross (1986) describes "a feedback system between the adult's Laius and the son's Oedipus constellations" (p. 124).[5]

In this retelling of the central version of the classical psychoanalytic narrative of conflict and deception, the parental (paternal) motives (and

---

[5]Laius, Oedipus's father, abandoned and mutilated his son in an effort to abort the prophesy that, as punishment for his rape of a young boy, he would be murdered and replaced by his own son.

accompanying deceptions) counterbalance the impulses, fantasies, and defensive distortions customarily attributed to the Oedipal child. There is a clear approximation in this retelling to what we have depicted as the *evolutionary narrative of conflict*:

> two individuals rather than isolated cauldrons of impulses toward constant objects . . . may best capture the unfolding, dovetailing and often unconscious communication of object relations. (Munder-Ross, 1986, p. 139)

Though much closer to the evolutionary narrative, this kind of classical revision still remains embedded in a version of drive theory and is narrower than the evolutionary model in its depiction of the meaning of the conflict between father and son. (See discussion at end of Chapter Five.) The emphasis on the universality of paternal aggression—in the sense of hostile, destructively envious wishes—seems to be an attempt to capture and characterize the fact that fathers will be geared toward the promotion of their own interests even, at times, in ways that may be destructive to their sons. Although violent wishes and destructive envy may be *parts* of such an overall paternal bias, they are, nevertheless, not likely to be the only or even primary form this bias may take. Hostile urges and envy are only parts of a much larger clash of interests that is simultaneously intertwined with truly overlapping interests. In this sense, paternal hostile aggression would represent what is essentially one inner, affective means used to effect larger, more complex relational ends. The danger lies in focusing on the supposedly universal urges and affects outside the adaptive context that shaped them and in which they continue to function.

A similar criticism can be directed at Winnicott's (1950) emphasis on parental (countertransferential) *hate* and ruthlessness as essential ingredients in the negotiation of genuine forms of relating. The issue is one of negotiating competing and overlapping interests. In the context of such negotiations, aggressive affects of all kinds may serve as functional signals and motivations. We return to this important theoretical question in our discussion of endogenous drives in Chapter Eight and developmental negotiation in Chapter Nine. For now we shall leave our attempt to revise the notion of what is average, expectable, and good-enough in the universal developmental setting with the following considerations.

There is a profound, essential difference in what is in the best interest of any two individuals. The human experience is indeed tragic, but the tragedy is the need to come to terms with what, along with our mortality, is the most basic narcissistic injury of all: because we are members of a sexually reproducing evolved species, nobody loves us as much as we love ourselves. In spite of the fact that much parental failure can be attributed to parental narcissistic defects, such parental pathology may simply "add (a terrible and damaging) insult to (inevitable, though not necessarily pathology producing) injury" (Kriegman & Slavin, 1990).

Even considering the overlap in interests between parent and child, there are, of course, huge variations in the degree to which parents (and caregiving environments as a whole) are able to recognize and respond to the child's unique self-interest, and this, to be sure, will profoundly affect the child's capacity to manage and integrate those intrinsic, universal tensions and inner divisions that we are depicting as an inherent part of relational reality itself. Though a crucial clinical issue, for the moment suffice it to say that the ways and the amount that these normative, universal conflicts influence the development of patterns we call "psychopathological," may be more of an open question than is usually recognized, particularly by retrospective analyses in which problems existing in the caretaking environment during early experience are assumed to be the primary factor in shaping later pathology (see Mitchell, 1988).

It may make a real difference in our eventual understanding of pathology whether we think about the normal relational world as inherently conflicted with only divergent, clashing interests, as inherently mutualistic with basically convergent interests, or as a world of intrinsic mutuality as well as divergent interests, deceptions, and motivated subjective biases (all of which become ambiguous because individuals engage in powerful deceptions to further their own inclusive self-interest). Might the dilemmas and conflicts encountered in the normal course of development be sufficiently complex, confusingly ambiguous, at times wrenchingly painful—and only partially soluble—to frequently lead to subjective distortions of reality and problematic behavioral patterns? Do not such universal, relational dilemmas remain, for many individuals, very deeply and elusively entwined with other facets of psychopathology?

CHAPTER SEVEN

The Paradoxical Challenge of Human Adaptation: Constructing a Self in a Biased, Deceptive Relational World

> [An individual] is the person most interested in his own well-being. The interest which any other person...can have in it is trifling compared with that which he himself has.... [W]ith respect to his own feelings or circumstances, the most ordinary man or woman has means of knowledge immeasurably surpassing those that can be possessed by anyone else.
> —JOHN STUART MILL, *On Liberty*

> I am a slave [to the Other] to the degree that my being is dependent at the center on a freedom which is not mine... insofar as I am the instrument of possibilities which are not my possibilities, whose pure presence beyond my being I cannot even glimpse...*I am in danger.*
> —JEAN PAUL SARTRE, *Being and Nothingness*

Creating a Self: Paradoxes of Human Psychological Evolution

We have now outlined our case for the existence of an inherent mixture of genuine mutuality, profound conflict, and strategic deception in the "evolutionary relational matrix"—the ancestral (and current) relational environment to which our evolved deep structure is adapted. Given this picture of the setting for which we are psychologically designed, what are the major, predictable psychological challenges posed by such a relational environment for the maintenance and enhancement of our inclusive fitness? Over the course of development, what must the evolved psychological deep structure of the child (and the adult) enable us to accomplish in the relational world? In short, what is the central adaptive dilemma of the human child?

The Overall Human Developmental Strategy

As a species, we have "sacrificed" certain kinds of hard-wired, innate knowledge about ourselves and the world. In exchange, we developed the capacity to construct a map of the self and the object world based on our ongoing experience of ourselves, the world's response to us, and the complex interplay that results as these phenomena become intertwined with each other as well as with our reactions to them. All of this initially takes place throughout an extremely prolonged period of childhood dependency. We are born with what the great evolutionist Ernst Mayr (1974) has called an "open program." We are "hard-wired" with a program that permits (and requires) us to learn much about ourselves and the world through interactional patterns. Innate wiring lies behind the very capacity to observe, organize, and make inferences about our experience. We make use of the information provided by our inner states and by the relational world in order to come to conclusions about who we are, where we fit in our particular familial and cultural scheme, and how to assess and promote our own interests. In the course of the time during which this quintessentially human adaptive strategy took shape—step by step through mutation, genetic recombination, and natural selection—those individuals designed to develop and learn most effectively in this complex, interactive manner had greater reproductive success than their compatriots. The highly dependent, slow to develop, but highly flexible, symbol-using child of today was favored by natural selection, that is, was the relatively "fit" child of this evolutionary narrative (La Barre, 1954; Konner, 1982).

In the course of human development that follows from such a design, most intersubjective transactions are mediated by language and other forms of symbolic communication. Much that is communicated and learned about reality (including crucial realities about the self) is not rooted in direct observation but, rather, comes transmitted through a complex system of signifiers—signs, symbols, and icons (Langer, 1942). What is communicated is a highly encoded version of reality, one that is inevitably infused with a great deal of "extra information," or meta-communication (Bateson, 1972) added by the sender. Through language and other forms of symbolic communication, the human child is able to efficiently construct a map of the world that far exceeds (and differs in quality from) anything that could be created from his or her direct experience (Konner, 1982). At the same time, this unique feature of human symbolic communication—its liberation from the need for direct observation—greatly amplifies or potentiates the power of the parental generation, as well as the child, to create *biased* versions of reality and to communicate them in strategic ways in the course of development.

Looking at our psyche as an evolved adaptive organ, we see several basic design problems inherent in the particular challenges faced first by the child in adapting to this symbolic medium within the family, and then by the adult in what we believe is a negotiation with a largely symbolically constructed culture. First, the child should be expected to be highly capable of (and intensely interested in) carefully monitoring the parental environment and evaluating its capacity and willingness to invest in the child's future. Children are likely to have been shaped by natural selection to try to maximize the amount and quality of investment (i.e., genuine interest, love, constructive guidance) they receive from their families. Equally, they are likely to have been designed to try to minimize the costs of such investment in terms of the inevitable tradeoffs—the attached strings and biases inherent in familial influences or "molding," as Trivers (1974) put it. The complicated affective and cognitive components necessary for these capacities may even have been critical to survival during the crucial, phylogenetically formative years—the many millennia during which our psyche evolved in its ancestral environment.[1] Above the need to maximize the chances of its own *survival*, it is likely that strong selection pressures throughout our evolutionary history have shaped the intuitive and emotional capacities of the child to pursue a strategy designed to *enhance* its own ends. The child's strategy is designed not simply to safeguard its own survival, but beyond this, to maximally promote its own aims, including its own relative advantage within the kin setting.

But what do we imply when we speak about these aims as though they exist somehow prior to and independent of the relational context that shapes them? If the human self is designed to develop within a context of shared, learned meanings, what aspects of the self and its meanings can we speak of as "essential" or "innate" as distinct from "relationally constructed?" How do universal, evolved human capacities relate to both unique individual givens and the operation of vitally

---

[1]The few studies that have actually looked at the relationship of different infant behaviors to survival have tended to confirm that, under precarious subsistence conditions, survival rates are significantly higher for infants who are seen as actively demanding care (crying more, protesting, etc.) (Blurton Jones & da Costa, 1987). Indeed, it is considered possible that under the conditions in which we evolved, infants who are able to maximize investment in themselves to the extent of delaying the birth of siblings, dramatically enhance their own chances of surviving (Blurton Jones & da Costa, 1987) (see also the box on p. 65).

needed (yet biased) influences and meanings constructed and provided by the relational world? We started to touch on this issue in Chapter Four, where we began to view the self in terms that transcended its phenotype, that is, our "gene's-eye" perspective in which we saw the boundaries of the self as being, in a meaningful way, distinct from and not coincident with the skin boundary of the body. We shall now elaborate on how this crucial issue weaves itself into our overall evolutionary–adaptive vantage point on the design requirements, so to speak, for a viable individual human psyche. In Chapter Eight, we begin to relate this vantage point in far greater detail to the problems that confront us as we attempt to critique and revise psychoanalytic models.

The Innate, Unique Aspects of the Self

Increasing evidence from a sophisticated new generation of well-controlled twin studies (see Farber, 1981; Tellegen et al., 1988), as well as observational studies of innate temperamental differences (see Brazelton, 1969; Green, Bax & Tsitsikis, 1989), lends enormous weight to the view that each child is born with a unique configuration of emotional and intellectual characteristics, dispositions, inclinations, vulnerabilities, and resiliencies (Neubauer & Neubauer, 1990). Consider the following:

> [Identical twin] girls were separated in infancy and raised apart by different adoptive parents. . . .
> When the twins were two and a half years old, the adoptive mother was asked a variety of questions. Everything was fine with Shauna, she indicated, except for her eating habits. "The girl is impossible. Won't touch anything I give her. No mashed potatoes, no bananas. Nothing without cinnamon. Everything has to have cinnamon on it. I'm really at my wit's end with her about this. We fight at every meal. She wants cinnamon on everything!"
> In the house of the second twin, far away from the first, no eating problem was mentioned at all by the other mother. "Ellen eats well," she said, adding after a moment: "As a matter of fact, as long as I put cinnamon on her food she'll eat anything." (Neubauer & Neubauer, 1990, p. 20)

This is an excellent example of the complex intertwining of innate aspects of the self (in this case, the preference for cinnamon) and the *meaning* these innate features may come to take on in different environmental (intersubjective) contexts. It is undoubtedly more likely that Shauna (rather than Ellen) will one day describe the struggles with her mother around self-definition (with possible oral imagery and memories) to her analyst. In an environment that is only partially attuned to an individual's interests, the developing self faces an immense problem in constructing subjective meanings that authentically express core aspects of its uniqueness.

Consider another example:

> Identical twin men, now age thirty, were separated at birth and raised in different countries by their respective adoptive parents. Both kept their lives neat—neat to the point of pathology. Their clothes were preened, appointments met precisely on time, hands scrubbed regularly to a raw, red color. When the first was asked why he felt the need to be so clean, his answer was plain.
>
> "My mother. When I was growing up she always kept the house perfectly ordered. She insisted on every little thing returned to its proper place, the clocks—we had dozens of clocks—each set to the same noonday chime. She insisted on this, you see. I learned from her. What else could I do?"
>
> The man's identical twin, just as much a perfectionist with soap and water, explained his own behavior this way: "The reason is quite simple. I'm reacting to my mother, who was an absolute slob." (Neubauer & Neubauer, 1990, pp. 20–21)

The obsessive–compulsive character styles of both men *might* be derived from the sources to which they attributed them. On the other hand, the fact that they are identical twins suggests the possibility that their similar character styles were in some way related to certain characteristics that each possessed that were, to some degree, independent of the environment in which they were raised. The fact that the twins' post hoc explanations of their compulsivity sound like confabulations that are complicated self-deceptions is, of course, interesting in itself. And it should give us some pause in evaluating our clinical inferences that may, in some cases, have the same quality of creating meanings because we have a need to make the unknown explicable. Yet, these "compulsive twins" may have responded, each in his own innate way, to extremes in maternal response, whether that extreme was compulsivity or absolute disregard for order. The issue of how innate individual dispositions interact with the intersubjective surround is, of course, extraordinarily complicated. We are simply emphasizing here that each child brings something uniquely his or her own to every interpersonal interaction that then shapes further development.[2] Often, these specific differences are highly relevant to the type of response one receives from the external world.

---

[2] See the related discussion on page 77.

In their report of many studies of identical twins reared apart, Neubauer and Neubauer (1990) concluded that "when there was either a low or high level of engageability in one infant, it *was shared* by the identical twin" (p. 45); in "comparably nourishing families," social smiles appeared in these infants, "a smile for a smile—within days of each other" (p. 60); and "a delay or difficulty in one twin's toilet training was almost always accompanied by a similar delay in his co-twin" (p. 61).

It is of utmost importance in the development of the individual self how one's unique features mesh with or conflict with the relational surround and how others then respond to the preferences for cinnamon or orderliness, or to one's high or low engageability. Such predilections may become fraught with pathological conflict; alternatively they may simply become interesting aspects defining the individual, or become important adaptive, functional features that can be incorporated into a healthy self. The complex interaction of *individual uniqueness* with the *specific relational world* in which one develops is what is most important.[3]

Existing studies of innate influences on personality now almost invariably take an "interactionist" perspective on the way innate, individual differences vary in their expression and meaning in different environments. However, the importance and nature of the *universal* aspects of human deep structure that are "designed" to process innate, *individual* inclinations in light of specific environmental conditions and meanings are rarely, if ever, depicted or described. Virtually all classical psychoanalytic models give a nod to "constitutional" differences. Freud repeatedly tried to include a recognition of innate individual characteristics:

> Each ego is endowed from the first with individual dispositions and trends [and] before the ego has come into existence the lines of development, trends and reactions that it will later exhibit are already laid down for it. (Freud, 1937, p. 209)

---

[3] Also note the fact that environmental influences begin before birth. Identical twins whose placentas are attached differently and therefore may obtain different amounts of nutrients in utero may have significantly different birth weights. While infants, in general, seem to show more of an interest in the human rather than the nonhuman environment:

> this orientation was often correlated with birth weight. Lower-weight twins focused more on the human presence, whereas the heavier ones interacted with a wider range of stimuli, including inanimate objects. One possible explanation for this phenomenon, though we should stress it is tentative, is that from an evolutionary standpoint it might be beneficial for a weaker infant to focus on people, who can provide the food and nurturing objects cannot. (Neubauer & Neubauer, 1990, p. 48)

Yet, with the exception of relatively vague references to the "true" (Winnicott, 1960a) or "nuclear" (Kohut, 1984) self in relational models, the process by which the child achieves some kind of access to the elements of his or her innate individuality—and becomes able to articulate it in environmentally relevant terms—has not been explored [see Bollas's (1989) notion of "personal idiom" for an interesting exception to which we shall return in Chapter Eight]. In other words, there has been little thought given to understanding how our psychodynamic architecture may be designed in some way to know, to adequately take account of, and, to the extent possible, promote inborn individuality and uniqueness.

It should be clear that we certainly do not mean to imply that the fully developed self, in its ultimate psychosocial form, is (in any psychoanalytically meaningful sense) "present at birth." What is present is some type of unique inner configuration that will, in interaction with the environment, become realized as that self. It is little more than a truism to say that given the same set of external developmental contingencies, no two individuals will experience them or react to them in the same way; although this fact is regularly overlooked in many retrospective clinical discussions. What is less obvious is that while the environment must powerfully influence the development of the self—as humans we can develop in no other way—the social environment *must not* define selfhood beyond certain limits. In fact, no adaptive design fitted to our relational world with its inherent biases and competing interests would regularly "allow" the social environment to have effects beyond these limits.[4]

For this reason, the evolutionary model leads us to seriously question the more extreme relational position put forth by theorists such as Mitchell (1988) and Atwood and Stolorow (1984). Reflecting a set of assumptions that are characteristically made by many theorists operating within the relational narrative, Mitchell (1988) asks:

> Why are relations with others the very stuff of human experience? . . . Why are our earliest relationships with others so crucial that we are actually composed of these relationships—"precipitates," as Freud (1923) put it, of our earliest attachments?" (p. 20)

As noted above (Chapter Two), Mitchell posits that the human psyche has no other a priori feature than the need for relatedness itself. While

---

[4]This conception of the need for a psychological design that checks social influences echoes Rappaport's (1960) notion of the way in which the drives guarantee a certain "autonomy from the superego"; or, as Eagle (1984), more recently put it, the instincts serve to resist the individual's "enslavement by society."

a reasonably good-enough environment, serving adequate selfobject functions, provides indispensable nourishment for the development of the self, the environment is not ever a "neutral" party or wholly an "ally": it is invested in having the individual develop what is, in effect, a somewhat distorted sense of self—a picture that is biased toward others' self interests. Thus, we believe that our innate, a priori inner preparation for functioning in such a world *must* include far more than attachment needs that lead to self–other schemas created from "precipitates of our earliest attachments." A self designed for the human relational world must be prepared to engage in an extraordinarily complex set of developmental strategies that serve, in part, to defend against having one's interests usurped by others. Such evolved strategies must go beyond the universal need for self-differentiation and self-demarcation noted by Atwood & Stolorow (1984). They must include motivations that lead one to maximize the utility of one's ties to others (to use others for one's own vital ends), in addition to simply creating desires for attachment and self differentiation.

Existential Conflict and Psychoanalytic Intersubjectivity

The branch of current psychoanalytic phenomenology that is emerging from the tradition of self psychology (Atwood & Stolorow, 1984; Stolorow & Lachmann, 1980; & Stolorow et al., 1987) has brought an extremely lucid conceptual focus to the universal issue of how we construct and regulate our subjective worlds in face of the enormous defining power (and power to thwart adequately self-interested self-definition) inherent in the environment. Through the focus on the phenomenology of subjective experience and its relation to reality found in the work of several major, modernist existential philosophers (primarily Husserl, Heidegger, and Sartre), the intersubjectivists build a foundation for examining the validity of an individual's subjectivity and its vulnerability to assault from the external world. However, from our point of view, the intersubjectivists tend to explain away a crucial dimension of the human adaptive dilemma as it is powerfully captured in the existentialist focus on the phenomenology of alienation: the universal conflicts between ourselves and others that we each must face in constructing and maintaining a viable, cohesive, and vital self. Instead, they go on to pursue a psychoanalytic theoretical strategy that essentially reduces this central, universal dilemma to psychopathology.

Where Heidegger and Sartre saw ontology (the essential nature of being in the larger, human world), the intersubjectivists see pathology (the failure of being as a function of invalidation by others). Pathology is, inevitably, a function of failing and inadequate environments. Let us look a bit more closely at the extraordinarily well-articulated view of the existential nature of reality and human subjectivity that emerges from the work of the intersubjectivists. We shall see how a central, human adaptive dilemma contained in that same existential viewpoint appears to be lost in the intersubjectivists' ultimate assimilation of this issue into what we consider to be a version of the relational narrative.

Atwood and Stolorow (1984) point out:

> In Heidegger's thought ... being-in-the-world is inherently and indissociably a being-with-others ... which has the property of alienating Dasein [or the mode of being belonging to a person] from its own true self. This is because Dasein's concern for others supposedly includes a constant care as to the way one differs from them. Such care gives rise to a tendency to become like others, to allow them to define who and what one should be. (p. 18)

> Dasein, as everyday Being-with-one-another, stands in *subjection* to Others. It itself *is* not; its Being has been taken away by the Others. Dasein's everyday possibilities of Being are for the others to dispose of as they please.... Being-with-one-another dissolves one's own Dasein completely into the kind of Being of "the Others." (Heidegger, 1927, p. 164, quoted in Atwood & Stolorow, 1984, p. 19)

Atwood and Stolorow (1984) go on to describe a closely related theme in the work of Sartre:

> The consequences of the "objectness" of being-for-others includes a severe threat to the continued life of the for-itself as an autonomous center of freedom. When a person comes under the gaze of another, he grasps the Other as a freedom that constitutes a world of meanings and possibilities around itself. This understanding may then extend to a sudden recognition that he is himself in the process of being articulated within the structures of that alien world, which threatens to displace his own and absorb him into pure objectness. He senses a foreign outline being imposed upon his nothingness, and without knowing what this outline is he feels himself being stripped of his subjectivity and transformed into an object. (pp. 27–28)

> I am a slave [to the Other] to the degree that my being is dependent at the center on a freedom which is not mine ... insofar as I am the instrument of possibilities which are not my possibilities, whose pure presence beyond my being I cannot even glimpse, and which deny my transcendence in order to constitute me as a means to ends of which I am ignorant—*I am in danger*. This danger is not an accident but the permanent structure of my being-for-others. (Sartre, 1943, p. 28, quoted in Atwood & Stolorow, p. 28)

The starkness—and, from our point of view, the one-sidedness—of this modernist view of individual alienation is clearly recognized by the psychoanalytic intersubjectivists. In Sartre's play *No Exit* (1947),

> Human relationships are ... pictured as never-ending battles between competing subjectivities struggling to strip each other of freedom and reduce each other to objects. (Atwood & Stolorow, 1984, pp. 28–29)

In reaction to what is essentially a pure narrative of interpersonal conflict and deception (purer and harsher than Freud or Hobbes), the intersubjectivists move radically and totally in the other direction, that is, to an assimilation of the existentialist's vision of the interpersonal conflicts and dangers into the psychoanalytic relational narrative. In doing so, however, they relegate this aspect of the existentialist vision to the status of an aberrant, pathological state of being in a failed environmental context. First, they (Atwood & Stolorow, 1984) argue that Heidegger's emphasis on "the tendency for man to mistakenly view himself in terms of that which he is not" (p. 22) and the human existential potential which includes "the possibility of becoming depersonalized and estranged from their own natures" (p. 22) is unnatural. They feel that Heidegger locates

> this alienation in man's ontological constitution. ... Such a postulate unnaturally magnifies the significance of this one human possibility to the exclusion of all others, and confers upon the resulting vision of mankind a very specific limiting focus. (p. 23)

Unquestionably, Heidegger presents a limited and limiting view of the human condition although, as Freud frequently noted, unpleasant conclusions cannot be construed as false simply by dint of their unpleasantness. Yet Atwood and Stolorow (1984) go further. They present as pathological just those aspects of Heidegger's thought that are most difficult to encompass in the relational narrative. They say that Heidegger's work needs to be read from a psychobiographical perspective in which it can be seen as

> a fascinating descriptive study of human self-estrangement. The ontology of Dasein may then be understood as a *symbol* of an anguished struggle for individuality and grounded authenticity in a world where one is in perpetual danger of absorption in the pressures and influences of the social milieu. (p. 23)

They manage Sartre, at least by implication, in the same manner, suggesting that Sartre's view that "the subjective being of the individual is perpetually threatened by the objectivating, engulfing power of alien consciousness" (Atwood & Stolorow, 1984, p. 29) is a view into which "deep insight . . . can be gained from a reading" of one of Sartre's autobiographical works. In responding to Sartre, however, we believe they reveal a self psychological bias:

> Sartre's treatment of social relationships does not include the possibility of being empathically understood in such a manner that one's sense of self is mirrored and enhanced rather than ensnared and degraded. This omission is of great significance, for once the experience of such empathy is introduced into the structure of being-for-others . . . [s]ocial life ceases to be a battleground of competing subjectivities locked in a life-and-death struggle to annihilate one another. The relationship to the other becomes instead a realm of experience in which one's personal selfhood can rest secure, indeed, in which it can be powerfully affirmed. (p. 29)

In our view, Atwood and Stolorow (1984) essentially tame the existential vision, purging it of *inherent* conflict so that it can be used as a foundation for a penomenological–self psychological version of the relational narrative. Atwood and Stolorow (1984) fully capture the side of the psyche emphasized in the relational narrative—inherent developmental striving and the primacy of mutuality and relatedness. However, it is rare to see this sensitivity combined with the full recognition of the ineradicable tensions between the individual and others that has been more clearly portrayed in the classical paradigm. In our experience, Winnicott's work taken as a whole approximates this exquisite inconsistency. This quintessential relational theorist ("There is no such thing as a baby, only a nursing couple"—see Chapter Three) at other points in his work addresses the existential dilemma inherent in human adaptation:

> I suggest that in health there is a core to the personality . . . that . . . never communicates with the world of perceived objects, and that the individual person knows that it must never be communicated with or be influenced by external reality. Although healthy persons communicate and enjoy communicating, the other fact is equally true that *each individual is an isolate, permanently non-communicating, permanently unknown, in fact unfound.* (1963b, p. 187, emphasis in original).

Using the World to Know Who You Are

From the evolutionary perspective, the child is highly dependent on direct investment and response from an environment that only partially shares the child's self-interest. The only chance for the child to develop an internalized map of reality—including a well-structured accurate guide to the relational world—is to have a way of dealing adequately with the inherent, often covert, biases in parental views of reality: the child must attempt to correct or compensate for these biases so that it can use parental presentations of reality to define and build crucial aspects of its own "self." This includes, of course, the complex ways in which the innate, individual givens we referred to above are actualized into a psychologically meaningful self. In short, the human child depends on self-interested parental figures not only to survive, learn and grow, but also, as objects to internalize in the very process of forming a self—a self that includes a useful, intuitive sense of its own unique interests.

We are saying that while, to a very large degree, children must be seen as actively reading, evaluating, and interpreting the parental environment in terms of the features that matter to their self-interest, the fact is that the child must use this same relational environment as the source of much of its core sense of meaning. In other words, while an enormous amount of intricately elaborated information (about itself and the world) comes to be accepted by the child as if it were emanating from within itself (its internalizations), the child must simultaneously take into account the fact that a significant part of what is presented and experienced as "real" has, in fact, been transmitted through its parents' (and others) biased vision of reality.

Faced with this paradox, how is the child able to construct, maintain, and promote a sufficiently accurate and unbiased sense of its own self-interest throughout the developmental process? How would one "design" a child to be, as much as possible, an active strategist of his or her own development? Such a child would need the complex capacity to do the following:

1. Monitor and evaluate its own unique, configuration of individual innate features, strengths, weaknesses, and propensities; and
2. Develop a viable self, actualized through the internalization of aspects of interactive relationships, while
3. Simultaneously looking out for its own specific interests and guarding against biased influence.

An evolved deep structure that is designed to be a strategy for knowing and regulating the immensely complicated balance of interests in one's personal relational universe faces its most horrendously complicated task in dealing with the developmental experience of the family. Defining and enhancing the true interests of the self (and trying to assess the true interests of close kin) are extraordinarily complex and difficult tasks. The existence of a high degree of overlapping interests combined with the inevitable divergences and conflicts of interest immensely heightens the subtlety and real ambiguity of judging how to form an internal representation of this configuration—the synthesis of which can be conceived of as one's working model of "inclusive self-interest." Add to this the fact that deception (and, as we shall discuss more fully later, self-deception) is an integral part of the normal, adaptive strategy of parent and child. The child must, therefore, be prepared, through inner design, to make inferences from what can be known by way of experience and deductive logic, as well as to intuitively (innately) arrive at certain inferences about relational reality that may be only partially knowable from individual experience.[5]

The Paradoxes of Relatedness

The picture of the developing psyche as guided by complex inner structures that are adapted to the conflict, deception, and ambiguity in the good-enough environment gives wider meaning to Winnicott's (1958) observations about the paradoxical relationship between the need for privacy and aloneness as internal experiences in order to achieve a true sense of "externality" and relatedness to others.

Many of Winnicott's metaphors—"the capacity to be alone in the presence of the other" (1958), the notion of "transitional relatedness" (1951)—stress the crucial need to create a relatively private inner world that is sufficiently buffered from influential others not only to survive the inevitable impingements of the relational world but, more significantly, to serve as a precondition for risking a full connection with and optimally "using" the external world to create the self and fully participate in one's culture.

The dilemmas and paradoxes described by Winnicott as the contours of individual experience echo Heidegger's and Sartre's insights into human alienation, and closely resemble the actual adaptive chal-

---

[5]Quite arguably, this complicated adaptive challenge may be a large part of the reason that psychoanalysis exists; the adaptive challenge may account for the complexity as well as the vulnerability of the inner psychological system designed to manage these tasks.

lenges that, in our view, we faced as a species in the course of our phylogenetic evolution. In order for our whole human adaptive program—with its prolonged dependency and heavy reliance on symbolic communication—to have been favored by natural selection, the evolving child must have had ways of constructing a personal identity that could maintain high levels of investment from others while controlling, to some extent, the inherent costs associated with such a social accommodation. A child must allow itself to identify with or internalize major components of the parents' psychosocial identities without sacrificing its own long-term interests. A child must possess reliable ways of dealing with the inherent parent-offspring conflict and ambiguity of developmental relationships as well as monitoring deceptions—deliberate deceptions in addition to unwitting, well-meaning ones.

It is often ambiguous in Winnicott's work whether he views phenomena such as the development of the false self and its sequestering and sheltering functions as integral features of normal development or as pathological, compliant formations in response to a poorly attuned environment (as the word "false" suggests). Discussions of the "capacity to be alone in the presence of another" (1958) and to relate in "transitional" ways (1951) seem to refer to intrinsic disparities and universal tensions between the inner realities lived by the individual (child) and those held by objects in the normal relational environment. Although more ambiguous, there is clearly, too, the implication that the emergence of a false self may occur frequently in normal as well as pathological development (1960a). The emphasis on the universality of such apparent contradictions and tensions in the normal environment is compatible with the evolutionary view that a deep structural capacity evolved to monitor and regulate such contradictions and tensions.

A Universal Adaptive Dilemma for the Human Child in the Average, Expectable Environment: A Summary

Let us summarize the "design problem" facing the human child. First, we need to keep in mind a few facts concerning human adaptation:

1. Our prolonged childhood entails a period of extreme dependency and immaturity in which an enormous amount of innate, hard-wired responsiveness has been "sacrificed" in the course of human evolution. Natural selection has replaced fixed action patterns with flexibility in the child's capacity to socially or interactionally construct an identity. Identity, in turn, serves crucial overarching functions in regulating behavior. A plasticity in the psychosocial representation of "inclusive self-interest"

can thus be exquisitely attuned to the complex, unique, and changing realities of the family and larger sociocultural world into which the child is born (La Barre 1954, Fox 1989).
2. During the long developmental period of parent–child interaction (as well as the interactions between the child and the wider social environment) most interpersonal transactions are mediated by language and other forms of symbolic communication. Much that is communicated and learned is not rooted in direct observation but, rather, comes through symbolic communication. Thus, the child is able to construct a map of the world that far exceeds (and differs in quality from) anything that could be created from direct experience (Konner, 1982).
3. Yet, as numerous theorists have increasingly noted (Mitchell, 1985; Trivers, 1985), deception is a pervasive, universal intrinsic feature of all animal communication. In the pursuit of their own inclusive fitness, organisms do not simply communicate in order to convey truth to others, but, rather, to convey tailored images of both self and environmental realities: to hide certain features and selectively accentuate those that they want others to perceive. Thus the unique feature of human symbolic communication—its liberation from the need for direct observation—greatly amplifies or potentiates the power both to convey realities accurately and to hide them.

With its prolonged childhood, learning, and language in mind, we have tried to formulate the central adaptive dilemma of the human child, that is, the essence of what it must accomplish psychologically in order to enhance its own inclusive fitness. Biologically, we know the child must maximize the amount of investment (time, interest, love, guidance) provided by the parental environment. This is critical to basic survival and any further maximization of fitness. Psychologically, we know that children must incorporate from their parents whatever is in their own interest to learn (about themselves and the world) as this is transmitted through the parents' visions of reality.

The child is almost totally dependent on direct investment from an environment that only partially shares the child's self-interest. More problematic is the fact that the only chance for the child to develop an internalized map of reality—including a well-structured accurate guide to the relational world—is to have a way of dealing adequately with the inherent, often covert, biases in parental views of reality: children must correct or compensate for these biases so that they can use their parents' presentation of reality to define and build crucial aspects of their own selves. In short, the human child depends on self-interested parental figures not only to survive, learn and grow, but also as objects to internalize in the very process of forming a self—a self that includes a well-defined sense of his or her own interest. How, then, is the child able to maintain and promote a sufficiently unbiased sense of his or her own self-interest throughout the developmental process? The evolutionist asks: "How did the human child evolve the basic structural, dynamic capacity, as it were, to accomplish this task? Given such a universally biased relational environment, how would one "design" children to be capable of developing a viable self internalized through interactive relationships, while simultaneously looking out for their own specific interests?

In the next chapter we examine various psychoanalytic concepts that may represent pieces of the "design solution." We suggest that the psychoanalytic concepts of endogenous drives, the true self, and repression provide basic clues as to the nature of the complex solution to this adaptive problem. We attempt to reframe several psychoanalytic issues in terms of a revised psychoanalytic narrative—a new narrative that is conceived as rooted in the philosophical synthesis of the classical, individualistic narrative of conflict and deception with the collectivist, relational narrative of mutuality and relatedness. Viewed as metaphors rather than as literally accepted structural/technical concepts, repression, the true self, and endogenous instinctual drives may be seen as signifying crucial elements in an evolutionary "design solution." We shall continue to wend our way through various analytic models, deconstructing metaphors from several versions of both classical and relational analytic traditions each of which may be interpreted as contributing concepts that suggest design elements necessary for the overall adaptive structure of the psyche.

PART III

Intrapsychic Dynamics and the Self System as Evolved Adaptations

CHAPTER EIGHT

Blind Mechanisms with Built-in Adaptive Vision: Repression, Endogenous Drives, and the True Self

> ... [N]o generation is able to conceal any of its more important mental processes from its successor. For psycho-analysis has shown us that everyone possesses in his unconscious mental activity an apparatus which enables him to interpret other people's reactions, that is, to undo the distortions which other people have imposed on the expression of their feelings.
> —SIGMUND FREUD, *Totem and Taboo*

In this chapter we focus on three processes that may have evolved in order to manage the complex challenges of constructing an individual identity in the average, expectable, good-enough environment as it existed for hundreds of thousands of generations of our ancestors: (1) the process of "repression"; (2) the so-called "endogenous" (or "instinctual drive") aspects of our motivations; and (3) the dynamic function of the "true self." In this chapter and the next, we shall look at the way in which these psychoanalytic constructs may actually connote certain evolved adaptations that function to make possible a *developmental "negotiation process" by which one's inclusive self interest comes to be actualized and through which the experience of self and others is organized and continually revised throughout the life cycle.*

Repression: Missing but Not Lost

In Freudian theory, the mechanism of repression serves as an energically driven, intrapsychic mechanism in which conscious awareness is diverted from certain aims, affects, or images of self and others. Into the realm of conscious meanings go those wishes and aims that are more congruent with what is acceptable in the family. This includes both universal taboos (on incest and violence) and beliefs that are compatible with particular parental identities and views of reality. In the classical psychoanalytic view, the energic dynamics of repression—cathexis and countercathexis—ensure that repressed *asocial* desires are always potentially ready to—indeed, are usually pushing to—return to awareness (Freud, 1915b; Fenichel, 1945).

More broadly defined, the concept of repression is incontestably at the heart of all of the narratives, classical and relational, within psychoanalytic theory. Yet, because analytic theorists have been caught up in the defense or critique of the literal meanings of the classical, hydraulic, mechanical system, they have not been free to explore the full range of its meanings as a functional metaphor of relations between parts, or versions, of the self. We can define repression in the generic, almost phenomenological fashion alluded to by Greenberg and Mitchell (1983): namely, a state in which something is missing in a person's "experience of him- or herself." Some "crucial dimension of meaning is absent in his account of his own experience" (p. 15). As Freud (1915b) put it, it is in *ein andere Platz*—another place. Missing, but not necessarily lost forever.

When we take this "other place" absolutely figuratively, without specifying the content of what is "missing" from focal awareness, we remain on fairly consensual psychoanalytic ground. Although the actual term "repression," per se, is not always emphasized in all psychoanalytic traditions (the notion of split-off, or disavowed, aspects of the psyche is more compatible with many nonclassical approaches)[1] the underlying view that there is some important set of needs or perceptions missing from the central conscious personality is, almost by

---

[1]While some self psychologists, following Kohut (1971), may focus more on "healing" splits in the self, we believe this definition of repression still applies. In Kohut's terminology a horizontal split corresponds to repression: that which is "below the line" is unavailable to conscious awareness. However, both sides of the vertical split (above the line) can at different times be in consciousness. That is, the grandiose self can dominate conscious experience at one moment, while at another point in time overwhelming sense of self-depreciation is all that exists. While both sets of feelings (or affect states) can enter consciousness, the psyche cannot "integrate" them, for example, cannot simultaneously have them in conscious experience (at which point each would presumably "defuse" the other). Both the horizontal and vertical splits are consistent with the way we are defining repression: at any point in time a significant content of the psyche, or aspect of the self, is "missing." There is also disagreement over whether the phenomenon of diverting awareness is best characterized by the term "repression" (Modell, 1989). We have used the term "repression" throughout this book because of its very broad connotation of a universal, innate process of regulating awareness.

definition, a universal psychoanalytic observation. Accordingly, the clinical task of the analyst—from the classical drive theorist to the object relations theorist, the interpersonalist to the self psychologist—always includes a significant effort to identify and understand that "missing dimension," to bring it into greater awareness and, thereby, in the context of the treatment relationship, promote greater continuity and integration within the psyche (Schafer, 1983).

In this more metaphorical light, the phenomena described in the classical model of repression can be viewed as an innate process that insures that conscious awareness will be diverted from certain aspects of the child's experience (aims, affects, or images of self and others) when these aspects are in conflict with other experiences as well as less congruent with parental views. If we do not interpret the content of what is repressed as derivative of blind, instinctual, mechanical forces but, rather, as representing aspects of the child's own version of reality (including its own needs and wishes), it can be said that the classical model of repression points to the fact that certain constructions of subjective experience must be regularly and normally "dissociated," sequestered, or compartmentalized separately from others. Into the realm of conscious meanings go those wishes and aims that are overtly more congruent with the family (including parental identities and views of reality). At the same time, however, the dynamic conception of repression describes a process that insures that many of the child's aims and views that are less congruent with parental views will not be lost as potential guidelines for the pursuit of the child's own interests. Unlike a nondynamic model in which behavioral responses are learned (brought into existence anew) and unlearned (extinguished, eliminated), the process of repression ensures that these aims, as well as related self and object representations, can be "put away" out of consciousness, but held in reserve, as it were, "pressing" to return. These repressed elements of the self can be allowed to reenter the child's repertoire when conditions change and the need for the repression is lessened, thus making the retrieval of the repressed something that, potentially, is in the child's long-range interest (Slavin, 1974, 1985, 1990).

Repression as a Deceptive Tactic and Developmental Strategy

Repression has been described (Slavin, 1974; Slavin & Slavin, 1976; Alexander, 1979; Trivers, 1976a, 1985; Nesse, 1990) as a mechanism by which the individual engages in a complex form of self-deception in order to deceive others into greater investment and reciprocity than would obtain if one's "true" intentions were known. Since deceit is fundamental to animal communication:

there must be strong selection to spot deception and this ought, in turn, to select for a degree of self-deception, rendering some facts and motives unconscious so as not to betray—by the subtle signs of self knowledge—the deception being practiced. Thus, the conventional view that natural selection favors nervous systems which produce ever more accurate pictures of the world must be a very naive view of mental evolution. (Trivers, 1976a, p. vi)

The self- (or inner) deception presumably makes the social (or external) deception far more believable by convincing others that what is overtly expressed represents the true experiential state and intentions of the individual who is engaged in the self-deceptive strategy (i.e., that they really believe it and mean it).[2] However, such a strategy exists at potentially considerable costs to the actor in divorcing him or her, even temporarily, from a more complete or accurate perception of reality (Trivers, 1985). The "cost" of repression—the structural distortion of reality and the "loss of crucial dimensions of meaning"—is quite real. Traditionally, it is this "cost" that overwhelmingly impresses the clinician in considering the workings of repression, and has created the common tendency to equate repression and psychopathology.

Although the overt, behavioral level description of repression as a "self-deception in the service of deception" has, no doubt, some validity, this conceptualization does not adequately explain how ultimately the benefits of having the capacity to repress might far outweigh the costs entailed; nor does the behavioral emphasis on short-term tactical deceptions convey several of the important human developmental and dynamic functions of repression that we have discussed.[3] We need a broader conception of repression, one that is compatible with these deceptive functions but also captures the full range of adaptive functions which repression may have evolved to serve. Such a view needs to embrace those aspects of repression that sequester and safeguard individual interests in a way that, indeed, makes it possible for the individual to become truly an integral part of the relational world, that is, those functions that, while safeguarding individual interests, also promote empathy, mutual identification, and cooperative action.

In terms of the "design problem" we are considering, repression

---

[2]See the discussion of the function of guilt (in Appendix C, p. 298) in convincing others that one is a reliable reciprocal altruist. Guilt can lead to a very convincing experience of shame and remorse—convincing to others because they can perceive that it is an actual, real experience. In addition, of course, guilt can lead to the repression of some of the more selfish aspects of the self.

[3]For a related discussion, see Appendix D, "The Confusion of Proximal and Distal Causes."

represents an evolved strategy, part of an overall adaptive design, that allows the child to identify with, to optimally internalize major components of the parents' psychosocial identities without sacrificing the child's own long-term interests. Repression permits the child to "plan" for—to be more fully equipped to deal with—future developmental contingencies some of which are predictable parts of the life cycle, others of which are not. What is crucial in this regard is that the process of repression actually ensures that aspects or versions of the child's self will never be irretrievably lost. Rather, they will be "put away" out of consciousness, but held in reserve, "pressing to return," not in a blind mechanical sense but functionally available under altered relational conditions.

From the evolutionary–adaptive perspective, what may be the most critical aspect of the classical concept of repression is that it signifies or alerts us to the fact that at the very center of psychic organization lies an organized inner system that enables us to use a diversion of awareness, an alteration of consciousness—in effect an act of "intentional" or strategic self-deception—while simultaneously preserving the possibility of access to temporarily unacceptable but potentially vital versions of the self.[4] Repression is a central aspect of our deep structure, one that makes it possible that a certain innate skepticism can exist in the child, a skepticism vis-a-vis the whole developmental process; this capacity allows the child to *temporarily* shape itself to fit into the family (or other social environment) while preserving (as an existing unconscious potential) parts of the self that serve as a check on "over-socialization" (Wrong, 1963) into the family culture.

Paradoxically, this dynamically accessible reserve of "consciously missing but not lost" identity elements actually permits the child to risk being far more open to, and influenced by, the parental environment. In Winnicott's (1965) sense of "the use of an object" and Kohut's (1984) sense of the "selfobject" (as a means of self-structuralization), the relational world can be "used" far more readily than would otherwise be possible. Thus, the child can avoid the potential costs to fitness incurred by a behavioral overconformity and internal overaccommodation to a psychosocial world biased toward parental as well as other social, or

---

[4]The dynamic process we refer to as repression thus can be seen as a strategically designed inner "force" (vs. a blind, mechanistic force in the classical model) as well as a subjective "state," *both* of which can be compatible with a relational perspective. In response to the blind, mechanistic, overly physicalistic view of repression in the classical model, Mitchell (1988) argued that repression must be viewed as a state rather than a force in order to be compatible with the relational perspective. The evolutionary adaptive view points the way to a relational model that can include both of these conceptions of repression.

group interests. And, given the intrinsic relational conflicts of interest within the family itself, pressures toward such a biased outcome are virtually always part of normative socialization pressures. This, we should remember, is prior to whatever further distortions of a child's ability to pursue his or her own self-interest may be introduced by parental pathology. Repression can only work adaptively, however, as an aspect of an overall evolved system that includes (1) powerful, universal, affective signals—motives and desires—that are innately linked to the fulfillment of one's inclusive self-interests, and (2) an intuitive sense of the degree to which one's vital, essential nature—including both the particular characteristics unique to a specific individual and the manifestations in the individual of universal, species-wide characteristics—is able to fit within a given set of social forms. We shall now explore how notions of "instinctual drives" and the "true self" may be viewed as elements of such an adaptive system.

RAGE AND THE REPRESSION OF INNATE ALTRUISM

While altruistic behavior is ultimately *self interested* (in that it ultimately benefits one's own genes), there is no reason to believe that genuinely altruistic motivations cannot exist at the level of the proximal mechanisms (i.e., the affective signals and perceptions that actually provide the inner push to motivate human action) (see Chapter Four and Appendix D). No doubt, it is sometimes necessary to repress one's more self-interested aims and egocentric aspects of identity in order to act in accord with such *internal* altruistic intent as opposed to only needing to repress selfish aims when there is pressure from the environment to do so. Just so, in other situations, it may be necessary to repress one's altruistic motives in order to effectively pursue and promote one's own more selfish interests when such self-promotion may be injurious to others.

In the clinical setting, it may be important to keep in mind that some problematic overaccommodations to familial needs may not be motivated by parental pathology, selfishness on the part of parents, abandonment, etc. There are times when the perceived needs of others directly move some children to repress important aspects of their selves in order to be true to other of their *own* (more altruistic) aims. Consider those cases in which rage—rather than being a drive that must be tamed—appears to be an attempt to keep in check an overdeveloped capacity for altruistic concern and compassion.

For example, there are patients who experience immense rage at their mothers who are perceived as painfully intrusive. Such patients may also find themselves vulnerable to acute disappointment and rage in many of their other relationships. In some of these cases, prolonged genetic exploration fails to confirm very much in the mother's overt behavior that ever corresponds with the patient's rage toward her.

This could support some version of a more classical view that such patients are struggling with untamed, innate sadism. However, the clinical data sometimes point to a radically different explanation. For some such patients, the most compelling memories and reconstructions point to the patient as being geared toward wanting to give to others, often in total, self-abnegating ways. This can be accompanied by a proneness to confuse his or her own needs with those of highly valued others.

From a self psychological standpoint, such a patient's failure to develop a sufficiently demarcated, defined self is likely to be interpreted as a *reaction*—an essentially learned response—to the threatened withdrawal of interest or approval by parental objects when the patient's self was not tailored to parental needs. The patient's extreme, often dissociated, outbursts of narcissistic rage would be seen as responses to traumatic repetitions of this kind of selfobject failure.

Yet, such a longstanding tendency to overidentify strongly with the needs of others may not primarily denote a learned accommodation to parental needs. This tendency may connote an innate feature of the psyche. A tendency to function, from the earliest point in development with what is essentially an extreme version of normal, altruistic, mutualistic aims may represent an individual, *innate* variation on a basic human adaptive program. A child with such innate cognitive and affective tendencies might well experience an even moderately intrusive mother (one with a tendency toward self-centered ways of relating to the child's easily overexpanded self-boundaries) as a pronounced subjective threat. The threat would, indeed, be a real one for a child whose altruistic tendencies were so easily aroused. Such a child might well sense the degree to which his or her self-structure and motivational tendencies were especially vulnerable to an even relatively normal degree of parental bias, and normal attempts by the parent to use the child as a selfobject. For such a child the mobilization and maintenance of extreme rage might well be an adaptive attempt to manage the real danger to self posed by such a parent. Essentially giving and caring children whose altruistic motivations may make them poorly able to protect their self-interest may, indeed, be placed in significant jeopardy.

For almost all of our patients, a milder version of this conflict is often a central issue. That is, contrary to the classical perspective, what often needs to be repressed is one's innate altruism in order to allow the individual to pursue his or her self-interest. Frequently we see anger and rage functioning to enable individuals to remain unaware of their altruistic motives which would render them less effective in promoting their overall interests. For most people, anger is necessary, at times, to enable adaptively selfish pursuits that would be "crippled" by the simultaneous action of the individual's genuine caring, mutualistic motivations.

Endogenous Drives: A "Different Drummer"

In the Freudian tradition, the concept of instinctual drives as an endogenous source of motivation[5] not only has several different historical

---

[5]Simply because motives can be meaningfully seen as "endogenous" there is no reason to equate them, as is often erroneously done, with what is "animal" or "biological" in our
(continued)

meanings (Rappaport, 1960), but can, like the construct of repression, be usefully read and deconstructed as a rich metaphorical element in a revised narrative of human nature (Slavin, 1990; Slavin & Kriegman, 1990). Indeed, we know that Freud (1933) recognized that "the theory of the instincts is so to say our mythology" (p. 95) and, as Bateson (1972) has pointed out, the concept of instinct itself explains nothing, but points to something that does need explaining.

As a part of the classical metapsychological edifice, libidinal and aggressive drives represent the emphasis on deriving motives—indeed ultimately all psychic structures—from physiological tension states, essentially bodily states (Greenberg & Mitchell, 1983). The concept of the "id" carries with it the sense of a linkage with a primordial level of impulse, of being "lived by forces beyond our control" (Groddeck, 1949), that somehow precedes the origins of the self. Within the larger theoretical superstructure of the classical paradigm, the metaphor of the "id" can thus be interpreted as implying the existence of an animalistic or "bestial" (vs. human) motivation on top of which a mantle of tamed, mutualistic, relationship-seeking, civilized (ego) motives are overlaid (see Mitchell, 1988). The "pleasure principle" gives motivational primacy to the subjective experience of pleasure or tension reduction in a fashion that begins, at least, as a closed, one-person psychological system (Modell, 1984). Yet, the classical motivational apparatus—drives, id, and pleasure principle—may also be viewed as overarching metaphors that fit within a conflict-based *relational* narrative.[6]

Let us reconceptualize the drives (as they are classically conceived)

---

nature in the sense of an opposition to something that is presumably nonbiological, or "cultural." From an evolutionary perspective this is a completely false, misleading dichotomy. In his metapsychological work, Freud overly concretized, or physicalized the notion of the origins of the drives, yet seemed to realize that even physiologically based motivational entities were likely to have evolved in response to phylogenetic, social selection pressures. See also the earlier discussion (Chapter Two, pp. 34–42).

[6]Fairbairn's attempt to recast libido theory in his own object relational terms ("I would prefer to say it is the individual in his libidinal capacity [and not libido] that is object seeking" [quoted in Greenberg & Mitchell, 1983, p. 158]) entailed a shift away from the classical view of drive discharge as the fundamental organizing feature of the psyche toward one in which libidinal drives functioned as the means by which holistically conceived, primary, relational goals were pursued. Our position resembles Fairbairn's in some respects, that is, that classical theory confuses the means and the ends. In this sense, the evolutionary view is closer to Fairbairn's than to Bowlby's (1969) attachment theory in which a system of primary object-seeking is believed to contradict and obviate the need for drive dynamics. Ultimately, however, the evolutionary perspective takes a far broader view of the "primary" motivational goals (and conflicts) of the individual than did Fairbairn (1952), whose model of motivation appears simply to consist of an innate need to establish (and then regulate) a meaningful closeness and contact with objects.

in terms of their relational functions, that is, their role in regulating the process by which the individual negotiates with the environment in forming a psychosocial identity or self. How could a socially or relationally constructed self—a self that, to a significant degree, is built out of biased familial (cultural) meanings—be counted on to generate the pursuit of a whole range of passionately individual aims? How did the human child take the enormous historical "risk" of "sacrificing" a huge amount of innate, reflexive adaptedness that was relatively immune to external (biased) influence? How did we risk the potential costs of evolving such a long, complex program of development? Especially how did we risk relying on a program that depends on symbolic, deception-prone interactions with powerfully needed objects—objects whose own interests will inevitably lead them to try to shape our identity toward what is beneficial to them?

The notion of endogenous or instinctual drives similar to those central to the classical narrative signifies a mechanism that guarantees access to some types of motivation that arise from nonrelational sources and are, in a sense, totally dedicated to the promotion of our individual interests (Slavin, 1986). The single, unique, functional feature of such "endogenous drives"—conceived as a type of motivation rooted in our bodily nature—is the way in which they serve as a guarantee against having one's genetic self-interest usurped by the social influence of important others, notably those serving as models for introjection and identification. As Wrong (1963) pointed out, theories that do not adequately address the question of how individuals are equipped with motives that are not entirely social in nature, will always present us with an "oversocialized image of man," an incomplete and unrealistic picture of human motivation. From this viewpoint, one of the most striking features of the classical metaphor of instinctual drives, the id and the pleasure principle, is the way in which such entities serve as guidelines, or sources of authority, that powerfully shape the experience of inner needs and the construction of subjective aims and desires. They signify the existence of some type of motivation that arises from sources that not only lie outside relational influences, per se, but are destined to exist in some inherent tension with relational influences and as a check on them. As inner imperatives, such endogenous entities, forces, or principles can be, in a sense, dedicated to the advocacy and promotion of our individual interests.

If we did not have some aspect of our motives that functioned in this way, it is likely that the child—whose conception of his or her own self-interest is inevitably highly dependent on information received from biased parental figures—would be so highly influenced by the powerfully articulated views and needs of the caretaking environment that his or her self-interest would regularly be usurped by the social influence of important others, notably those serving as models for introjection and identification. Given the substantial conflict and deception in the ordinary parental environment that faced the evolving child who, over vast stretches of evolutionary time, had sacrificed innate behaviors to develop an "open program" of receptivity to social learning (Mayr, 1974), we would expect natural selection to have endowed us with powerful, complementary, motivational structures that serve to signal and guide us independently from relational influences.

Some form of endogenous motivation—in this sense like Freud's conception of an id rooted in ancient imperatives that are poised to counterbalance socially constructed versions of the self—is quite likely to have evolved as a safeguard for self-interest. Although desire may only occur in a relational context (Mitchell, 1988), the adaptive design of our motives must fulfill what we interpret as the core relational *function* of the classical drives—the "different drummer" function if you will—captured within the metaphors of the id, the instincts, and the pleasure principle. In this sense, the "drives" are not "outmoded and implausible... phylogenetic vestiges" (Mitchell, 1988, p. 81) for they are seen as filling a crucial, current day function. Indeed, it is possible that classical drive theory has survived as well as it has, despite the manifold, well-documented conceptual and clinical problems associated with it (e.g., Klein, 1976; Holt, 1976) because, at a metaphorical level the theory retains a persuasive power in capturing an aspect of our deep structure that we intuitively sense is a functional necessity for dealing with problems inherent in negotiating development within a biased, conflictual, deception-prone relational world.

The core function of such a principle of motivation is to serve as an imperative, gut signal of self-interest: a kind of test, a reference point in relation to which the complex variety of other, relationally derived, more socially responsive motives would have to stand.[7] Such endogenous motives are in essence one unique source of adaptively relevant information operating within the larger functional organization of the aims of the individual, information that has a certain degree of inherent immunity from the vicissitudes of relationships. There is absolutely no implication that such motives represent more than *one* type of adaptive signal, *one* way of giving meaning to relational experience. No doubt, many other types of motivational signal also exist—some of which arise

---

[7]For example, there are patients whose passions for certain unavailable objects, rather than being an indication of neurotic "fixation," sometimes serve as the only conscious indicators of needs that are, indeed, in danger of being submerged in the choice of more available, secure attachments.

fairly directly from internal (endogenous) sources and others that are patterned more directly on relational needs and internalized interactions.

The Projective Anticipation of Conflict and Deception

The Kleinian version of "endogenous" motivation depicts the child as driven—by the inborn, destructive urges of the death instinct and "innate preconceptions" (Bion, 1962a, 1962b) about objects—to create a highly conflictual, internal, relational scenario (see the discussions on pp. 73 and 123). Quite independent of the actual behavior that the child observes or in any way directly experiences from his or her primary caretakers, the child will begin to construct a map of the relational world in which (part) objects compete for basic resources, clash in their aims, and engulf one another's identities. According to Klein, this inner mapping of intrinsic, relational competition and conflict that begins in the "paranoid–schizoid position" is what organizes the child's deepest level of relating, what guides and motivates the child's basic transactions with his or her real objects. To some extent, this conflictual scenario is tempered in the progression to the "depressive position," an innate, maturational capacity to appreciate the interdependence of good and bad objects. The depressive position essentially signals the child's capacity to painfully negotiate a compromise of his or her own interests so that conflict, while unavoidable, need not result in the destruction of loved objects or the self. Relatively whole, multifaceted objects begin to appear in the child's inner map of the relational world (Segal, 1964; Grotstein, 1985; Ogden, 1990).

As with the id and the libidinal drives in the Freudian system, the Kleinian conception of innate instinctual motivation can be viewed as an overarching metaphor, aspects of which point to, or signify, crucial functions that should be innately structured into the human child's adaptive capacity to anticipate universal, relational realities. Though cast, like Freud's metaphors of instinctuality in very physical, bodily terms (in Klein's case as an elaborate drama of interpenetrating or engulfing bodily parts) the Kleinian metaphor depicts the child's ability to "know" (in the sense of anticipate and internally represent) that, over and above what can be effectively gleaned from its caregivers' communications, there is a pervasive dimension of competing interests and deceptive communication running through the relational world.

In short, through its innately unfolding projective–introjective creation of a conflictual inner map of the relational world, the child is prepared to intuitively know a level of covert parental (and other) motivations—perhaps even vigilantly attend to seemingly negligible environmental cues—that would otherwise be impenetrably obscure and easily disguised from the child's awareness. If the "paranoid-schizoid" position evolved to detect and predict conflicts of interest inaccessible to the child's direct observations, the "depressive position" may represent the earliest developmental form of the capacity to recognize, however sadly, the imperative need to achieve a negotiated solution to these conflicts—conflicts in which the object is preserved and one's own needs tempered. Through evolved, deep structure, these conflicts are now innately appreciated and anticipated by each individual in the course of their development. Winnicott's view of the child's "innate guilt" and the recognition of the existence of what has been called "original morality" (Samuels, 1990) fits well within the evolutionary perspective.

Klein formulated her innate relational scenario (and many Kleinians may tend to apply it clinically) as operating largely independently of the actual behavior of objects (the breast will be experienced as withholding and destructive regardless of how attuned and appropriately responsive the mother is). Though projections such as this may, as we noted, serve to anticipate covertly conflictual dimensions of relating, it is most unlikely that they would ever be cut off from inferences based on actual interactive experience in the dichotomous way and to the degree portrayed in the Kleinian model. Rather, such anticipatory intuitive knowledge—innate preconceptions (Bion, 1962b) about the relational world—is likely to operate in a complementary fashion that counterbalances inferences based on the real behavior of early objects.

Repeated ancestral experiences of conflict and deception in development comprised the selection pressures that shaped an inner *anticipation* of conflict that is juxtaposed with interactive experience, to *supplement* actual experience, and correct what is learned from it in a way that equips the child with a strategy for assessing the object world in order to maximally promote its own interests. As such, the revision of anticipated scenarios using actual experience becomes the kind of problem-solving capacity, "epigenetic rule" (Alexander, 1990), or "Darwinian algorithm" (Tooby & Cosmides, 1990) that contemporary evolutionists see as the products of natural selection in the social environment. Blind endogenous forces functioning in a manner that is relatively unaffected by actual relational events and serving as the primary human motivations—be they Kleinian projective–introjections or Freudian libidinal and aggressive instincts—ultimately connote a far different process, one that has little functional or adaptive logic inherent in it from an evolutionary–adaptive point of view (Kriegman & Slavin, 1989).

Melanie Klein's (as well as Freud's) overall metaphor of the innate deep structure of motivation emphasizes a certain quality in the "aggressive" component of endogenous motives that deserves some discussion. From an evolutionary–adaptive perspective, the aggressive component of these aims can be viewed as essentially "self-assertive" in nature rather than emanating from a central reservoir of inherently destructive or sadistic energies as the Freudian, and particularly the Kleinian, metapsychologies would have it. From an evolutionary perspective, the existence of an overriding destructive drive energy that requires discharge in the way these models depict would be a costly, unnecessary, and thus unlikely way for natural selection to have designed us (Kriegman & Slavin, 1989).

Our view may seem to resemble the familiar notion that aggression is a response to frustration or a more contemporary "systems" view (e.g., Stechler & Halton, 1987) that healthy self-assertiveness becomes "contaminated" with destructive rage in an unresponsive environment. Yet a closer examination of such views shows the distinction between the evolutionary perspective and what we believe is a more relational bias. For example, Lichtenberg (1989) notes that for a child's normal development:

> *the caregiver must empathically accept* that it is to the child's advantage . . . to formulate his or her own agenda and even at times to have a vigorously aversive reaction to interference with that agenda. (p. 174, emphasis added)

In this statement, as well as in Lichtenberg's larger discussion of the "aversive motivational system" (1989), the notion that the fundamental aims of the child and caregiver contain significant and profound interpersonal conflicts is largely absent. However, the built-in human capacity to experience the world in a self-interested manner, and to act on this experience, will lead, in the thoroughly "good-enough" environment, from earliest infancy onward, to certain major clashes between our aims and those of even our most intimate objects. Further, the proximal mechanisms that equip us to deal with a biased, self-interested relational world may well include the capacity to mobilize, under certain conditions, urges and desires to harm others in the pursuit of our own advantage. The interpersonally adaptive rage (Kriegman & Slavin, 1990) that frequently ensues as a result of conflict within this universal relational matrix could then lend itself readily to interpretation (in our opinion, a misinterpretation) by the outside observer as emanating from a central fund of innate, intrapsychically determined destructive aggression.

We would expect natural selection to have favored those individuals who could call rage into play whenever it would be adaptive to do

NARCISSISTIC RAGE: SELF PSYCHOLOGICAL VERSUS EVOLUTIONARY CONCEPTIONS

In an attempt to deal specifically with the issue of aggressive drives, Kohut (1984) tried to differentiate between

> aggressions directed at objects (who stand in the way of cherished goals) and those directed at selfobjects (who have damaged the self). (p. 137)

The former

> cease as soon as the objects cease to be obstacles or as soon as we have reached our goal. In addition, these aggressions do not produce psychopathology. . . . Narcissistic rage, on the other hand, cannot be satisfied via successful action against the offender—the injury lingers and so does the rage. It is narcissistic rage in childhood . . . that does indeed play a significant role in the genesis of pathology. (p. 138)

The goal-directed oedipal competition or sibling rivalry that Kohut acknowledged but claimed *does not* produce pathology (1984, pp. 137–138) is given no clear motivational status or source *within the self psychological system*. To find such motivations we must rely on the complementary perspective provided by the traditional paradigm. This is the "complementary strategy" that Kohut (1977) had proposed. However, the traditional paradigm that Kohut must turn to in his complementary strategy explains such motivations using the conception of drives that Kohut (1982) referred to as "a vague and insipid biological concept" (p. 401). In this, and other ways, the complementary strategy breaks down (Kriegman & Slavin, 1990).

Two of Kohut's claims are worthy of further comment in this context. First, he claims that object-directed aggressiveness and the conflicts it engenders

> e.g., guilt or an unstable equilibrium between currents of fondness and anger . . . are not constitutive of psychopathology, however severe they might be, but [are] part and parcel of normal human experience. (1984, p. 138)

We would suggest that such conflicts (both intra- and interpsychic) may actually play an important role in *engendering* some of the selfobject failures that Kohut claims are the actual cause of psychopathology. For example, is there not likely to be an interaction between inner love–hate conflicts and subsequent responses by the selfobject? The resulting patterns of inner conflict and selfobject failure may then be organized into pathological structures.

Second, Kohut claims that narcissistic rage "cannot be satisfied via successful action against the offender." This suggests that narcissistic rage (as in Kohut's (1972) example of Captain Ahab in *Moby Dick*), has no other function than reconstituting the injured self (i.e., that it can be understood purely in intrapsychic terms). Can it be that there is no adaptive *interpersonal* function for narcissistic rage? In Kohut's purely intrapsychic model of the function of narcissistic rage, we are reminded of similar attempts in traditional theory to define processes such as "ego mastery" solely in intrapsychic terms, that is, as the mastery of dangerous instinctual pressure

(Bibring, 1943). What is neglected in both types of overemphasis on purely internal processes is a recognition of their potential adaptive functions in the relational environment. It is unlikely that narcissistic rage has no adaptive interpersonal function, as Kohut (1984) suggests. Simply by applying our basic adaptive perspective we are led to ask how such a universal feature that has enormous *interpersonal* consequences could have evolved. Immediately, we must consider it unlikely that such a prominent feature of interpersonal relations has only intrapsychic functions. Going beyond our basic adaptive perspective, we have sketched a more specific view of the relational world in which an individual is frequently either attempting to influence others, or ready to respond lest others should attempt to selfishly use the individual. While narcissistic rage can be quite adaptive in such a relational world, self psychological theory lacks a formulation of any inherently self-interest-promoting motivations that are destructive to others, except in those situations where the self has been threatened or injured (Kriegman & Slavin, 1990).

so. Consistent with relational narratives that explain anger and destructiveness as reactions to the environment (e.g., Kohut, 1972), rage may be called for when the self is threatened. However, it may also be brought into play in situations in which the threat is mild or even nonexistent, when the interpersonal consequences of the expression of rage are likely to be advantageous. Note that similar frustrations, injuries, or threats to the self may initiate a rage response in one setting but not in another; for example, a child who throws an enraged temper tantrum with one parent, who responds by relieving the frustration, but not with the other, who responds by ignoring the tantrum. A conceptualization of rage as serving purely intrapsychic functions—protecting and preventing further dissolution of a fragmenting self (see the box on narcissistic rage on p. 168)—simply cannot account for many complex human interactions. We are biological organisms that have evolved to actualize our own self-interest, even if at times, when functioning in accordance with our design, we intentionally and unintentionally inflict injury on others.

All of our patients have to come to terms with the fact that, typically, in conflicts of interest with others, when functioning in a *healthy* manner in accordance with one's design, an enormous, self-centered, and righteous rage actually can be quite *easily* aroused. If one sees this simply as a *response* to the environment, which it invariably is, one can be experienced as denying a powerfully felt reality about our patients; a reality that could be stated by them as follows:

> I can *easily* be moved to actively and aggressively promote my own interests in the form of a frighteningly destructive rage. *That is me.* At

times I want to promote my interests over your very existence, if necessary. Along with my equally integral mutualistic wishes for empathic connectedness to affirming selfobjects (and my desire to play such a role for others), I can be imperious and destructive. Any attempt to deny the full extent of my struggle with this inherent conflict at the core of my being, in my deepest motives—e.g., by implying that I am merely responding to the environment (as if this imperious *destructive capacity* somehow exists or derives from sources or events outside of me)—will ultimately be met with a sense that I am deceiving you and/or myself.

We must keep in mind that in the evolutionary perspective, the whole notion of inherent destructiveness versus environmental "frustration" and "unresponsiveness" is altered by the recognition that frustrating conflicts of interest and biased responsiveness are thoroughly expectable features encountered by an innately self-interested, self-promoting individual developing in the nonpathological, normally biased object world. Aims that begin simply and benignly enough as self-assertion will, at times, take on significantly destructive and hostile qualities in an environment of inherently competing interests.

From this perspective we see the child as an active agent struggling to grasp and represent his or her own inclusive self-interest, and to influence others to act in the child's best interest even if it is at the other's expense. That is, some significant failures of the self–selfobject milieu may be due to the conflict between the child's healthy attempts to meet its needs and the parents' own healthy nonoverlapping interests (i.e., interests that are not shared with the child). Although the source of such failure may at times lie primarily in parental pathology, more frequently the failure is likely to derive from the interaction of parental pathology, constitutional factors in the child (Fajardo, 1988; Shane, 1988) and/or parent, the inherent conflict between the interests of parent and child, and external stresses impinging on the self–selfobject unit. Even some of the most disturbed parents will at times function as adequate selfobjects to their children. Likewise, even the most healthy parents will at times traumatically fail their children. What accounts for these variations in response are the various factors that interact with the dimension of parental health/pathology to yield a specific level of "environmental" success or failure. To focus on parental pathology as primary may, in some cases, simply be erroneous.

We believe that Freud and Klein resorted to constructs such as libidinal energy and the death instinct in part because they never took the conceptual step of explicitly defining the universal challenges of the relational world, for example, the need to have a functional, adaptive solution to the problem of intrinsic relational conflict and deception. An

THE VITAL CAPACITY FOR INNATE DESTRUCTIVENESS AS AN INTRINSIC ASPECT OF THE SELF: A CLINICAL EXAMPLE

Ms. B, a 35-year-old, intermittently psychotic woman, had been hospitalized more than 30 times over the past 17 years. She harbored a deeply felt delusion that came from a sadistic set of hallucinated voices who said, "You ate your mother." (Her mother died in childbirth when Ms. B was 1 year old. When she was 3 her father contracted a fatal disease and she was placed in a children's home.)

Occasionally, when discussing the four or five incidents of truly destructive rage (for example, pulling her baby sister's hair, attacking an attendant who was trying to forcibly medicate and seclude her) the patient would dramatically regress when her therapist expressed what seemed to be an empathic view of her desperately self-protective rage, that it was a response to the terribly brutal and damaging mistreatment she had experienced; and that her rage was a response to her feeling as though she were being annihilated and destroyed, and would disappear forever.

While this explanation actually was experienced as "correct," that is, the patient fully acknowledged and remembered the terror repeatedly engendered by mistreatment at the hands of others (beatings, sexual abuse, locked-door seclusions, forced medications, etc.) that, indeed, made her vividly feel that she would, in fact, disappear forever or die, she became quite anxious and accused the therapist of being "an attorney for the defense." Contrasted with the therapist's explanation was her primitive, sadistic psychotic inner voice that claimed she was an evil murderer. This tenaciously held negative view of herself—what would be traditionally called a primitive, sadistic superego—was understood by her therapist as a way of maintaining a shred of identity and relatedness, however painful; as her need to hold onto "bad objects" (echoes of numerous frustrated and angry caretakers), because a "bad object is better than none."

In fact, innumerable associations and memories indicated that the persecutory voices were an internalization of the relatively uncaring succession of frustrated and angry children's home workers, the psychotic fundamentalist couple who adopted her at age 9, and the long series of psychotherapists and inpatient mental health workers who treated (and mistreated) her. All would become frustrated and withdraw—and/or enraged and attack—when Ms. B failed to improve, regressed, or made infantile demands on them. Thus, Ms. B appeared to be clinging to a sense of herself as the devouring, destructive child in lieu of her greater fear that the persecutory voices (that, in her life, were the only possible internalization of a sense of human community) would also abandon her to face the empty void alone. There was sufficient data, indeed, quite experience-near data, to support the view that her tenaciously maintained "sadistic superego" was experienced as the only alternative to a total abandonment and utter psychic death. Yet even when made with full respect for her need to hold onto the "bad object," the "bad object is better than none" interpretation invariably led to marked despairing panic along with a desperate yearning to return to her former analyst who, she was convinced, was profoundly afraid of her!

It soon became clear that at least this former therapist respected the enormity of her rage and destructive potential. He would be careful and not let her destroy

> him, while her new therapist was experienced as "frivolous" in regard to the danger posed by her destructive rage. Further work clarified the fact that she felt that she *needed* to be punished because the "attorney for the defense" would encourage her fragmented, overwhelmed self to come together as a whole, undestroyed person whose destructive potential could then be unleashed. While she was eventually able to question whether she actually "ate her mother," she was terrified that her very real rage would, in fact, lead her to kill someone.
>
> When the therapist acknowledged that the hateful rage within Ms. B was an integral part of her that, indeed, needed to be controlled, and when the therapist warily joined her in carefully watching for such in the treatment relationship, Ms. B was able to feel calmer, safer, and gradually reconstitute a sense of self that was not totally enveloped in fragmenting terror. In this state she was able to begin to *angrily* recall and mourn the enormous, devastating narcissistic and life-threatening injuries and abuse that characterized her life; injuries that, she could now see, had left her desperately holding onto a miserably painful set of persecutory voices that she needed because of a complete absence of any but the most rudimentary self-structures with which to ward off chronic psychotic fragmentation.
>
> While the earlier explanations that were offered were unquestionably correct, with enormous experience-near data to support them, they were *unempathic* for they did not acknowledge the degree to which her rage response was felt *as a part of her*; as a deep, innate capacity that arose from the deepest part of herself; from her "gut."

explicit understanding of the nature of the relational world would have necessitated the development of a model of a motivational system that can anticipate covert conflict and aggressively sustain individual interests (including reproductive interests), especially in face of the simultaneous need to construct an identity built on internalized social and symbolic interactions. Thus, the real functions connoted by the metaphors of blind, mechanistic sexual and aggressive forces in classical concepts of drive and instinct need to be included in our model of the psyche as forces that somehow compel individuals to act, often in opposition to the social context, in the pursuit of selfish (asocial) reproductive aims.

Innate Affects and Expectations

Stolorow et al. (1987), as well as many others (e.g., Basch, 1984),[8] have argued that we need to do away with drives and replace them with the

---

[8] Unlike Basch, Stolorow does not attempt to integrate psychoanalytic knowledge with other scientific fields. Defining psychoanalysis as a hermeneutic, intersubjectively derived field, he questions whether it should be considered a science. Thus, while Basch and Stolorow appear to be in agreement, profound differences underlie similar statements about drives (Basch, personal communication, 1988).

notion of "affects." Stolorow (1983) criticizes drive theory and Kohut because "it was never Kohut's intent to do away with drives" (1986, p. 41). Although the term "drive" can be translated into a more experiential form, Stolorow believes that even a revised version of the drive concept "carries so much semantic baggage" that it will inevitably burden us with its blind, mechanistic, energic connotations (personal communication, 1989). Stolorow's solution is to redefine whatever subjective emotional experiences are encompassed by the concept of drives in terms of "affect states."

There is no evolutionary reason why innate, evolved motivational principles that function as the "different drummer" in the Freudian system and the projective anticipation of relational conflict and deception in the Kleinian model need to be viewed as "instinctual drives" rather than cast in the language of "innate affects," and "complex, inner, signal functions" as is increasingly common in psychoanalytic discourse (Lichtenberg, 1989; Tompkins, 1962, 1963; Bowlby, 1969). However, many translations of the "drive" concept into relational terms (e.g., Loewald, 1980) slide into the relational narrative and thus do not fully capture the crucial functional properties we are referring to; most "affect" theories (e.g., Tompkins, 1962, 1963, Stolorow, 1986) or attachment theory (Bowlby, 1969) do not emphasize the functional need for a strategic design that operates as a critical anticipation of, and check on, relational influences. For example, for Stolorow (1986), "affect states" connote a source of motivation that is shaped and regulated entirely within the infant–caregiver (the self–selfobject) intersubjective system. This is in contrast to the sense in which "drive" motivations arise *endogenously* within the individual.

Ultimately, the issue may reside in whether we find sufficient evidence for a distinct class of motivations that primarily arise from innate sources, that is, motivations that are largely independent of the shaping and regulating influences of the relational world. Not only do we believe that such evidence exists, but the very research that "affect theorists" turn to for validation of their ideas contains some of the most convincing evidence supporting innate, affective motivations.

Numerous authors have turned to the work of Tomkins (1962, 1963) in order to use his conception of affective motivational systems based on his infant observations. For example, Lichtenberg (1989) uses Tomkins's work to support his affect theory—part of his larger motivational theory—which includes "innate programs for the categorical affects of enjoyment, interest, surprise, distress, anger, fear, shame, and disgust [that] exist at birth" (p. 260). Lichtenberg (1989) describes the motor responses and facial expressions linked to the different affects suggesting, following Tomkins, that "each affect is an entity (a brain program)" (p. 269). For Tomkins events trigger affects. But Lichtenberg goes further, linking these innate affect programs to his five motivational systems which are:

> (1) the need for psychic regulation of physiological requirements, (2) the need for attachment-affiliation, (3) the need for exploration and assertion, (4) the need to react aversively through antagonism or withdrawal, and (5) the need for sensual enjoyment and sexual excitement. (p. 1)

Lichtenberg links the innate brain programs called affects to his innate motivational systems:

> Each motivational system refers to innate programs of perceptual–action responses built around basic needs. The needs are recurrent. Each innate program will trigger particular affective responses. (p. 269)

Thus, for example, the affective experience of excitement can be triggered by events occurring when the sensual–sexual system is dominant, by the sight of the mother when the attachment motivational system is dominant, and even by mathematical understanding when the exploratory–assertive system is dominant (Lichtenberg, 1989, pp. 270–271).

Even though Lichtenberg focuses on the development of a psychoanalytic system of motivations other than simply sex and aggression—something that has been sorely needed in psychoanalytic theory—it is our sense that his motivational systems cannot fully encompass the inherent conflict in the relational world. Despite his suggestion that there is a need for some conflictual struggle (following Wolf, 1980; also see p. 167), we believe he continues to see *problematic* conflicts as resulting from parental inability to adequately meet the child's needs (personal communication, 1989). In our view, the notion of innate, affective motivational programs needs to be taken further than it has; it must include evolved, affective adaptations to the full range of conflict inherent in the relational world.[9]

Given the normal child's need to deal with the normal bias and deception in the normal environment, we could never be designed in a

---

[9] In addition, we note that affects must be understood as operating within specific relational contexts. Affects may not be relatively simple mechanistic reactions that are then further shaped by learning (experience) in a straightforward manner. Fairly complex specificity and subtlety can be innately structured into affects to serve specific interpersonal functions. This issue is discussed in greater detail in Appendix E, "Specific, Functional, Affective, Motivational Systems from an Evolutionary Perspective," p. 305.

way that permits the claim that "motivations arise solely from *lived experience*" (Lichtenberg, 1989, p. 2). The challenge is to incorporate certain of the key relational connotations of the Freudian and Kleinian models—that is, the relative resistance to socialization and the innate anticipation of conflict and deception—without the narrow and inaccurate implications of a somatically derived, physicalistic libido theory, the death instinct, and the tremendous overemphasis on the primacy of selfish and destructive motives that overlooks equally innate, primary, altruistic, and mutualistic motives.

The True Self: A Biased Set of Innate Dispositions

Winnicott's idea of the "true self" (1960a) represents a psychoanalytic concept that, along with the broader, adaptive view of repression and the "different drummer" function of certain endogenous drive signals, may contribute to understanding how the child attempts to solve some of the thorny relational dilemmas it faces. The concept of the true self may be understood in terms of the evolutionary narrative in which the child's innate individuality must be able to find expression in the creation of an adult self—a self that can use an intrinsically ambiguous, biased, and deceptive environment to create a viable sense of its own self interest.

Rather vaguely defined in a variety of ways—such as "that inherited potential with which each person starts out in life" (Bollas, 1989, p. 213), or "the source of personal impulses" (Bollas, 1989, p. 43)—the notion of a true self continues to be a powerful, evocative metaphor. It connotes the aspect of the self that is, in some sense, individuated from the outset of life. In addition to the *universal* needs, wishes, and other sources of personal satisfaction that each individual experiences anew as a vital part of his or her own self, the true self also includes those aspects of one's self that are *uniquely* one's own. Thus, the true self ultimately stands for *that unique constellation of universal and individual characteristics, the maximization (full actualization) of which is known in each individual's experience through a sense of vitality, aliveness, meshing or fit within a relational context.* This "fit" is experienced when there is a sense that the relational world recognizes and provides opportunities for the elaboration of the deep structural aspects of the true self into effective action or more directly conscious, verbalizable pictures of the self and self/other schemas.

Focusing on those aspects of the true self that are unique to the individual, Bollas (1989) has taken the poetic license freely granted by

Winnicott's metaphor and further defined the true self as a "genetically biased set of dispositions" (p. 9). Like Winnicott, Bollas acknowledges yet leaves implicit the powerful biological implications of this concept. He seems to be referring to a unique set of individual givens, or predispositions to interpret and respond to experience in certain ways. Such predispositions do not exist simply as a static collection of "inherited traits" or "behavioral inclinations," but rather as a preexisting inner foundation or vantage point from which one reads interactive experiences with the environment in light of their implicit, individual meaning.

Recall that in Chapters Three and Seven we discussed the view that each individual varies considerably in his or her innate propensities and character. Because of this variation, part of the *universal* adaptive psychological design of the human psyche must include a process by which the unique *individual* attributes, which could not have been predicted prior to birth and subsequent discovery by the individual's developing self, can be "known," integrated, and maximally utilized in pursuit of inclusive fitness. As we have repeatedly stressed, the way in which both universal human strivings and innate individual features are integrated into a developing unique self is certainly not a matter of indifference to either the parental surround or the child whose interests inevitably differ from those of all other players. Because of this, we believe that the idea of the "true self" may signify a dimension of our overall adaptive design—our evolved deep structure—that serves to provide us with an absolutely critical source of information about our individual interests. As such, it is a crucial inner guideline that, as Bollas (1989) notes, exists "before object relating" and, indeed, makes effective object relating possible. Viewed in this way, the meaning of the "true self" goes far beyond the more limited phenomenological connotation of an experienced "authenticity" about certain experiential states or versions of self-experience. Indeed, a subjective sense of "authenticity," "realness," "genuineness," or "naturalness" may well function as some of the evolved signals that one's own core self-interest has been engaged.

Similar capacities to those implied by the "true self" are implicit in Erikson's (1956) view of the process by which a "psychosocial identity" is successfully achieved: through the utilization of a capacity to intuitively distinguish how well different versions of the self, roles, and interactions correspond with one's innate endowment. Kohut's (1984) "nuclear self" and "nuclear program for the self" point to similar adaptive capacities and processes. The "true self" is presumably less effectively expressed when we observe premature "identity foreclosures" (Erikson, 1956) or, as Winnicott sometimes called it, "false self" organizations, that is, crystallizations of self-experience that are overly attuned to constructions of reality deceptively offered by a biased relational environment.

The existence of some form of implicit, intuitive capacity to "know" certain basic givens about oneself—Bollas (1989) includes this in his notion of the "unthought known"—represents an aspect of the crucial ability to evaluate the "goodness of fit" between the configuration of innate, individual givens (including both universal strivings and an individual's specific strengths, vulnerabilities, dispositions, and sensitivities) and the way these primordial givens are viewed and responded to by the familial environment. The adaptive capacities represented by the "true self" may be as important in the child's initial assessment of what is genuine or authentic about *others* as it is in making such determinations about oneself. For example, true self experience may provide innate anchors, or reference points that can be used to assess the validity of the attempts of others to mirror (or claim they are mirroring) the child's experience as well as assisting the child in "reading" the more covert or hidden aspects of others' identities.

The "true self" as an adaptive capacity includes the "unthought dispositional knowledge" (Bollas, 1989) that enables us to evaluate the "goodness of fit" between one's own innate characteristics and the "versions" of the self offered by the environment. It carries with it a certain sense of urgency about listening to what these "inner whisperings" (Barash, 1979) may signify (and, if needed, to be moved powerfully by them as a crucial piece of adaptively relevant information). In our view, a design element such as this became a critical, functional necessity for a species in which our subjective experience of self is, to a large extent, constructed from a world of ambiguously deceptive social interactions that are inevitably somewhat biased toward the interests of others.

"Repression" and endogenous "instinctual drives" have traditionally been conceived as blind, mechanistic processes, and the "true self" has remained an evocative, if somewhat mystical notion. Yet, all three concepts can be deconstructed as elements of the universal narrative of conflict within a highly social species. A narrative of conflict weaves through classical psychoanalytic models and aspects of Winnicott's work.[10] Viewed as evolved, adaptive aspects of human nature, the

---

[10]Winnicott's thinking ranges all over the psychoanalytic map. The extreme relational position is found in the denial of innate individuality associated with his statement, "[T]here is no such thing as a baby, only a nursing couple" (see Chapter Three). The true self/false self distinction, however, as discussed at the end of Chapter Seven, totters between a characterization of a universal tension rooted in the inherent conflict between

(continued)

broadly adaptive process of repression, the different drummer functions of certain endogenous, driven motivations, and the intuitive sense of psychosocial fit connoted by the true self can all be understood as parts of an interlocking system of inner adaptations to the dilemmas and paradoxical tensions in our ancestral relational world. Such a reinterpretation by no means validates the specific mechanistic and physicalistic reductions of repression and drives that have been generated by classical psychoanalytic theorists. In our view, these theorists groped toward ways of representing the essential, innate, functional principles that needed to be incorporated into our psyche as it opened itself, as it were, to the evolutionarily unprecedented challenge and opportunity of leaving behind simpler, reflexive responsiveness in order to replace such with the adaptive advantage of flexibility inherent in the construction of an identity from materials (meanings) provided by an inevitably self-interested and somewhat deceptive environment. Chapter Nine represents a more detailed view of the intrapsychic and relational process (largely seen as a "negotiation" process) by which the self is constructed, challenged, and continually revised throughout the life cycle.

---

inner and outer realities (a more classical view) and a characterization of false self as a defensive response to an environment that has failed to live up to the fully mutualistic interactions that are normal, expectable aspects of the relational world (a more relational view).

CHAPTER NINE

Negotiating and Re-negotiating the Self: Transference as an Evolved Capacity to Promote Change (With a Special Focus on Adolescence)

> [H]uman beings are not born once and for all on the day their mothers give birth to them . . . life obliges them over and over again to give birth to themselves.
> —G. G. MARQUEZ, *Love in the Time of Cholera*

> [E]very human benefit and enjoyment, every virtue and every prudent act . . . is founded on compromise and barter.
> —EDMUND BURKE, second speech on conciliation with America

The Provisional Organization of the (Divided) Self

Mr. T is a 66-year-old man whose life long ago became one of cautious creature comfort focused on TV, napping, and quiet dining with his wife. Earlier the couple had enjoyed dancing and traveling. Following a minor heart attack, all such activities were experienced by Mr. T as overtaxing, dangerous, and uninteresting, much to his wife's consternation. A period of intense mourning following his wife's death completely destabilized Mr. T, making him "dysfunctional," and causing him to regress to a relatively helpless, emotional, childlike position. Following the intense period of mourning, as he began to cope again, he remained in a highly anxious but lively state. In this state, once again he went out, socialized, traveled, went dancing, and became emotionally alive in a successful search for a new wife. Aspects of his self that had been set aside as unnecessary in his relational world, and had been experienced as dangerous, were recaptured in a powerful adaptive–regressive revival. Through the process of mourning, these elements could then be reincorporated into a reorganized self.

In this chapter, we suggest that from birth to death, all individuals are centrally engaged in the negotiation and continuing renegotiation of identities that remain, to some extent, *provisional*, or contingent upon an *intricately negotiated fit* between their unique individual interests and the complex networks of converging and overlapping interests in the relational world. In countless cases, like that of Mr. T above, we observe a capacity to retrieve and reaccentuate relatively repressed parts of one's inner experience, including vital aspects of the "true self" (as we have defined it), and to make use of this "reserve" when relational conditions change (particularly when external investment from others decreases or is threatened).

Repression, Regression, and Transference: A Coordinated Set of Evolved, Adaptive Processes for Negotiating Change throughout the Life Cycle

An evolutionary perspective on the lifelong process of "negotiating the self" must begin with the recognition of the universal tendency on the part of parents to confuse their own interests with those of their children. The universal overlap of interests within the family is reflected in a natural fluidity and blurring of interests; it is unclear whose interests are actually represented and promoted in the course of many crucial developmental exchanges. The far greater experience and knowledge of parents is often the only accessible source of vital information regarding crucial aspects of their children's larger, longer-term interests. Thus genuine ambiguity may frequently envelop a web of valid information that is interwoven with misinformation (from the child's point of view) that derives from the continuing operation of parental, subjective bias. This bias may remain hidden from the child and, indeed, from conscious awareness in the parents themselves.

Over and above the complications introduced by the ambiguities of overlapping interests and the greater realistic knowledge possessed by parents, there is the regular operation of more motivated forms of deception and self-deception practiced by all parents and all offspring in the service of promoting their own interests. When all the normal advantages of parental control over investment and cognitive skill at fashioning and using symbolic communication are brought to bear on the negotiation of a view of reality, the child, (despite its complex, evolved capacity to counter this reality) is likely to develop a perspective that, in some ways, is poorly aligned with its core self interest. In this sense virtually all of us follow a developmental course that, to some degree, leaves us somewhat alienated from our own interests.

Thus, as hard as it may be for us to accept, every human child is, in a significant sense, deceived, confused, and led astray by the normal, developmental negotiation process. Growing up in a thoroughly good-enough family brings with it the internalization of a picture of oneself and the world that is inevitably less than optimally attuned to the individual's long-term interests. The child's capacity for repression is an adaptation to this dilemma. Similarly, the intuitions and affects concerning what is "true" or "authentic" about oneself—the signals of one's degree of alignment with one's true self-interest—are the evolved adaptations to the inevitability of this developmental alienation. Repression and so-called false self defenses are crucial adaptive ways of simultaneously accommodating to, while resisting and tempering, parental influences. Yet, repression and the false self are costly compromises with the parental world and must be accompanied by a capacity to realign oneself in a more genuinely self-interested way when there is an opportunity to do so safely. When the alienation and self-deception appear to reach a certain point and the realignment process appears to be stymied or malfunctioning, we say that we are dealing with psychopathology.[1]

What are the ongoing events that may require the existence of such an innate capacity to renegotiate "inclusive self-interest?" Of enormous adaptive significance from an evolutionary point of view (and usually subjectively experienced equally powerfully) are the common, unpredictable changes in the amount and kind of investment that the child receives from parents and other kin—for example, the birth of siblings, changes in the health and emotional state of parents, as well as the myriad ways in which family fortunes and dynamics can drastically change over time. Life's small blessings, deprivations, and disasters were always quite uncertain throughout our long evolutionary past; a good deal of this uncertainty continues to exist today even in our privileged, postindustrial settings where we still see substantial shifts in the quality and attunement of most interpersonal environments at different periods or phases of development (Erikson, 1968; Lidz, 1963).

---

[1]We frequently also find the term "pathology" (mis)applied to fundamentally adaptive struggles to resist and undo such alienation. This will be discussed more fully later.

Indeed, the evolutionary biological theory of parent–offspring conflict predicts that the resurfacing of conflict over normal parental attempts to "mold" their offspring—the resurfacing, that is, of more individually self-interested motives in the child—will occur under relational conditions that have been altered whenever (1) conditions of direct parental investment decrease or threaten to decrease (Trivers, 1974); or (2) the parents' control over resources and view of reality becomes less closely aligned with the child's interests than it was when the repression took place (Slavin, 1985). Or, in Winnicott's terms:

> The equivalent of the False Self in normal development is that which can develop in the child into a social manner, something which is adaptable. In health this social manner represents a compromise. At the same time, in health, the compromise ceases to become allowable when the issues become crucial. When this happens, the True Self is able to override the compliant self. Clinically this constitutes the recurring problem of adolescence. (Winnicott, 1960b, p. 150)

Thus the capacity to retrieve and reaccentuate parts or versions of the self can serve as a real, effective means of renegotiating inclusive self-interest. The broad adaptive regulation of awareness that includes the process of repression may serve to reversibly suspend access to parts of one's inner experience—indeed, to vital aspects of the self—and to make use of this "reserve" at a later point in the process of reconfiguring one's *inner* life, as well as in the ongoing negotiations/renegotiations designed to effect changes in the *environment*. Repression thus serves as a major fulcrum for ongoing adaptation and dynamic change throughout the entire life cycle. Consider those times when one's own identity is called on to complement another's, or to compensate for the loss of an other: marriage, divorce, death of a spouse. In all of these normal crises in relationships, one can well make use of that reserve of alternative identity elements and repressed wishes that natural selection has enabled the child to bequeath to the adult (Slavin, 1985).

From childhood onward, individuals need some form of experiential (lived, vital) access to repressed layers of experience (versions of self and others, fantasies, wishes, and affects) that can serve to reduce the *normative alienation* that characterizes human development. We shall depict the overall developmental process as deriving from an adaptive system built around repression, "developmental transferences," regression, and ongoing revision. This system, we suggest, is designed to deal with universal alienation and threatening subjective states of confusion by providing a process by which individuals can continuously evaluate and alter the basic structure of how they experience, articulate, and promote their interests. In the context of this universal alienation, the innate "tendency to regression," as Winnicott (1959) put it, "is . . . seen as part of a capacity to bring about self cure" (p. 128).

The human capacity for self-revision—for negotiating change—can thus be understood as rooted in the workings of repression. Repression in this view is defined as a normal, universal process that limits our awareness of multiple, competing self-other schemas. While limiting our access to such schemas may be functionally adaptive, these same schemas are potentially vital for both ongoing and future use in the relational negotiation process that starts at the beginning of life. As we discussed in Chapter Eight, the evolved capacity for repression makes transferences both possible (by creating a rich reserve of less conscious potentials) and necessary (so that these potentials will be moved into experience) when the ever-changing balance of overlapping and competing interests indicates that it may be advantageous to tap into the reserve of adaptive information carried by repressed versions of the self.

Originally, transference was conceived by Freud (1900) in topographic terms as the "transfer" from the system unconscious to the system conscious of some memory or impulse. Later, the term came to be utilized clinically to refer to the "transfer" of earlier "fixations" (in classical models, e.g., Fenichel, 1941) or "patterns of relating" (in relational models, see Mitchell, 1988) to the current situation in a manner that colors and distorts the perception of present realities. Still later, the question of "distortion" in transference was challenged (Stolorow et al., 1987; Gill & Hoffman, 1982) and numerous modifications and adumbrations were added along with the notion of countertransference.[2] Later, we shall discuss the clinical aspects of the crucial question of how valid (vs. distorted) transferential experiences are viewed. Clinically observed transferences, however, have often been recognized as only one, limited manifestation of the process by which our experience of any given relationship or situation is enriched (and complicated) by a host of meanings derived from other (often earlier) experiences and imaginative constructions. Freud (1925) recognized that

> Transference is a universal phenomenon of the human mind . . . and it in fact dominates the whole of each person's relation to his human environment. (p. 42)

---

[2]It might be wise to use a new term that does not have these connotations. However, like a brand name that is so successful that it enters the language as the word that refers to the entire class of objects, "transference" has come to be the word that almost all psychoanalytic clinicians use to refer to some of the most important, problematic, interesting, and dynamically alive aspects of the treatment interaction. Rather than invent a new word, we believe we can use the evolutionary perspective to develop a new understanding of the phenomena to which the word "transference" now refers.

The analytic situation, as Hoffer (1956) noted, "is a variant of human transference relationship."

Recently, J. Slavin (1990) has pointed out that clinical transferences are, perhaps, *most* usefully understood as vehicles for constructing the therapeutic relationship in ways that will elicit or provoke responses from the therapist that are needed by the patient in order to pursue his or her developmental agenda. Quoting Bird (1971), Slavin expands the ego psychological view of transference to one in which transference is seen as an essentially adaptive, interpersonal process that operates by intrapsychic means:

> Transference must be regarded as one of the ego's principal structures, a very special, very powerful, and possibly even a very basic ego apparatus. Most remarkable is the closeness of its relationship to the drives. This closeness, amounting to almost an alliance with the drives, may make it possible . . . to think of transference as being the ego's main antirepressive device. Such antirepressive action, so clearly exemplified by the usefulness of transference in analysis, may be seen as the power which in a general sense endows the ego with its crucial capacity to evoke, maintain, and put to use the past-in-the-present. (Bird, quoted in J. Slavin, 1990, p. 9)

Consider the experience of Basil, a middle-aged professional man who, at the onset of treatment, rather spontaneously experienced an activation of his capacity for transferential experience:

> Basil began to report an alternation between daydreams about scenes from the past and moments of living far more vividly and consciously in the present. This signaled a powerful activation of what we mean by the universal capacity for transferential experience or a heightening of the transferential dimension in ordinary experience. The sense of "vividness" in experiencing the present was made possible because important, neglected, background dimensions of experience—without which current experience was missing something vital—had been brought to the fore. He described feeling "much more aware of how things look when I'm walking down the road . . . buildings, landscapes, the sky . . . and much more conscious of myself being there, in it." This vivid sense of the present felt to him, simultaneously, "more like it was when I was a child . . . more the way things are . . . less inhibited." New memories of places and atmospheres from the past would then return. Moments of lightness and clarity alternated with an unnerving sense of "walking in a circle . . . a recurring confusion."
>
> He then dreamt that he "had to be in three places," had left something in each, and was traveling between them. One place he

connected with the town of his birth (in a foreign country), the second to the United States, the third, simply to "another place": a nameless, third destination that finally led, by association, to the line, "Who is the third who walks always beside you?" Basil knew this to be a reference by T. S. Eliot[3] to the hallucinatory companion that some travelers report "seeing" on very long, arduous treks, an allusion to a guiding spirit that enables one to endure arduous, spiritual journeys.

Though elicited by the analytic situation, Basil's experience illustrates the activation of the universal capacity for transferential experience: the evolved, intrapsychic mechanism that would allow him to begin the intrapsychic aspect of creating a "potential space," as Winnicott (1951) put it. In Basil's transference, he construed this space as a "third place" connected almost simultaneously with the past and the present along with the therapist's "presence" as a somewhat unreal or ethereal, but crucial, guiding force.[4]

The specific meanings of the intermittent memories that revived themselves in Basil's mind during this period remained entirely secondary to the "adaptive regressive process" (Kris, 1952)[5] of moving into and out of the transferential space announced by his dream. The evocation of the evolved capacity to shuttle adaptively relevant (formerly repressed) information between the past and the present was signaled in this case by Basil's experience of past "atmospheres" commingling with the present; it was not a sense of *living in the past* but rather a heightened, subjective sense of an intensified, *enlivened present*.

In evolutionary terms, the experiences we label "transferences" can best be understood as rooted in an evolved, universal deep structural capacity designed to permit us to make use of the relational environment in order to evaluate and organize the experience of the self—its internalized self and object schemes that represent and regulate all

---

[3]From *The Wasteland*.

[4]The pervasive real–unreal, sometimes uncanny forms in which transferential experience is encoded explains, in part, why we often label it a distortion. The issue of "distortion" versus "truth" is a complex one that we shall discuss later.

[5]As we have noted at many points in this volume, in regard to a wide range of issues (see discussion of Hartmann in Chapter Two, pp. 42–45; Blos, Chapter Two, pp. 195–198; and later in this chapter), the ego psychological emphasis on the adaptive aspects of the regressive side of transferential phenomena (Kris, 1952)—while suggesting a valuable direction in which a truly functional–adaptive analysis needed to move—ended up creating many obstacles to such an adaptive understanding by preserving classical assumptions about the drive-based origins and dynamics of conflict. The ego psychologists often simply added adaptive considerations to classical assumptions as though adaptation occurred in a realm separate from the main thrust of psychic operations.

interactions. Throughout the life cycle, an ongoing process takes place in which the need for an adaptive "regressive renegotiation" of specific aspects of one's self-interest is continually evaluated. In this process, the kinds of experiences and ways of relating we call "transferential" play a crucial role in this undoing of repression and the provisional psychic organization repression helps maintain. *From an evolutionary perspective, we experience transferences in order to actively experiment with virtually all new situations as demands and opportunities for the reorganization of intrapsychic structure.*

In this process of exploring and negotiating with the relational world, two components of transference play crucial functions. First, transference brings prior "learned" expectations into current relationships. This aspect of transference is adaptive (functional) in that it brings organized, learned expectations to bear in novel situations for which the individual would otherwise be totally unprepared. However, if we focus on this aspect alone, as is frequently done, we can conclude that transference maintains older self/other schemas rather than being a process that brings about adaptive change. The notion that transference represents a return to the past and a stubborn maintenance of the status quo (resistance) is only possible when we restrict our focus to this one aspect of transferential experience. We suggest that in addition to bringing the past into the present, transferential experience includes a constant testing and probing of the environment to determine whether it matches these older expectations. In exploring (and challenging) the environment, transference becomes a negotiation process for the modification of both the self and others. In addition to bringing learned expectations to bear on new situations, transference is a system for *changing* such expectations. This conception of transference explains how coloring the present with the past (which may well include some distortion of current reality) may be an adaptive process (i.e., bringing learned expectations to novel situations). The specific "colorings" may only be changeable through lengthy, arduous, and difficult negotiations with the relational world, for example, adolescence or psychoanalysis. Thus, it may appear that transferences, with their distortion and coloring of reality, are malfunctioning resistances to adaptive change. Rather, transferential experience may be an adaptive process composed of two parts: (1) the application of learning to novel situations and (2) a process for bringing what was learned more vividly into current experience and carefully modifying what was learned.

In contrast to classical, topographic views of the realm of the repressed (the unconscious) that emphasize a certain sealed-off, world-of-its-own quality, our clinical observations as well as evolutionary theory are more in accord with the growing relational view that the normal "boundary between conscious and unconscious mental content is actually more permeable, shifting and indistinct" (Mitchell, 1988; p. 261); that there is, in effect, a "fluid, ever shifting boundary" between what is conscious and what is not (Stolorow & Atwood, 1984, p. 368).

In the evolutionary view, the *heightened fluidity* of the relationship between conscious and unconscious configurations in transferential experience is actually an integral feature of a functional system "designed" to synthesize aspects of self-structure in a *provisional* way. Throughout the life cycle, an ongoing process takes place in which the need for an adaptive, "regressive renegotiation" of specific aspects of one's self-interest is continually evaluated. In such "regressive renegotiation," relatively repressed versions of the self (including alternative relational scenarios embodying wishes, fears, memories, ideals, and fantasies) are revived and brought to the experience of new situations in an attempt to both assess and influence the environment. As Modell (1990) describes the inner goal, it is to revise psychic structure, to "retranscribe" the initial ways we have categorized (internalized) formative experiences in order to accommodate the changing relationship between oneself and the environment.

Repression (ongoing storage), regression (activation and retrieval), and transference (the experimental re-evoking, reliving and comparing of old solutions with a new tested and probed reality) may thus operate in an intricately coordinated way from childhood onward. They may operate as elements in an evolved system for creating the "interrelated and interpenetrating" relationship between conscious and unconscious life that Loewald (1965) suggests is an inherently creative function of much clinically observed transference repetition. As we noted in Chapter Seven, repressed schemas, especially affects, can be seen existing in the terms Modell (1990) outlines, that is, as "a potential, awaiting activation in current experience" (p. 65). In our view, the shifting boundary between what is conscious and what is not always remains highly contingent on the continuously constructed, probed, tested, evaluated, and re-evaluated fit between one's self-interest and the environment.

Intrapsychically, these interrelated functions repeatedly serve to heighten the subjective experience of the "quality of fit" between inner individual needs, old structuralized solutions (self-definitions and ways of being in the world), and the current texture of relational reality. These inner stucturalized solutions, built in part around repression and expressed through transferences, were, to begin with, the result of compromises with the relational world. Many aspects of the individual's whole outlook on life are involved in this process. Of central psychoanalytic concern is the continual dissolution and resynthesis of those aspects of self-structure that regulate the ambiguously changing balance between directly "selfish" and "mutualistic" aims.

The Construction and Probing of Relational Contexts

The evolved system for self-revision that makes use of transference differs significantly from the intrapsychic process that is emphasized in classical accounts of analytic transferences. Despite Freud's clear statement that "transference is a universal phenomenon of the human mind ... and in fact dominates the whole of each person's relations with his human environment" (1925, p. 42), his major discussions of transferential phenomena (1912, 1913, 1914b, 1915c) were geared, in our view, to protecting the analyst from becoming deceived by the recognized power of transferential experience to draw one into the web of defensive self-deceptions that transferences may carry with them (Szasz, 1963). As Gill and Hoffman (1982), Stolorow et al. (1987), and others have noted, this defense of the analyst's identity and normal subjective bias was rarely recognized as such within the classical paradigm. Indeed, attempts to defend the analyst's own subjective reality became encoded in the overall classical vision in which transference is seen as rooted in drive-based pathological distortions that are projected or displaced upon the analyst (Brandchaft & Stolorow, 1990). The fact that transference-based interpretations became seen as the most valuable means for entering a patient's inner world and effecting analytic change (Strachey, 1934)[6] did not fundamentally broaden the limiting emphasis on transference as distortion inherent in the classical view.

We believe that elements of compulsive repetition and the struggle for mastery that can be discerned in the developmental process are more akin to the "new beginning" that Balint (1965) speaks of or the "retranscription of the past" (Modell, 1990) than to the far narrower conception of traumatic mastery in the Freudian conception of the "repetition compulsion" (see also Kriegman & Slavin, 1989). As an evolved, adaptive process, repetitive attempts to engender the interactive conditions in which new structure can be created are not seen simply (and misleadingly) as directed toward the restoration of a preexisting equilibrium (Bibring, 1943) or the correction of instinctually based distortions on the part of the child (Blos, 1967, 1972, Fenichel, 1945).

---

[6]Of course, this is what has always been known to differentiate analysis from nonanalytic therapies. In addition, the evolutionary view suggests that psychoanalysis has been engaged in some fashion with the basic program for change that is an integral part of the psyche. Only psychoanalysis, despite its many variations, has remained resolute in attempting to maximally tap into this design feature of the psyche.

These attempts at effecting "inclusively self-interested changes" in the self and others are only "regressive" in the sense that they serve to retrieve previously less acceptable versions of the self and to undermine provisional, interpersonally more acceptable self-organizations and relational arrangements. Regressive renegotiation may, indeed, also place pressure on the external surround to make accommodating changes. The retrieval of previously repressed versions of self are commonly seen in Freudian, Kleinian, and ego psychological terms from the adult (parental) perspective—and termed "regression," with the connotation of being infantile or immature and of "distorting reality." In self psychological, intersubjective terms, regressive retrieval is seen from the child's perspective—and viewed as resulting from a form of "environmental failure." As we view it, the retrieval of previously repressed versions of the self (which may have accompanying "selfish" relational demands and more "primary process" qualities) may well serve the crucial function of compelling a degree of accommodation, or renegotiated modus vivendi, by the environment, thus attaining a more optimal expression of the child's (or patient's) self-interest.

In this sense the regression we are describing is not a literal return to the past, but is similar to Winnicott's (1954) suggestion that regression entails a gradual shedding of the false self. While regressive retrieval may be functioning quite adaptively, what is retrieved may be labeled by the surrounding environment "regressive" in the pejorative sense of selfish, immature, or pathologically attached to earlier objects or identifications. This is to be expected as this is the biased viewpoint of the surround that may well be invested in seeing these aspects of the self remain repressed. In addition, regression can be chaotic and problematic to the individual retrieving the repressed information. Unrepressed impulses, affects, and resultant behaviors can be quite inconsistent, confusing, and unpleasant, especially until the renegotiation with the environment yields a new repressive solution and stability of self-structures. This can easily cause problems to those interacting with the regressed individual.[7]

In this view of the ongoing, renegotiation process, there is no particular emphasis on the fact that in retrieved, previously repressed relational schemas, reality is, in some sense, "distorted," or that there is less validity to the child's demands, that is, that they are less realistic

---

[7]Consider Miller's (1981) "gifted child" who, in analysis, slowly struggles to overcome severe constriction only to go through a phase in which he or she is often enraged. The same may be true of many traumatized individuals who massively repressed significant parts of their experience, for example, sexually abused patients and Adult Children of Alcoholics (as well as patients who have been sexually abused by former therapists) who confront their abusers after retrieval of repressed memories of the abuse.

representations of the child's interests than were the existing compromise arrangements. We must remember that, from an evolutionary perspective, all parties to the dialectic of development, the child as well as his or her parents, are naturally biased toward constructing reality in accord with their own interests. All engage in degrees of deception, particularly the coloration of reality by their own biased "true self" and "endogenous motives" that are, themselves, part of a complex adaptation designed to anticipate relational conflict and deception and to promote their own inclusive self-interest.[8]

In our view, as an evolved, adaptive capacity shaped by eons of phylogenetic experience, transferential experiences were likely to have been designed, precisely, to deal with the recurrent relational dilemmas posed by development and change throughout the life cycle in the normally conflictual, deceptive, good-enough environment—the environment that our universal intrapsychic dynamics evolved to fit. The capacity for transferential experience is thus likely to have become heightened and refined over evolutionary time as a subjective vehicle for the crucial "probing" of the object world necessary to monitor, evaluate, and alter crucial aspects of the fit between one's inclusive self-interest and relational reality. This "fit" is experienced as one monitors the ongoing sense of the balance of overlapping and competing interests within the network of one's relationships. Individuals may have, in fact, been designed to use a transferential engagement with many life situations in exactly the way several theorists have described patients as using the transference in analysis: to probe the analyst's subjectivity, interpret countertransference, and evaluate his or her willingness and ability to be influenced by the patient (Tower, 1956; Hoffman, 1983; J. Slavin, 1990; Modell, 1990). Through such transferences our psyche experiences many new situations as opportunities to invite, persuade, or compel our objects to enter the process of revising, renegotiating, and reorganizing intrapsychic structure and its complex patterns of interaction with the relational world.

Sometimes this reorganizing process takes the form of inducing

---

[8]This bias in perception includes our attempts as psychoanalysts, as well as other social scientists, to formulate and understand human development. As each individual tends to identify him- or herself to a greater or lesser degree with either parents/clinicians or children/patients, one or the other of these biases will be present in their perceptions, values, and beliefs. Such biases may account for some of the dichotomizing/polarizing debates between various psychoanalytic positions; for example, there may be a tendency toward child bias in self psychology and the parental bias in the classical view. From an evolutionary perspective, these are simply two inevitably competing viewpoints that must be expected given the genetic basis for parent–offspring conflict that has shaped the development of our psyche.

others to play out reciprocal internalized roles such as "enactments," "symbolic actualizations" (Modell, 1990), and the sometimes uncanny interpersonal experiences that are sometimes explained in terms of the concept of "projective identification" (Klein, 1946; Ogden, 1990; Bion, 1962a). Whether or not one likes the term "projective identification" or many of its connotations,[9] we can see some of the phenomena it purports to explain as specific intersubjective vehicles, or means, by which the overarching process of transference operates.

The activation of the capacity for *transference in the analyst*[10] makes it possible for the interpersonally evocative side of the patient's transference to work as it is designed to do: to fully elicit and probe an enormous range of potential responses from the relational world through the analyst. As the child animates the transitional object and the adolescent forms a semi-illusory set of beliefs about the world beyond the family, the patient uses the relatively undefined analyst to create a "working illusion," a form of transitional relatedness (Modell, 1990; Winnicott, 1951) that functions as a vehicle for probing relational reality and revising the self. In effect, the capacity for transference in both patient and analyst creates a system of expanded resonance and mutual influence. This system has, from time to time, been recognized by analysts in many traditions (Tower, 1956; Langs, 1978; Levenson, 1983; Hoffman, 1983). In our view, transference evolved as a capacity that is designed to work in a highly reciprocal fashion—that is, in an interactive context that evokes our natural capacity for reciprocal altruism (see Chapter Four)—within a network of social relatedness.

We discuss the complex interplay between the analyst's and patient's transferences in the next two chapters when we address the importance of the fact that our patients are designed to change in *far more reciprocal contexts* than the analytic situation can ever be. For now, we simply want to underline that the range of phenomena we call transference can be seen as an expression of a larger, complex, evolved strategy for presenting new objects with an array of relational challenges and opportunities. The responses to these challenges are read and interpreted for implications regarding the crucial inner process of creating and enhancing (and then recreating) a sense of one's inclusive

---

[9]As we shall discuss later on, a disturbing tendency exists for therapists to use the concept of projective identification as a way of disowning the therapist's own transferential responses by claiming that they have somehow been "put inside" one by the patient. This tendency ends up overemphasizing the distortions and deceptions in transference in a manner similar to that used by early classical theorists, for example, Fenichel (1945) and Glover (1956) (see also Stolorow et al., 1987, for a critique of this concept).

[10]Because our emphasis is on transference as a general human capacity, we often use the term for the analyst's experience rather than "countertransference."

self-interest. Behind this process is our evolved system for the creation and continuing modification (through new challenges, probes, and compromises) of the crucial balance of competing and overlapping interests.

Adolescent Change: A Developmental Case in Point

> I had to learn to tell the difference between me and my mother's son.
> —Harry Stack Sullivan

We have evolved the capacity for adaptive developmental transferences, reenactments, and the renegotiation of the self because of life's myriad, unpredictable vicissitudes as well as the major, expectable developmental changes in the degree and quality of parental investment built into every child's life cycle. The major instance of such expectable change is adolescence, and it provides an exceptionally clear case in point of how the evolved capacity for regressive renegotiation might operate. Through our discussion of adolescence, we hope to show that viewing the process of developmental negotiation in terms of underlying, evolved adaptive design fits with the "evolutionary narrative" as we have defined and developed it thus far. This developmental, evolutionary narrative incorporates elements of the classical narrative (stressing inherently conflictual motives, dividedness at the core of individual psychic structures, deception, and self-deception) and the relational narrative (stressing the social construction of the psyche, mutuality, and a more unified inner psychic structure), while differing in crucial respects from both perspectives.

A prolonged childhood in the context of primarily very close kin relations is, no doubt, an excellent environment in which to develop. Although we have repeatedly emphasized the inherently competing interests, natural subjective biases, and deceptions in the thoroughly good-enough family, we must also remember (see the discussion of kin altruism beginning on p. 86) that the natural overlap of interests between parents and their children, their "flesh of my flesh" adaptive alliance, should yield the highest ongoing levels of innately programmed attunement to each other's needs found in any relational context. Indeed, no one but parents and close kin (whose own "inclusive fitness" is frequently increased by altruistic actions toward the child) could conceivably be counted on, in any reliable way, to be psychologically organized so as to experience their own self (including even the highly individual biases of the "true self") as ultimately enhanced by the major, long-term, and arduous investment necessary to ensure a human child's development (Kriegman, 1988, 1990).

There is, however, a radical shift for the child at adolescence. From an evolutionary perspective, the biological change that takes place at adolescence consists not only of the well-known physiological maturation of the body, the brain, and the capacity for reproduction. Adolescence is also marked by an equally fundamental, equally "biological" shift from an adaptation to a relational context composed primarily of interactions between close kin toward an adaptation to a relational world composed primarily of unrelated individuals who are bound together principally by ties of reciprocity and exchange.[11] We would thus expect that a complex, evolved program exists by which those elements of individual identity developed in the context of parental and other close kin attachments can be re-evaluated and renegotiated in the context of the radically different relational conditions outside the nuclear family.

Several aspects of adolescent change, particularly the paradoxical mixtures of regression and developmental progression in adolescence that have long fascinated psychoanalytic theorists, may be understood in these broader, adaptive terms. As a process guided by a complex set of evolved deep structural adaptations, adolescent change can be seen as revolving around (1) the prior negotiation of a "provisional self-structure" in childhood; (2) the increasing failure of this provisional self-structure to represent the true inclusive self-interest as the adolescent child begins to leave the family; (3) the "regressive" creation of a transitional state of heightened "developmental transferences" and "enactments" in adolescence; leading to (4) the renegotiation of self-structure, including a new internal working model of the balance of competing and reciprocal interests within the wider relational world. Viewed in this fashion, adolescent change can be understood as a process that embraces many of the disparate phenomena that have been described separately, and often in rather contradictory terms, within the classical and relational narratives.[12]

---

[11]There are, of course, vast differences (between social classes, cultures, and historical eras) in the degree to which the adolescent is expected to move away from the nuclear and extended family in the process of establishing an adult life. There is, nevertheless, a relative movement away from primary relationships with close kin toward more distant and unrelated individuals (for mating and exchange relationships) in virtually every human social setting. Biologically, there is a vast difference in the meaning of relationships with different degrees of relatedness (Trivers, 1985).

[12]Many aspects of the comparative psychoanalytic view of adolescent change presented

(continued)

The Classical Narrative of Adolescence

The classical narrative of conflict in adolescence was initially characterized by Freud (1905) in his brief, entirely unsystematized, remarks on adolescence when he noted:

> one of the most significant, but also one of the most painful psychical achievements of the pubertal period ... is detachment from parental authority, a process that alone makes possible the opposition, which is so important for the progress of civilization, between the new generation and the old. (p. 227)

Yet, when the classical psychoanalytic theory of adolescence became systematized by Anna Freud (1936, 1958), the emphasis on inner conflict and turmoil was no longer envisioned as a genuine clash of subjectivities between parent and child. Instead, it was viewed as a clash between the newly intensified, bodily based endogenous drives of the pubescent child and the child's relatively "weak" ego. The "upsurge of the drives" was assumed to amplify repressed incestuous and aggressive (parricidal) wishes that regularly "overwhelm" the adolescent ego causing a regression to "primal defenses." These primal defenses (asceticism, intellectualization, flight into relationships, and displacement) were essentially elemental self-deceptions by which adolescents disguise and contain the overwhelmingly selfish, asocial impulses that comprise their primary motives. The whole defensive system was viewed as working primarily to "restore the inner psychic equilibrium" that existed prior to pubertal bodily change. Ultimately there was little in the classical view of adolescent change that could explain the occurrence of nondefensive, *progressive* developmental change.

In fact, the attempt to cast the classical narrative in systematic, theoretical, intrapsychic terms almost completely eliminated any sense of interpersonal goals and purposes behind adolescent conflict, that is, the freeing of oneself from inner and outer authority that controlled the earlier accommodation to the family that Freud noted. Devoid of such purpose or ultimate relational function: (1) the drives become enshrined as the prime movers and creators of psychic structure (outside of any larger interpersonal, developmental process) as though the ultimate human maturational goal consisted of coming to terms with these im-

---

in the following section were developed in collaboration with Dr. Jonathan Slavin whose own writing in this general area (J. Slavin, 1990) is not yet published. Though Jonathan Slavin shares the name of one of the authors, they are not kin; in evolutionary terms they are genetically distinct.

mutable somatic forces; (2) the whole narrative becomes slanted toward the parental point of view (with little appreciation of the biases inherent in parental authority that Freud recognized) as though the adolescent is struggling almost entirely with his or her own impossible, uncivilized desires; and (3) the entire adolescent developmental process becomes virtually equated with a pathological process, that is, as a structural failure in which heightened impulses burst through inadequate, defensive structures resulting in regressive acting out while distortions of reality are perpetuated by defensive operations geared not toward change but toward a restoration of a former status quo.[13]

In certain respects, the evolutionary perspective on the adolescent process—the retrieval of adaptively repressed alternative selves (or self-other schemas)—echoes aspects of what, in the hands of Blos (1967, 1972), become the ego psychological revision of Anna Freud's attempt to systematically spell out a classical narrative of adolescence. Drawing heavily on the work of Hartmann, Jacobson, and Mahler, Blos attempted to revise the classical view of adolescent change while remaining, like the other ego psychologists, true to the basic assumptions of drive theory and the underlying individualism and emphasis on conflict and deception in the classical narrative.

In brief, Blos appears to have realized that perhaps the weakest aspect of Anna Freud's drive based codification of the classical narrative of adolescence lay in the implication that the adolescent, like the neurotic, remains forever caught in an endless succession of defensive efforts to disguise true motives while covertly attempting to realize them through transferences; and that all of this forms part of the pursuit of the elusive goal of returning to a lost state of inner equilibrium. Though Blos never actually states this, he appears to realize that Anna Freud's vision has, perhaps unwittingly, led us back to an incredibly stark version of the classical narrative: a vision of individualistic striving for narrow self interest in a fashion that remains quite alienated and divided. Such an untamed vision was far from the sensibility of the ego

---

[13]The clinical need to distinguish between developmental and pathological processes, though clearly recognized by Anna Freud, produced only the most minimal accommodation of the classical model: namely, the adoption of the notion that the adolescent makes use of a new intrapsychic maneuver known as "removal" (Katan, 1937) by means of which early conflictual attachments are moved into an arena outside of the family. Unlike the basically conservative dynamics of Anna Freud's "primal defenses," removal somehow permits the adolescent to make genuinely new attachments to figures outside the family breaking the virtually endless process of defensively disguised, repetitive displacements of incestuous parental ties. The concept of removal remained a terminological *deus ex machina* by which the fact of progressive development could be accounted for and allows the classical drive conflict dynamics of adolescence to be largely retained.

psychological movement, loyal as it tried to be in many respects to Freudian thought.

Blos's solution was to create the view that adolescents possessed a built-in capacity to "cure" this developmental neurosis, leave childhood behind, and resolve the alienation: to become, in effect, a genuine individuated adult who fits comfortably into the realities of the adult world as portrayed in the classical narrative. The adolescent "self-cure" as Blos termed it, turned around a postulated phase-specific capacity to actually "court regression," to make use of the upsurge of drives in order to reexperience the specific individual "traumata of childhood" that create the instinctual fixations, overweening attachments, and irrational fears that, in Blos's view, are the actual sources of the pathology-like repetitions and enactments of adolescence.

For Blos, adolescents are not "ill" or alienated due to a universal condition of asocial desires or "an innate antagonism between the id and ego" as they are for Anna Freud. Rather, they are ill from a childhood disorder—the "infantile neurosis"—the instinctually driven, cognitively immature child's misreading of itself and the family. It is an illness that leaves a residue of repressed distortions of reality that, through regression, can be revived and revised in adolescence, that is, "played out" on a new "social stage." With the adolescent's vastly improved cognitive–symbolic ability and what Blos claims are the potentially less ambivalent, less antagonistic instincts of the adolescent at play, the childhood traumata can be "abreacted," corrected, and neutralized into an autonomous, integrated adult character. This "second individuation process"[14] creates what Blos—in the traditional psychoanalytic appeal to evolutionary authority (see Chapter Two, p. 48)—believes is a virtually unprecedented evolutionary form: human adult character structure. "We can view character formation," says Blos, "in an evolutionary perspective and contemplate it as a closed system . . . " (1979, p. 191). As a result of the adolescent "self-cure," adult character is seen as essentially "transcending" the social matrix. The child, on the other hand, was viewed as totally, dependently, and ambivalently "embedded within" the social matrix.

From our own evolutionary perspective, we can see that Blos has moved toward placing the process of adolescent change in a functional–adaptive context. No longer are the instinctual drives accepted as immutable antagonists of the ego (and the family); instead, they are viewed as playing a role within a larger, coordinated maturational system designed to promote separation and individuation. No longer

---

[14] A reference to the first separation–individuation process of early childhood as described by Mahler et al. (1975).

are the developmental phenomena of regression and the revival of repressed aspects of the self seen simply as a breakdown countered by massive defensive operations; instead, the reexperiencing of feelings and fantasies that had been repressed is seen as a crucial way of bringing the past (and its internalized representations) into better alignment with the present, allowing the creation of genuinely new attachments (as opposed to disguised repetitions of a fixated-on past). In short, Blos comes close at points to the perspective on the developmental process we have outlined. But, like the radical differences between a true contemporary evolutionary perspective and the attempts by Hartmann to create a biological adaptationist point of view in psychoanalysis there are major differences between Blos's ego psychological views and what we believe to be the fundamental paradigmatic shift required and made possible by the evolutionary narrative.

Blos constructed a way of reframing the adolescent process in a fashion that made genuine (as opposed to disguised and essentially repetitive) change appear possible. But he did so in a way that remained basically loyal to the classical vision—the classical narrative of human nature. And it is this "paradigmatic loyalty" (Greenberg and Mitchell, 1983) that ultimately weakens the explanatory power of his model. The classical narrative is fundamentally rooted in a view of our species as highly individualistic, divided creatures engaged in deception and self-deception because we are driven to pursue our self-interests in ways that will inevitably, antagonistically clash with social reality. Almost complete fidelity to this narrative led Anna Freud to generate a vision of an essentially endlessly alienated adolescent. How did Blos remain paradigmatically loyal, yet find a way to cure the divided, self-deceptive child? Primarily, it appears, by putting all of the inherent conflict, asociality, and self-deception and dividedness into the child. The *child*, in Blos's account, turns out to embody the classical vision of the human condition. The child's perceptions, cognitively immature to begin with, are colored and distorted by narrowly self-interested instinctual needs. The child is the asocial being whose object relations are characteristically "ambivalent" with an "intrinsically antithetical affective nature." The adult, by extreme contrast, "transcends" childhood and overcomes this alienated state through the adolescent self cure (Slavin, in preparation).

The extreme dichotomization of a transcendent adulthood—"character takes shape like a phoenix rising from the ashes" (1979, p. 176)—and an instinctual, distorting childhood seems to be an inevitable consequence of Blos's attempt to find a way to "civilize" the child envisioned in terms of the classical narrative. The repressed traumata of the infantile neurosis that the adolescent "self-cures" are viewed as virtually entirely due to the child's developmentally skewed way of experiencing adult realities. Parent–offspring conflict, in short, is seen almost entirely from the parental point of view, more so, even, than in Anna Freud's essentially pre-ego psychological model in which the classical narrative of human nature is more consistently applied to both child and adult. The transcendent, individuated adult in Blos's (1979) ego psychological adaptationist view is removed, as well, from most of the significant embeddedness in the relational matrix that was characteristic of the child. The child's self is in a highly contingent, dependent relationship with the parental environment. The adult's character becomes a "closed system" (p. 191) in which "one feels at home in one's character or, mutatis mutandis, one's character is one's home..." (p. 189).

From an evolutionary standpoint, Blos has to invent new dynamic processes (like phase-specific adolescent regression) and restrict these processes, arbitrarily it would seem, to one developmental period rather than viewing the adolescent process as one major instance of an evolved human process. He restricts his fascinating idea that the "infinitely more resourceful" (p. 484) adolescent ego "courts" the repressed to engage in an "obligatory" (p. 180) regression in the service of development by limiting the realm of the repressed to the child's drive-based distortions, rather than viewing the repressed as a function of a complex two-sided conflict built into the parent–offspring relationship. Finally, and most expectably, Blos's "paradigmatic loyalty" to the classical narrative leaves him not only with a rather idealized "neutral" adult removed from the relational matrix, but with a vision of both the child's and the adult's psyche that contains no inherent, mutualistic motives, no innate connection to genetically or reciprocally related others. As we have argued in previous chapters, this other primary side of our motivational system, and the relational embeddedness of the self (the "gene's-eye view," Chapter Four) to which it is integrally linked, are essential features of the evolutionary perspective. In psychoanalysis, these features have only been integral to the "relational narrative." And, as far as the adolescent developmental negotiation process is concerned, it is only Erikson who has systematically told the story of adolescent change from within this narrative structure.

The Relational Narrative of Adolescence

Although Sullivan dealt extensively with adolescence, Erikson's model is by far the most systematic treatment of adolescent change emanating from the relational narrative. Breaking far more decisively than Blos with the classical tradition (though, like Winnicott, he did it very diplomatically and in his own unique idiom) Erikson (1956, 1964, 1968) in effect questioned the whole notion that childhood experience was normally characterized by instinctually based distortions of reality that invariably had to be revised and even "cured." He questioned, as well, the almost exclusive emphasis on intrapsychic processes as the means by which adaptive change is brought about during adolescence. In contrast to these views, Erikson depicted the child (and the parent) as an intrinsically social creature, whose relationship with the object world is innately far more mutualistic than in the classical and ego psychological visions. Throughout the life cycle, the creation of an inner identity and the process of developmental change are fundamentally reciprocal, contextual matters, inextricably intertwined with patterns of recognition and response from the real, external relational setting.

In leaving his or her preliminary place within the family, the adolescent's central tasks, in Erikson's view, are to establish a sense of continuity between early, familial experience and the wider social world; and to create a sense of cohesion between disparate "identity elements" (versions of self and others) that have coexisted, often in contradictory ways, within the child's psyche (Erikson, 1956). For Erikson, the adolescent is thus not out primarily to "correct" his or her own instinctually driven distortions of reality (though this may be part of the process), but, rather, to discover what is most genuinely true and valid about early experience—often in opposition to accepted familial and social views—and to maximize the expression of self/other schemas that incorporate these vital truths into an adult "identity" (Shapiro, 1969).

In a mutual and reciprocal developmental movement, the adolescent and the adult world essentially cooperate to create a transitional period (the "psychosocial moratorium"), an experimental, semi-real state, one that is similar to Winnicott's (1971) "potential space," in which development can optimally take place (Slavin, in preparation). Thus, adolescent "regression," in Erikson's view, consists primarily of the adolescent's internal and enacted attempts to deal with the pressures of the adult, relational world (family, peers, and institutions). These pressures consist largely of sanctions and rewards that play on adolescent fears and need for approval. A major psychological danger to the individual, in Erikson's view, comes from a "pre-mature foreclosure of identity" (Erikson, 1964). By this Erikson meant a reconfiguration of the self that is influenced too strongly by the need to simply "fit in" rather than a self-structure that genuinely and fully embraces the true, unique scope of individual personality and explores the limits of its expression beyond the bounds of the family.

Adolescent Change in the Evolutionary Narrative

Viewing the process of adolescent change as guided by a system of evolved, strategic adaptations begins with an emphasis on the interactive nature of identity formation within the family. In good part, developmental "truth" is defined in a relativistic and socially constructed fashion. Unlike Blos's ego psychological model, the child's perspective is not seen as more inherently distorted by its own instinctuality and cognitive immaturity than is the equally biased, even if more knowledgeable and mature, perspective of the adult, parental world. Yet, the evolutionary assumption that the child is normally adapted to deal with significant amounts of conflict, bias, and deception within the fully good-enough environment is distinctly different from the almost exclusive emphasis on normal mutuality and interdependence between the generations in Erikson's way of viewing development.

The reemergence of what has traditionally been considered repressed urges and self/other schemas accompanying puberty represents the first part of the adolescent, developmental renegotiation process. From the biased adult perspective, the process may appear like a "breakdown" of the hitherto "civilized" child due to the "upsurge" of some new, physical, instinctual, driving force, that is, the formulation of the classical model (A. Freud, 1936, 1958). Yet, to the extent that the adaptive system guiding this developmental movement has been extensively shaped by social selection pressures operating on hundreds of thousands of generations of adolescents, it is a breakdown in only the most narrow, limited sense. It is an initial stage in the process of renegotiating the self, a process that has been "prepared for" by the existence of a reserve of potentially viable alternative ways of experiencing the self and others assured by the operation of "repression" in childhood.

Through the regressive retrieval of self/other schemas that were repressed through prior negotiations carried out in the family, the ingredients are gathered for a complex, internal and interpersonal renegotiation of a "new compromise" between individual interests and the new possibilities of the larger environment. Conceived as a special instance of the basic developmental negotiation/renegotiation process we outlined at the beginning of this chapter, the adolescent "regression" affords access to the imperative, inner signals of "endogenous motives," and the biased, innate, and individual givens of the "true self." These innate inputs afford crucial information about what needs to be included in the pursuit of individual inclusive self-interest in the relational world outside the family. A "developmental transference" is engendered—a heightened, fluid, "transitional" state, a "potential space" (Winnicott, 1951), accompanied by a widened range of experimental, interactive scenarios in the relational world beyond the family.

Given this inner state—the essence, we believe, of adaptive levels of "identity diffusion" as Erikson (1956) put it—the adolescent developmental transference serves as the basis for a massively reinitiated negotiation process, one that is not simply an "externalization with the aim of creating a new and perfect need gratifying world" as Blos (1979, p. 414) says, referring to an almost exclusively intrapsychic drama "played out" on the stage of the new social environment. Far more broadly, the adolescent process is designed to serve as an adaptively crucial probing of the relational world beyond the family. "Transferred" expectations, rooted in the cost–benefit assumptions of kin relationships, the morality of exchange between closely related individuals, and other patterns of meaning and power within the family, are juxtaposed with the demands, character, and opportunities afforded by new objects and new, often reciprocally altruistic relationships.

The adolescent psyche is designed to experience new, largely non-kin role relationships and objects through a heightened developmental transference: a state in which the relational world (like the usable therapeutic relationship), including its ideals and its limits, is continuously and simultaneously experienced in terms of what is "old" (replicates familial configurations) and what is "new" (calls forth altered self/other schemas) (Greenberg, 1986).

Similar to Erikson's "psychosocial," essentially relational model of adolescent change, our perspective on the adolescent process as an evolved, adaptive strategy brings the environment—the familial environment as well as the wider social world—into an integral role in the process by which developmental change comes about. In sharp distinction to the classical and ego psychological traditions, Erikson contended that the adolescent psyche (in a spirit that almost exactly echoes the developmental picture drawn by Kohut and Winnicott) develops in a way that is inseparable from the interpersonal field; progressive development requires affirmation (or mirroring) by a world that offers hospitable roles, usable ideals, and ideologies.

Adolescent developmental transferences that entail elements of sought after "mirroring" and alter ego sharing of identity—the reflected sense that one's newly forming sense of self is, to some palpable degree, seen, appreciated, and shared—represent crucial ways of using the environment for help in giving new, viable, relational meaning to revived, repressed versions of the self, and, ultimately, for help in reorganizing the self into a form that does, indeed, capture enough of one's innate, unique "true self" to accurately reflect one's genuine self-interest. So, too, adolescent transference relationships that entail significant idealizations, of real, mythical, or historical figures, recreate the conditions for replacing some of the inherent mutuality of close kin relationships with new, shared identities, that is, shared, internalized schemas of overlapping interests and goals that will give legitimate expression to (or re-legitimate the renewed repression of) vital aspects of the true self and endogenous desire.

The adolescent cannot expect the non-kin world to share his or her self-interest with nearly the depth and reliability as did the family, though the degree of difference and many crucial details of how the differences operate will have to be resolved through intense, repeated, "transferential" experiences and their relative resolution. Although in most cases adolescent developmental transferences, and the renegotiation process of which they are a part, represent real opportunities for expanding and revising the self, we must remember that, in the evolutionary view of the average, expectable, good-enough environment, these interactions will operate within a relational world that, under the most receptive conditions, is still characterized by an ambiguous mixture of competing and shared interests and deceptive and self-deceptive portrayals of reality. Adolescents, like the child before them, are out, in part, to negotiate the optimal expression of their own interests, their own relative advantage; even more than the child, adolescents are strategically equipped to engage in many of the complexities of creating a world view tilted subjectively toward their own interests, that is, to engage in degrees of the deception (and inner self-deception) that are part and parcel of how the self has evolved to organize and present itself within the wider relational world.

From an evolutionary–adaptive perspective, transferential negotiations—play, experimentation, probing, and enactment—will entail the navigation of an intricate web of clashing interests and the deciphering of new forms of deception in the world beyond the family. This fact ultimately distinguishes the evolutionary view from the more mutualistic relational perspective portrayed by Erikson. Though there must certainly be a degree of genuine reciprocity and opportunity for the adolescent in the world beyond the family, such mutuality is still far from the portrayal of normal, intergenerational relationships as a process that can "only be viewed as the joint endeavor of adult egos to develop and maintain, through joint organization . . . a mutually supportive psychosocial equilibrium" (Erikson, 1968, p. 223). Erikson's characterization of the adolescent negotiation process as a joint, mutualistic, interdependent endeavor, "the persistent endeavor of the generations to join in the organizational effort of providing an integrated series of average expectable environments" (Erikson, 1968, p. 222), ultimately overemphasizes the mutualistic side of the relational environment and intergenerational relationships; and it thus underestimates the significance of inner evolved adaptations to the inherently conflictual and deceptive dilemmas of development in the context of other biased, self-interested adult egos.

The evolved quest to optimally realign inclusive self-interest during adolescence appears to entail an adaptively useful belief that the transformed self will not only give vastly greater expression to the unique characteristics of the "true self" and the fulfillment of the "endogenous desires," but will find a context in which to do so where one's interests coincide with the interests of others. Although the capacity to entertain this belief may serve as a critical motivation for change, it is, of course, only partially true. Like the "primary omnipotence" (Winnicott, 1951) of infancy—and, in our view, the operation of the "curative fantasy" (Ornstein, 1984) guiding the ongoing process of regression and renegotiation in treatment—the unrealizable aspects of the belief in self-transformation (the versions of self that cannot be actualized) and the unlimited possibilities inherent in a world that is experienced as overly mutualistic must to some degree be relinquished. For the most part, though, the wishes and identity elements that comprise these adolescent beliefs will again become the repressed aspects of a new compromise between individual potential and social possibility. They will then comprise a new reserve of identity elements held in storage for future possibilities for regressive renegotiation (e.g., falling in love, marriage, divorce, deaths, major changes in the larger social environment). The creation of a "new compromise," including the successful reorganization of repression, comes about, in part, through the painful, subjective process that has been called developmental mourning (May, 1978).

Inclusive self-interest, as provisionally mapped out, modeled, and partially internalized in a childhood identity, is thus reopened, and, in the course of experimental, probing negotiations with new objects, redefined into its adult form (Slavin & Slavin, 1976; Slavin, 1985, 1985a). The self-organization or "character consolidation" created in adolescence remains a relationally contingent *working compromise* balancing a complex web of older kin ties (and the internalized identities they engendered) and new reciprocal relationships. In clear distinction to Blos's ego psychological model, adult character does not become a new "closed system" that, somehow, removes us from the highly interdependent relationship with the relational world characteristic of the child and, as Blos (1979) puts it, "lower forms of life" (p. 191). The evolutionary perspective is closer to Erikson's, Kohut's, and Winnicott's views that there is a lifelong, maturing need for and use of selfobjects. Though relatively more independent than the child, adult self structure, achieved through renegotiation in adolescence, is still the product of an evolved internal system designed to monitor, represent, and regulate inclusive self-interest within the relational world. This "negotiated self" is sustained by (and thus must be highly sensitive to) the complex textures of interpersonal and intersubjective exchange—the canonical music of overlapping and competing interpersonal interests.

Multiple versions of the self exist within an overarching, synthetic structure of identity (this superordinate structure being the experience of a cohesive, relatively unfragmented, vigorous, stable, and unique self). This adaptively divided yet cohesive self along with the continuing potential for creating new versions of this self—a potential maintained by the repression that made possible the ongoing renegotiation process throughout childhood and adolescence—continue to be the innate design features of adult self-structure that function as vehicles by which future change may also take place.

The Adaptively Divided Character of the Self

As the product of an evolved strategy for reconciling competing and shared interests, the "negotiated self" probably cannot possess the degree of internal cohesion or unity frequently implied by concepts such as the "self" in the self psychological tradition, the "consolidated character" in Blos's ego psychological model, or "identity" in Erikson's framework. As these terms are used in most discussions, the idea of an individual "identity" or a "cohesive self" serves as an extremely valuable metaphor for the vital experience of relative wholeness, continuity, and cohesion in self-experience. Yet, as has often been noted (Mitchell, 1991; Meade, 1934) when we look within the psyche of well-put-together individuals, we actually see a "multiplicity of selves" or versions of the self coexisting within certain contours and patterns that, in sum, produce a sense of individuality, "I-ness," or "me-ness."

Indeed, if the process of developmental renegotiation is, in fact, rooted in the need to revise and reorganize our overall self-structure in accord with continuously shifting interests in the relational world, one can argue that we are quite *unlikely* to have evolved an inner system that fosters a state approaching the true cohesion or integration implied in at least certain metaphors of self (Kohut), identity (Erikson), or character

(Blos). In contrast to a design that included *multiple alternative* versions of the self's relationship to the world—versions that may be necessary for different contemporary social contexts in addition to the renegotiated versions of the self for future contexts—even a flexible version of a unitary self-organization would likely reduce the range of available adaptive self/object schemas.

Although the coexistence of "multiple versions of the self" (Mitchell, 1991) that we observe introspectively and clinically may thus represent crystallizations of different interactional schemes, this multiplicity may also signal the existence of an inner, functional limit on the process of self-integration. Evolved strategies probably operate through limits or templates within which overall self-organization takes on its range of individual forms. The kind of evolved system we would expect natural selection to have favored is a kind of ongoing tension between multiplicity and unity: a constellation of multiple versions of the self that was set up to be "purposely upsettable" by the kinds of recurrent, ancestral situations that signaled significant potential changes in the complex tapestry of competing and shared interests into which each human life is woven. What might seem to be simply the "practical difficulty" in achieving inner integration or cohesion (given the varied contexts of human experience and the limited capacities of the individual ego) becomes further understandable as an evolved strategic patterning, a divided deep structure, designed to increase functional, context-dependent flexibility and to facilitate adaptive change.

All complex evolved strategies and structures represent adaptive compromises and entail trade-offs (Mayr, 1974). The cost of our human strategy for structuring the self in a provisional fashion—around a sometimes precarious confederation of alternative self/other schemas—lies in the ever-present risk of states of relative disintegration, fragmentation, or identity diffusion. The maintenance of self-cohesion, as the self psychologists contend, should thus be one of the most central, ongoing activities of the psyche—possibly the most central or superordinate principal of human psychological activity.

The strivings of such an evolved "superordinate self" would emanate, however, not primarily from a fragmentation induced by trauma or environmental failure to fully provide its mirroring (selfobject) functions. Rather, the intrinsic strivings would emanate from the very design of the self-system. A powerful, innate inner criterion for the evaluation of the meaning of any event or relationship might well consist of that event's potential for enhancing cohesion or evoking some degree of dangerous fragmentation. Yet, the forces that upset the provisional organization of the self (and may bring on states of relative fragmentation) must extend beyond failures in empathic attunement within the relational environment.[15]

Specifically, the provisional organization of the self is *designed to be disrupted* by events, internal or external, that signal an altered balance in real or potential possibilities for optimizing inclusive self-interest. Fragmentation, regression, and identity diffusion are thus far more than reactive processes triggered by "failures" in the environment. In relation to the evolved, inner quest to maximize inclusive self-interest, external events may be experienced as failing to provide what one needs. But, at times, they may only fail in reference to an inner, adaptive optimizing process that is geared to the promotion of individual advantage. Similarly, in relation to the evolved need to maintain a sense of cohesion or integration in the experience of the self, states of relative fragmentation may be designed to occur when the maintenance of cohesiveness becomes secondary to the need to tap into the store of alternate identity elements because the current fit with the environment is poor. Although subjectively painful because they are experienced as "deficiencies" or "defects" in the structures that create or maintain a sense of wholeness, the inner loss of a sense of wholeness signals the operation of an evolved strategic process in which multiple self/other schemas are revived and reanimated in an attempt to revise self structure or to induce changes in the environment.

---

[15]Such failures may actually signal a failure of fit between the individual's needs and the environment, and thus the *fragmentation* such failures provoke may function as an *adaptive* call for a restructuring of the self. This is in contrast to the self psychological view in which fragmentation tends to become synonymous with a sense of adaptive failure or breakdown (even if the fragmentation is a reaction to the failure or absence of selfobjects). Although chronic fragmentation that is not part of the process of regressive renegotiation is quite a problematic, maladaptive condition, some fragmentation may be part of an adaptive process.

PART IV

Contemporary Evolutionary Theory and the Clinical Process: Conflict, Negotiation, and Influence in the "Good-Enough" Therapeutic Relationship

PRELUDE

Clinical Discovery, Comparative Psychoanalytic Narratives, and the Evolutionary Perspective

> The search for hidden function in what seemed to be grossly maladaptive traits has been among the most fruitful procedures of evolutionary science ever since Darwin.
> —R. D. ALEXANDER, *Darwinism and Human Affairs*

> If there is one lesson that I have learned during my life as an analyst, it is the lesson that what my patients tell me is likely to be true—that many times when I believed I was right and my patients were wrong, it turned out, though often only after a prolonged search, that *my* rightness was superficial whereas *their* rightness was profound.
> —HEINZ KOHUT, *How Does Analysis Cure?*

Throughout this volume, we have used the evolutionary perspective to help reconceptualize the fundamental underlying differences between the classical and relational paradigms. We have viewed the narratives that underlie these analytic paradigms as built around a set of radically different assumptions about the structure of the human psyche and the nature of the relational world: conflict versus mutuality; deceptive versus valid communication; endogenous versus environmental patterning of motivation; and individual versus social structure of mind. In this process, the rough outline of a new psychoanalytic paradigm has begun to emerge, one in which it may be possible to incorporate the existing dichotomies, or biases, of the classical and relational traditions into a more inclusive model. Such a new model could be based on an understanding of how the mixture of mutuality, conflict, and deception inherent in the relational world selected for, or shaped, the particular evolved design solution—the "semisocial" solution—that characterizes the deep structure of our psyche. We turn to this new model and the larger implications of the evolutionary perspective in

Chapter Twelve. In the next two chapters, we examine some of the clinical implications of the approach we have been presenting.

We began to develop an understanding of the clinical implications of the evolutionary perspective with a recognition of the deep tensions within the range of clinical stances, the views of the clinical process, that derive from divergent psychoanalytic paradigms. Once these paradigms have been glimpsed at the level of their basic assumptions and markedly differing clinical sensibilities, many forms of pragmatic integration (e.g., Pine, 1985) become more problematic. In our approach, we make use of the available range of existing clinical-level theories, as far as they go. Increasingly, we also rely on the functional–adaptive implications of the evolutionary perspective. Through trial and error, a continuing negotiation with our patients, and an effort to be taught by them, we have approached each unique analytic relationship and have arrived at our current (provisional) understanding of the analytic process.

In essence, we have worked to *heighten* our awareness of the tensions between incompatible models. As Hoffman (1987) notes, a comparative "perspectivist" stance "dereifies" specific, theory-bound conceptions, enhancing our attunement to our patients' experience. Within this comparative context, our clinical–theoretical perspective has grown in its own way and its own time. Perhaps especially, our perspective grew at those inevitable moments when we felt abandoned by virtually all the broader theoretical notions about the mind, the moments when we yielded to and embraced the particular internal and interpersonal imperatives of the clinical encounter. Slowly, our evolving clinical experience has worked its way back, upwards through the levels of theory. It has pervasively shaped our view of the broader implications of the evolutionary biological perspective for psychoanalysis. We have tried to fit together our understanding of existing psychoanalytic models with current knowledge of the evolutionary forces that shaped our species in a manner that is consistent with the clinical data. In this process, the separate and differing clinical experiences of each of the two authors were incessantly argued out. The dialectic of our differing views and personal inclinations (biases) led to a different and far more balanced, multifaceted view of the implications of the evolutionary perspective than could ever, conceivably, have occurred as the product of one individual's experience.

As time has gone by (nearly two decades) we have begun to create a praxis that is, de facto, increasingly informed by our sense of evolved deep structure and the negotiation process of self-revision for which we, and our patients, are functionally designed. We shall try to define our approach and de-construct the different comparative psychoanalytic models on which it is based back into the language of the evolutionary framework.[1] From this evolutionary perspective, we then attempt to reenter the psychoanalytic *clinical* setting bringing an altered way of making sense out of the analytic encounter.

Our aim is to illustrate a way of looking at the analytic clinical process[2] that includes an understanding of the *echoes of ancient relational imperatives encoded in the workings of innate, evolved psychological design in both participants in the analytic process—therapist and patient*. We move back and forth between relatively experiential levels and theoretical perspectives, looking at a range of clinical phenomena with an eye toward observing the evolved, adaptive threads that reveal universal patterns interwoven within the fabric of our patient's and our own individual, idiosyncratic experience and specific familial histories. We discuss several basic clinical phenomena—transference, countertransference, subjective truth and distortion, repair and repetition, empathy, interpretation, and the therapeutic alliance—as we have come to observe and understand these phenomena in our own work, that is, infused with our own lived sense of the evolved, innate dimensions (the ancient phylogenetic dimensions) of the relational and intrapsychic experiences these concepts describe. We discuss these phenomena in comparative psychoanalytic terms with reference to the clinical implications of the major existing contemporary analytic paradigms, making use of brief, illustrative examples drawn from our clinical experience, consultation, and the supervised work of our students.

---

[1] In the effort to articulate our views we shall not attempt to advance notions of technique per se. Our recurring impression is that questions of technique (in the sense of guidelines about what to do, or how to produce such and such a result) almost literally *disappear* in face of a sufficient understanding of "what is going on," as Sullivan would put it, in the lived clinical process. This natural flow from understanding to praxis must, of course, include an openness to what is going on internally within the subjectivities of therapist and patient as they both enact roles rooted not only in their own histories and unique, innate potentials but also in the history of human relating as encoded in the underlying universal design of the psyche.

[2] In speaking about the analtyic process and the analytic situation, we refer to a generic, transference-based form of treatment in which customary distinctions between analysis and psychotherapy have minimal significance. Our position on this issue is rooted in our conception of transference as a universal, adaptive psychological function that will be mobilized under the right relational conditions—conditions that we believe are quite independent of many of the traditional criteria that distinguish analysis and psychotherapy. Analysis is supposed to mean the establishment of a set of expectations and conditions that amplify this process, but in our own experience and that of many others (see Gill, 1984; J. Slavin, in press) many "analyses" do not actually bring this about while some "psychotherapies" do.

We focus in more detail on recent self psychological views such as "intersubjectivity" (Atwood & Stolorow, 1984; Stolorow et al., 1987) and the American interpersonalist tradition than we did in Chapters One through Nine where we were more concerned with evoking for the reader a starker, purer sense of the underlying tensions between the two historical–intellectual analytic paradigms—the classical and the relational. As we have noted, the diversity within the relational paradigm, which includes Winnicottian, Fairbairnian, interpersonalist, self psychological, and intersubjectivist versions, was not emphasized in order to keep our focus on the dialectical quality, the deeper tension between the classical and relational narratives. In our clinical work, the particular visions represented by the intersubjectivists and the interpersonal tradition (especially the modern synthesis of it, Mitchell, 1988; Ehrenberg, 1975; Bromberg, 1989; Hoffman, 1983, 1987; Levenson, 1983) turns out to require a closer look.

CHAPTER TEN

Transference, Resistance, and the Evolved Capacity for Creative Self-Revision

> The tendency to regression is now seen as part of the capacity of the individual to bring about self-cure. It gives an indication from the patient ... as to how the analyst should behave rather than how he should interpret.... [T]he clinical fact of self-cure through a process of regression ... is commonly met with outside psycho-analytic treatment ... [R]egression represents ... the hope that certain aspects of the environment which failed originally may be relived, with the environment this time succeeding instead of failing in its function of facilitating the inherited tendency in the individual to develop and mature.
> —D. W. WINNICOTT, *Is There a Psycho-Analytic Contribution to Psychiatric Classification?*

Change and growth in personality revolve around the workings of transference, an evolved human strategic capacity and, consequently, a crucial aspect of the therapeutic relationship. Though universal, and sometimes overwhelmingly obvious, the transferential dimension of experience is usually subtly interwoven with the overall texture of life. The meanings of transference, both within the clinical situation and beyond it, are complex; and they are very easily oversimplified or reduced to one or another salient part that is mistaken for the whole. Indeed, an evolutionary perspective alerts us to the complexity of the human motives involved in the theoretical enterprise of trying to develop an understanding of transference. We expect that at the core of the very effort to define transference, we shall encounter the powerful human need to detect and identify others' expectable distortions, to discern truth from illusion, to guard ourselves against the dangers of deception of us by others and, by us, toward ourselves. Indeed, as we shall argue, the effort to define and use the concept of transference can, itself, be understood as part of the complex human strategy for dealing with the ubiquity of deception and self deception in our patients and in ourselves.

The Comparative Psychoanalytic Context

The classical and ego psychological psychoanalytic traditions have viewed transference primarily in terms of its functions in concealing and distorting reality. These distorting, deceptive functions of transference are codified in the classical idea that repressed impulses generate transferences in order to gratify instinctual drives; the accompanying defensive operations needed to keep the impulses themselves out of awareness create complex, perceptual distortions (Fenichel, 1941). Ego psychologists, such as Blos (1979), emphasize the expression through transferences of the relative cognitive immaturity of the child's conclusions about the world. In addition, the interpersonalists have emphasized the ways in which transference serves to maintain a system of self deceptions and distortions that perpetuates itself through its power to elicit the ongoing participation of others in recreating old patterns and interpersonal dramas that seem to validate the distortions. They speak of the transferential distortions of "consensual reality" and the reproduction through transferences of known, familiar relational scenarios in order to maintain loyalty to archaic familial objects as well as the idiosyncratic picture of reality engendered by one's family experience (Levenson, 1983; Mitchell, 1988). In sum, the various expressions of the patient's transference in treatment (in symptoms, dreams, resistance) is often experienced, defined, and responded to as a deception and form of distortion.

Although adaptive in its ultimate aims, individual efforts at renegotiating the self and self-interest through a sometimes painful, costly "regressive renegotiation" will often encounter little acceptance from others. Indeed, such efforts may be experienced by others as highly ambiguous interpersonal communications—communications that contain deceptive messages and distortions of reality. As a group response to the apparently deceptive messages and distortions inherent in efforts to renegotiate the self, the analytic tradition generated a broad clinical strategy based, in part, on an extremely profound mistrust of the very appearance of repetition. The history of the influential clinical concept of the "repetition compulsion" and its pervasive use (Kriegman & Slavin, 1989) in classical, ego psychological, and interpersonal guises attests, in part, to the magnitude of the need for a collective belief system to help individual therapists cope with the stupefying complexity of clashing subjectivities, aims, and interests that become activated when the transference-based process of regressive renegotiation is mobilized and under way in the treatment relationship.

To some degree, each analyst's subjective world will be substantially biased by his or her personal and social identity. Every therapist needs to maintain a somewhat idealized view of him- or herself as a change agent within the therapeutic process; every therapist needs help in maintaining his or her therapeutic and personal identity in the face of the often powerful pushes and pulls of the patient's transference, that is, the patient's attempt to use the therapeutic relationship and the therapist for the patient's own (biased) ends. We believe that the classical notion of the "repetition compulsion" was born, in part, in order to support each analyst in particular and the analytic identity in general. The idea of the repetition compulsion codified the notion that a huge range of the patient's observed, transferential repetitions can be readily understood as reflecting the underlying motivational *goal* of repeating dysfunctional behavior. Elevated to the level of a universal human trait, the idea of the repetition compulsion ties together many of the story lines in the classical narrative. Drive theorists, ego psychologists, as well as many theoretical models that are otherwise "relational" in nature[1] derive part of their coherence from positing one or another version of a universal, inherent compulsion to repeat: an inherent tendency for the perceptual *distortions* in transferences to become concretized repetitively in ever-changing, deceptive guises (Freud, 1920; Bibring, 1943; Fenichel, 1945; Glover, 1955). This systematic repetition operates in the service of maintaining repression—of *not* knowing, and *not* changing. Though even classical writers such as Bird (1971) have noted a "paradoxical" way in which "transference may be the ego's main antirepressive device,"[2] by and large most analytic therapists accept the view (codified in the idea of the repetition compulsion) that transferential reality is "illusory."

The implication is that, somehow, by their very design, transferences and "projective identifications" find their predominant expression through compulsively repeated action and therapeutic resistance. As Hoffman (1991) points out, whether transference repetitions are seen more classically in one-person terms as primarily intrapsychic events

---

[1] Several object relational and interpersonal theorists maintain this "classical sensibility" while focusing on the interpersonal implications of the classical narrative (see Chapter One for a fuller reference to the way Kleinians and some interpersonalists give essentially classical accounts of repetition and distortion).
[2] See the discussion in Chapter Nine, pp. 183, 184.

THE DEATH INSTINCT, BIOLOGY, AND ADAPTATION

Freud based his conception of the repetition compulsion and many of his discussions of "the negative therapeutic reaction" on the death instinct. In this concept, the major "goal" of all life is death, a return to an inorganic state. This became the motivational force behind regular painful repetitions and masochistic, negative therapeutic reactions. In a recent paper (Kriegman & Slavin, 1989), we critiqued the concepts of so-called "repetition compulsion" and "negative therapeutic reaction" in psychoanalytic theory and clinical practice. Even when separated from the death instinct, an evolutionary analysis reveals serious theoretical problems with these concepts.

There is a fundamental principle of nature that does parallel Freud's (1920) death instinct: the Second Law of Thermodynamics, which governs all matter in its relentless movement toward a maximum state of entropy or disorganization. With a logical structure similar to the Second Law, Freud's death instinct became his rationale for the existence of the repetition compulsion as a universal human characteristic. However, the Second Law is not a biological principle and certainly not an "instinct." Indeed, biological forces as shaped by natural selection have found a way of *opposing* this natural tendency of matter. They form a "matter eddy" wherein the forces of nature are *organized*. Of course, they don't operate outside the Second Law, but the degree to which an organism is favored by natural selection is *precisely* the degree to which it can *stave off* entropic disintegration (maximum entropy in its own body) long enough to pass on its principles of organization (genes) to other viable new organisms (offspring) (Kriegman & Slavin, 1989).

Freud apparently confused a fundamental law of nature (the Second Law of Thermodynamics that governs inanimate matter) for which "purpose" and "function" are inappropriate terms, with biological processes for which an adaptive functional analysis is necessary (Mayr, 1983b). There is no reason whatsoever for organisms to be motivated to promote the Second Law of Thermodynamics. It has, and will for all eternity, do quite well on its own without compulsions to enhance it. Conceptualizations of repetitions based on the death instinct have no clear biological basis. In Freud's view, the tendency to repeat became its own motivation; an irreducible inherent quality of all instincts.

Freud (1920) felt that the concept of the death instinct as the ultimate explanation for repetition, resolved certain theoretical inconsistencies in the notion of libidinal discharge and the pleasure principle as the overarching motivational forces. He was, of course, correct in sensing the need for a far broader, overarching, motivational principle. But in trying to find it in the death instinct and the repetition compulsion, he left the realm of valid biological discourse. The death instinct is a biologically untenable notion. We would suggest that what is "beyond the pleasure principle" is *adaptation*. The organism is not designed to function to maximize pleasure (or minimize pain); it is designed to maximize its inclusive fitness. As we discuss in this chapter, the dynamic aspects of the psyche designed to promote inclusive fitness may necessarily make use of a substantial amount of repetitive action and interaction as well as generating painful affects and problematic interpersonal situations as the *means* toward the complicated process of self-revision.

projected onto the relationship or as two-person events that fundamentally occur within the relational field (Levenson, 1983), there is often the same sense that behind the deceptive transferential facade lies a truer, objective reality that is presumably knowable (and to some extent is known by the analyst). It is this reality that must, in turn, be revealed to an often reluctant, self-deceptive patient.

As we have defined it, a "pure" relational narrative gives an account of the human condition in which the inherent effort to know and communicate subjective truth is emphasized over individual efforts to conceal or distort aspects of reality (see Chapter One). In the relational narrative there exists an account of so-called transference distortions and compulsive repetitions that is philosophically equal and opposite, as it were, to more traditional versions. As we have noted in regard to several other issues, contemporary self psychology—particularly its philosophically more radical form "intersubjectivity theory" (as developed by Stolorow, Atwood, Brandchaft, Lachmann, and others)—represents the clearest manifestation of this far more "constructivist" and "collectivist" narrative of the human condition. Indeed, nowhere has the clash of psychoanalytic narratives been more profound in the last decade or two than in the sharply dissonant voice of the intersubjectivists in response to the question of deception and distortion in the therapeutic relationship.

From our evolutionary vantage point, we see the intersubjectivists as having created a kind of equal and opposite system of beliefs. In this view, the core definition of transference is changed to one that is compatible with a thoroughly subjectivist, relativist, constructivist view of reality. To begin with, transference is equated with the functioning of the universal, subjective "organizing activity of the psyche." No longer can we talk about transferential experience as "distorting" reality. Indeed, we cannot conceive of transference as a "distortion" because it is now defined as simply one individual's way of using his or her "pre-reflective, unconscious, organizing principles" to structure current sensations and experience in terms of existing (known) categories. Since this is the only manner in which any perception can occur, reality becomes an individual's construction; reality can have no relevant (or meaningful) existence independent of the subjective act of constructing perceptions. Thus, no perception can be (or needs to be) considered a distortion of reality; perception, itself, is the act of creating reality. The subjectivity of the patient, no matter how idiosyncratic it may be, is never described as self-deceptive. There is no knowledge of reality that is kept from awareness as a function of the patient's own innate strategy for creating a subjective world biased toward his or her own interests. The repressed or "unthought known," as Bollas (1989) puts it phenomenologically, is described in the intersubjectivist view entirely as a sanctuary from the pathological biases of others. In this system, the subjective worlds of analyst and patient can and, of course, do differ. But neither the patient's transferences, nor the analyst's for that matter, are actually geared in any primary way toward generating illusions or creating distortions, self-deceptions or deceptions. Both subjectivities are simply aimed at maintaining organized (nonfragmented, nonpsychotic) experience. There are no *inherent* biases in this process.

As a corollary, the patient is essentially never viewed as motivated in any significant way by a compulsion toward transferential or other symptomatic repetitions. Indeed, the enactment of traumatic repetitions is usually more "dreaded" (Ornstein, 1974)—and avoided if possible—than unconsciously courted; the quest for self-repair and self-revision is viewed as a more primary motivational goal than any form of attachment to the past, be it a Freudian "fixation," Kleinian or Fairbairnian "introjected bad object," an interpersonalist "loyalty," or inappropriate generalization from familial objects (Mitchell, 1988).

The intersubjectivists do not deny the existence of transferential and symptomatic repetitions. They simply do not respond to them as deceptions or self-deceptions (efforts to act, or enact without remembering or feeling). Prereflective organizing principles that unconsciously operate to organize conscious experience are seen as leading patients to attribute familiar rather than new or alternative meanings to unfamiliar intersubjective contexts (Stolorow & Lachmann, 1984; Stolorow et al., 1987). However, patients substantially prefer to engage with individuals who respond differently from old traumatizing objects, that is, who (in self psychological terms) will provide a self–selfobject relationship that allows the structure of his or her transferences to change, for new meanings to emerge (Ornstein, 1984). Those transferential repetitions that do inevitably occur in treatment are almost always seen as resulting from "disjunctions" between the analyst's subjectivity and that of the patient. While they are treated as functions of the collective unit of analyst and patient, they almost invariably tend to be discussed as technical failures on the part of the analyst to "de-center" and enter the patient's subjective universe (Stolorow et al., 1987).

From an evolutionary perspective, both the classical accounts of self deceptive and repetitive transference distortions and the relational narrative of valid, idiosyncratic meaning and communication describe intrinsic aspects of our evolved system for regulating transactions with the relational world. Because of this it is extremely important to develop an alternative (an antidote) to the mistrust of patients' communications that became encoded in the concept of the "repetition compulsion" (rooted in the doctrine of transference distortion). Yet shifting into the relational narrative (and its implicit philosophical–biological assumptions) in order to develop an alternative goes only so far. We still must grapple with the inherently confusing mixture of adaptively vital information revived and conveyed by transference and the (equally vital) efforts to bias (distort) information through those aspects of transference that serve as strategic self-deceptions and interpersonal deceptions in the therapeutic relationship. In practice, we believe that in recent years analysts have essentially had to choose between "working illusions," so to speak: the older objectivist stance in which transference is seen as a distortion of reality buttressed by the "repetition compulsion" or this more radical, subjectivist, constructivist view in which the patient is seen as striving primarily to avoid repetition and relate in ways that will generate a new, strengthened sense of him- or herself.

Modifications of these views have certainly been developed: for instance, the clinically crucial notion that repeated transference enactments must occur in the analytic relationship in order for unconscious patterns to become known (J. Slavin, 1990; Levenson, 1983); or that creative (reparative) repetitions are a healthy alternative to pathological, deceptive, static ones (Loewald, 1965). From an evolutionary perspective, we believe we can go further and explain why both the distorting and intrinsically valid aspects of repetitive transference enactments are integral adaptive features of the normal psyche in the normally constituted relational world.

Distortion and Repetition in the Context of an Evolved, Adaptive System

Distortion and Self-Interest

In the evolutionary view of the ongoing, transference-based renegotiation process, the question of transference-based "distortions" and repetitions is a very complex, multifaceted one.[3] Retrieved, previously repressed, relational schemas are not seen, as they are in the classical/ego psychological view, as inherently distorted derivatives of the child's instinctually saturated view of the world (e.g., Blos's view of the

---

[3]Though our critique of the concept of the repetition compulsion (and, clinically, the negative therapeutic reaction) is far more extensive and thorough in an earlier paper (Kriegman & Slavin, 1989) than in the present chapter, the view of the meanings of repetition is less well integrated with the larger evolutionary perspective, particularly the process of regressive revival of repressed versions of the self and its clinical implications. The reader will thus find that here, as on a number of issues, our views continue to "evolve."

infantile neurosis discussed in Chapter Nine, pp. 195–198). In contrast, we assume that there is an intrinsic validity to transference-based perceptions and demands. They are versions of the self; they are as "realistic" a representation of the individual's interests as were the existing unrepressed compromise schemas that were most salient in awareness. In this sense, transferences as well as transference regressions are often alternative *versions* rather than distortions.

In our view, the organization of perception and the subjective states designed to preserve and change this organization (such as transference) cannot be understood as if they were neutral, objective mechanisms for simply and accurately knowing "reality." These features of the psyche evolved, in part, as ways of constructing reality in a manner that is likely to enhance individual interests. All parties to the dialectic of development—the child as well as the parents, the patient as well as the analyst—are naturally biased toward a *net* construction of reality in accord with their own interests. All engage in an entirely normal degree of relative "distortion," rooted in the necessary tendency to read and define reality in terms of the individual's innate, adaptive predisposition toward self-promotion. As we discussed in Chapters Four through Seven, we are not designed simply to see reality "accurately," that is, as would be intersubjectively verifiable by a large group of individuals whose individual biases are averaged into a consensually "realistic" view.

Our biases are adaptively crucial. Transferences will inevitably contain and reveal some of this bias. This is part of their function as vehicles for bringing individual self-interest as prominently (though not as explicitly) as possible into the current experiential, interpersonal field. Transferences mix inner, formerly repressed truths (affects, associations, images) with ongoing perceptions. They create developmental states of "transitional relatedness." Although crucial to the dynamic of change, these states can be viewed as relative distortions: They include experimental accentuations of relatively more self interested versions of reality in contrast to less repressed, everyday, reasonable schemas. Such "strategic distortion" may be viewed as an integral part of an evolved, longer-term, adaptive design in which patients bring to awareness and use accentuated subjective states as *more effective* relational probes, more effective ways of eliciting a sense of the genuine responsiveness, true identities, and readiness for investment by others. As Weiss and Sampson (1986) have shown

> the coming to consciousness of a particular mental content . . . is typically controlled by the patient . . . [the patient] is primarily testing the therapist . . . [and] in treatment, the patient works in accordance

with certain unconscious plans to change . . . [pathogenic] beliefs. (p. xiii)

The vital "kernels of truth" encoded in transferences may be far less separable from the distorting vehicles—the metaphors, concrete enactments—that carry their interpersonal meanings than is implied in much of the self psychological literature.[4] Kohut (1984) and the self psychologists (e.g., Ornstein, 1984), and the intersubjectivists (Stolorow et al, 1987) accurately stress the inherent validity in the patient's subjective reality and the striving toward self-revision that is contained within these subjective truths and revealed when they are effectively "heard." But these models tend to ignore the vital exploratory, evaluative functions that may actually be served by the encoding of these valid truths as *evocative* "distortions."

A major function of distortion may be to "overstate one's case." Because others can be anticipated to be somewhat antagonistic to the fullest pursuit of one's personal agenda, the case may have to be overstated (sometimes even created from scratch in anticipation of conflict). This is what frequently characterizes couples therapy and the endless arguing (negotiating) that couples do. Thus, transferences are not simply intersubjective or interpersonal *probes*; very prominently, they are a means of creating *bargaining positions*. Self psychologists tend to ignore the bargaining functions of transference, emphasizing instead the probing for the other individual's alignment with one's interests. Because it is usually far more clinically effective to re-frame the relative distortions entailed in transference as efforts at self protection, the self psychologists tend to emphasize this aspect of transference. The intersubjectivists (Stolorow et al., 1987) provide a consistent philosophical justification for this emphasis (Atwood & Stolorow, 1984; also see our critique in Chapter Six). We believe, however, that it is important to understand the ways in which transferences may, in part, represent active strategies for actualizing one's self-interest (even if it is at the expense of the other).

In our view we have, in effect, been designed to ensure that vital configurations of impulses, aims, and fantasies are effectively held in reserve in order to be retrieved, reanimated, and reenacted through transferences that, in part, through their very form, evoke, test, and compel the renegotiation process. Transference-based attempts at effecting inclusively self-interested changes in the self and others are "re-

---

[4]Though as the self psychologists point out a *clinical* focus on *distortion* per se (and even worse, trying to "correct" it) may be like "missing the forest" (vital information encoded in the metaphors and enactments) "for the trees" (distortions). Yet a recognition of the distorting aspect of transference may be important both for the therapist, and, at the right time and in the right context, for the patient.

gressive" in the sense that they serve to retrieve versions of the self that were, at some point, experienced as not fostering (often as endangering) relational ties and identifications. While the regressive retrieval ultimately contains the seeds of an adaptive revision, what is retrieved is, in fact, frequently experienced more strongly by others as a perceptual distortion, a deception of others that somehow needs to be corrected and counteracted, often by emphasizing its self-deceptive features.[5]

When transferential states occur outside of ritualized "rites of passage" and other acceptable symbolic forms, they may be labeled by the surrounding environment as "regressive" in the pejorative sense of selfish, immature, or pathologically attached to earlier objects or identifications. Understandably, one of the main ways we protect the relative coherence of our own naturally biased views of reality from challenges by other biases is by detecting evidence of distortion by others, attempting to correct such biases, and trying to influence others to accept our biases as more accurate.[6] Moreover, others may be quite invested in seeing things remain repressed.[7] Regression can be chaotic

---

[5]The focus on the self-deceptive features of the patient's conflictual material may sometimes enable the analyst (or parent, teacher, colleague, etc.) to appear as more of a neutral, helping teacher, thus downplaying their own bias and attempt at exercising influence. See the discussion of Trivers' views on the biology of parental deception in Chapter Five, page 113, as well as Gill and Hoffman (1982) and Stolorow et al., (1987) in the analytic literature.

[6]While not characteristic of modern analytic practice, we know that in extreme (worst) cases, when the analyst attempts to explain those aspects of regressive renegotiation that the analyst finds most troublesome by utilizing notions of inherent, universal, maladaptive features (such as the repetition compulsion), a pejorative assault on the patient can result. The patient's failure to improve can be blamed on some inherent characteristic in the patient; the patient is thus responsible for his or her so-called negative therapeutic reactions (Kriegman & Slavin, 1989). The extreme that this can reach was described by Schmideberg (1970) who reported knowing "of two patients treated by leading analysts for twelve and twenty years respectively, who eventually were sent by their analyst for lobotomy" (p. 199).

[7]Frequently a complicated system of "disbelief" on the part of the environmental surround is engendered by the process of regressive renegotiation. These complications begin with efforts by others to "correct" what they perceive as the "regressed" individual's distortions. The individual may then struggle harder (especially if what is needed is experienced as vitally necessary) to obtain a response from others that is consistent with the individual's goals for this attempt at regressive renegotiation. This can lead to greater determination on the part of those others to correct what, due to the ensuing struggle, appear to be even clearer "distortions." The resultant escalating feedback pattern can lead to self-perpetuating, self-defeating struggles that are repeated over and over again in the course of an individual's life. Moreover, the experience of retrieving repressed schemas is often extremely tense and frightening (remember, they were repressed as part of a vital compromise with a specific relational world in which they were experienced as dangerous or destabilizing). A web of further self-deceptions and deceptions often grows to attenuate the inner tensions and counter the invalidating external

and problematic to the individual retrieving repressed self-knowledge. Regressive ways of experiencing are often grandiose, idealizing, dependent, unstable, and painful. As such they trigger evolved inner signals (such as anxiety, shame, guilt) which indicate that the line between an adaptive regressive retrieval and a more dangerous regression needs to be monitored and managed[8]; for example, dangerous regression may include serious social impropriety, hazardous acting out, or a loss of a sense of reality.

The Compulsion to Repeat versus the Capacity for Creative Repetition

As we noted previously, in the classical tradition, the persistence of symptoms, painful repetitions, transference reenactments, and the like are ultimately described as extended ways of reliving the past without consciously experiencing or knowing it. Most clinically observed forms of repetition are thus regarded not as persistent efforts at accomplishing something new but, rather, as deceptive strategies for maintaining or restoring something old, some prior state of equilibrium, or status quo ante (Bibring, 1943). The emphasis is thus on repetition as signaling self-deception, as a way of distorting meanings, of not remembering, not knowing, and not changing.

From the evolutionary perspective, such repetition may be more fully understood as reflecting a striving to retain and make use of certain older, highly subjective meanings for the self-protective and adaptively relevant information they contain, that is, for their use as developmental guidelines in testing and eliciting responses from a new environment. Encoded in these early "fixations" is a version of reality that contains substantial subjective truth. The acceptance and articulation of the truth contained in this version of reality can be particularly crucial in the process of redefining self-interest—slowly and cautiously relinquishing no-longer functional fixations—and thus adaptively reorganizing the self. There is always an effort to communicate such meanings in some form and, to the extent that the new environment is reliably different from the old one, to renegotiate their place in the structure of the self.

Even before our patients enter treatment, a process of "regressive

---

assumptions that a "crazy" process of distortion, self-deception, and deception is at the heart of the individual's overall transference-based strategy for renegotiating identity.

[8]See the discussion of regression on page 189 where we also discuss how conflict between the individual and the environment gives vastly different (biased) perceptions of the process of regression.

renegotiation" has usually already begun. The pain and symptoms that cause one to seek treatment can be viewed as an inner adaptive signal that the current "fit" between one's inclusive self-interest and the relational world is a poor one; it is not serving one's interests adequately. The pain and symptoms cause an adaptive, regressive longing and search for someone who can help, that is, a "selfobject" with whom one can engage in the process of renegotiation. Consider the case of Christina.

> Christina grew up in an extremely religious, Catholic family. She was always responsible for her two younger sisters (one of whom became a nun). The family was organized around industrious activity. Both mother and father worked two jobs. Yet, there was virtually no communication regarding family members' feelings: activity replaced any form of empathic communication. A vividly remembered, frightening sexual molestation when Christina was a child was never discussed (though it never reoccurred). The severely alcoholic uncle who was the offender was never removed from the house. Raging battles would ensue between the father (when he was drunk) and the mother over the presence of mother's brother in the house. But nothing was done no matter how disruptive the drunken uncle became before he died. Christina was to take care of her next younger sister, and the two children were dealt with as if they were one.
>
> As an adult, Christina frequently found herself feeling depressed and sought aid from an endless number of psychological self-help books. She became quite an expert in dealing with issues in these terms and received an advanced degree in counseling focused on personnel development, becoming an unusually successful employee assistance manager for a major corporation. She taught numerous different types of "improvement" courses designed to enhance employee functioning and self-esteem. Her students always gave her the highest ratings possible for an instructor. They found her inspiring.
>
> Yet, Christina remained depressed. In every setting she experienced herself as performing, presenting a "facade," never experiencing a sense of authenticity, never being herself. At the same time, she viewed her depression as the problem to be solved. There must be something wrong with her because she could not muster up the energy for enthusiastic activity. Thus, the search through endless self-help books.
>
> Had Christina encountered the mental health system at a typical entry point, this view might have found an empathic, but misguided, echo in providers who would have seen an intelligent and talented woman struggling against an unreasonable depression. In many treatment settings, efforts might have been made to remobilize her capacity for gratifying, productive activity, defining this simply as a useful set of defenses. Alternatively, her depression might well have been viewed primarily in

terms of the losses that Christina (like most patients) had experienced, or in terms of the shame and guilt that surrounded certain events in her history. To some degree, all of these factors may have played a role in Christina's difficulties, but there was meaning in her current distress itself.

An attunement to the adaptive processes we have been discussing prompted Christina's therapist to consider the possibility that her depression carried a critical adaptive message. Indeed, it was not difficult to gather evidence supporting this view.[9] The depression was, in many respects, the healthiest part of her life! It became increasingly clear that her life was not satisfactory because, subtly and repeatedly, she had made sacrifices that left her feeling empty and used. The depression was not a debilitation that blocked a healthy set of coping mechanisms. Being a "good girl," going to church, taking care of her sisters, constant industrious activity, etc., had all taken their toll, leaving her with an empty feeling that, at age 41, she had abandoned herself and those pursuits that truly excited her. In simple terms, the depression was a powerful sign emanating from within her that the existing "fit" between her needs and the external world was a poor one.

This view of Christina's depression as carrying important adaptive information was, in fact, experienced by her as a hopeful alternative to her view of herself as defective. In addition, it became the cornerstone of an analytic treatment that involved an extensive effort to recover those repressed parts of herself that had been denied expression. In effect, she and the therapist could then struggle to engage in the rediscovery, indeed, the creation of a sense of her genuine, "inclusive self-interest."

As Williams and Nesse (1991) have argued for a wide range of physical illnesses, we believe psychological symptoms (such as acting out, drug abuse, manic episodes, anxiety attacks, phobias, obsessions, etc.) can sometimes be advantageously viewed as complex adaptive processes. Rather than being seen as failures or breakdowns, they often contain encoded messages that indicate that one's self-structure and picture of the world are, in some significant measure, alienated from one's genuine interests. Much symptomatology may be usefully viewed

---

[9]Although other viewpoints would also conclude that her depression was a reasonable response to her environment (i.e., not unhealthy), we know of no other viewpoint that as consistently suggests that what are traditionally considered symptoms of illness might, in fact, be adaptive, healthy functioning. In this case, given the degree and chronicity of the depressive affect, which at times was paralleled by near delusional self-devaluating ideas, it is hard to imagine another viewpoint that could have as consistently led to viewing her depression as a sign of an essentially adaptive developmental striving.

as an internal strategy that serves both as a signal to oneself and as a process leading to engaging others in the attempt to obtain the relational conditions that may allow the renegotiation of one's identity. Christina's transference was, in good part, the reactivation of a thwarted search for parents who were genuinely allied with her interests, (who would protect her and recognize her feelings and inner experience).

Her depression turned out to prompt just this kind of transferential quest. A primary feature characterizing this patient was her absolute terror at the thought of engaging anyone in a real relationship. She was profoundly convinced that no one would be there, and "terror" is not too strong a word to use in describing her fear regarding treatment. Yet, despite this, the depression led her to follow a friend's suggestion and enter treatment. She, in fact, described her intense anxiety in just this way, and fully verbalized how terrifying it was to even begin to believe that she might actually *want* to have a significant relationship with her therapist in which she did more than just describe her problems without affect. Despite years of attempting to solve her emotional problems alone, which was something she had become convinced was necessary in her unresponsive family, her persistent depression forced her—despite considerable fear or what, in her case, could truly be called her dread to repeat (Ornstein, 1974)—to enter into a new, therapeutic relationship in which she could engage in the process of regressive renegotiation of her identity.

Repression confers, in effect, a capacity for future repetition, a capacity that may well take symptomatic forms and appear to be (or even need to operate by means of) a subjectively experienced compulsion. Unlike the "repetition compulsion," it does not fundamentally operate in the service of returning to an earlier state or restoring and maintaining an inner "equilibrium," in the sense of a status quo, but rather promotes adaptive change, growth, and developmental repair. Encoded within the seemingly irrational "fixations," "parataxic distortions," "projections," or "projective identifications" that often infuse powerful transferences in the therapeutic relationship there is a subjective version of reality that contains substantial, relevant truth. Such truth will be used as a renewed developmental guideline for testing and eliciting the possibilities for a reciprocal fit between the patient's subjective world and that of his or her analyst. Thus, our readiness to see transferences and repetitions (within and outside the therapeutic relationship) as distortions of reality often reflects the challenge they contain for the analyst's professional and personal reality. In the face of this challenge, a huge amount of therapeutic energy is typically spent in the articulation and defense of the therapist's subjective world, the therapist's identity.[10]

It is not without therapeutic value for therapists to show and assert their identity in response to patients' transferential probes. Indeed, as we discuss in the next chapter, such expression is an inevitable and essential aspect of the process by which patients continually assess the reciprocity and degree of alignment of interests in the treatment relationship. Yet, it is perhaps equally crucial for therapists to take this challenge to their reality as a vital sign that the patient is engaged in a particular form of adaptive striving: an effort not simply to communicate an idiosyncratic subjective truth and have it empathically heard, but, beyond this, to more fully evoke, define, and grasp aspects of their own reality that may be intuitively sensed in a confused, inarticulable form.

We assume that over and above all notions about the distortion and deception that are intertwined with all complex, vital communication, there is always an effort to know and communicate valid meanings in some form and—to the extent that the new environment is reliably different from the old one in ways that make it possible—to renegotiate their place in the structure of the self. Repression confers, in effect, a capacity (operating through transferences) for future repetition—a capacity that may well function, in part, through complicated interpersonal negotiations. At their core, the repetitions and distortions involved do not fundamentally operate in the service of returning to an earlier state or restoring and maintaining an inner "equilibrium" in the sense of a status quo, but rather, as in Loewald's (1965) conception of "creative repetition" or Balint's (1965) "new beginning" they function as a force for overall dynamic change, of promoting adaptive growth and developmental repair.

---

[10]Transferential experience and communication are likely to contain a significant amount of biased, deceptive information designed to convey self-interested, subjective truths. Thus, our common reaction to transferential experience as "distortions" may, in part, reflect the fact that transference may include a strategy for influencing others to see things from the individual's biased perspective. If biases inherent in subjective experience are, in fact, *adaptive biases,* they must remain invisible to the individual who holds them. If these biases are conveying adaptively relevant information that ought to be listened to (i.e., if holding such biases and using them in interactions with others enhances one's inclusive fitness) then the individual may need to believe that they are accurate reflections of the way the world is actually structured; the individual must not become aware of the biased nature of his or her subjectivity. By the same token, we are not only *able* to see the biased aspects of *others'* transferences, we are *designed to search for* and be sensitive to bias in others' perceptions because they may (and often do) conflict with what is advantageous for us.

Some Universal, Adaptive Meanings of Resistance

Comparative Views

Classically, resistance (like transference and compulsive repetition) has been seen as a function of the patient's need to maintain ongoing unconscious gratifications and attachments through transference distortions, that is, disguised, displaced repetitions for the Freudians (Freud, 1912, 1915c; Fenichel, 1941; Glover, 1955, 1956) and projections for the Kleinians (Racker, 1968). The analyst was presumed to maintain an essentially nondistorted, "neutral" objective stance that would ultimately help understand, define, and revise the subjective world of the patient. The ego psychological heirs to the classical paradigm widened and shifted the theoretical emphasis to the ways in which resistances reflect the patient's difficulties in the very establishment of a therapeutic transference—difficulties in trusting, tolerating ambiguity and inner conflict, and those that result from the need for characterological defenses (Kris, 1956; Greenson, 1965). Inner tensions were still assumed to arise as a function of the challenge to the powerful transference distortions and repetitions that would necessarily occur in the analytic process. The emphasis on the patient's capacity to "tolerate" the analytic challenge to deception and self-deception led to the development of the idea that a relatively nonconflictual part of the therapeutic relationship—the therapeutic or working alliance (Greenson, 1965; Zetzel, 1956)—had to be cultivated in order to create "an atmosphere of safety" (Sandler, 1960; Weiss & Sampson, 1986) in which subjective distortions could be more effectively and palatably challenged.

Although they are far less apt to see resistance as something the patient's psyche projects onto a presumably blank-screen analyst, the interpersonalists (Sullivan, 1953; Levenson, 1983) ultimately trace the origins of resistance to the patient's transference distortions of consensual social reality and/or their compulsive repetitive enactments of familiar (familial) relational assumptions and scenarios. The crucial difference from more classical and ego psychological views lies in the interpersonalist view of resistance as a phenomenon that exists, and is only really knowable, as a lived, so-called "two-person" process. From this perspective, the analyst must become part of the lived experience of transferential relational scenarios. There is no observation without participation: the analyst must, to some degree, participate in the distortions and repetitions in order to observe them. The patient's transference-based enactments are a resistance to entering into, creating, and fully experiencing a truly "new" relationship that is undistorted by archaic relational scenarios (Levenson, 1983). Thus actual enactments are an inherent part of the unfolding analytic process. These patterns can become intuitively known only through the analyst's participation, unintentional as it may be, in the construction and experience of the older relational patterns (Greenberg, 1986).

Resistance came to have radically different connotations for Winnicott (1959), Kohut (1984), and the intersubjectivists (Stolorow et al., 1987). In keeping with the view that transferences and symptomatic repetitions are primarily valid subjective efforts at reinitiating thwarted development, clinical discussions of resistance in this tradition focus on the failure of the "holding environment" (Winnicott, 1965), the analyst's empathic failures (Kohut & Wolf, 1978; Wolf, 1980; Ornstein & Ornstein, 1980), or the occurrence of "disjunctions" between the subjectivities of analyst and patient (Stolorow et al., 1987). The emphasis on the origin of these breaks is almost invariably placed on the failures of the analyst to sustain a sufficient awareness of, and responsiveness to the patient's experience of meanings and developmental striving. Each time such a subjective break (or "selfobject failure") occurs a feared repetition of the experience of thwarted or traumatically disrupted archaic longings is apt to emerge—a repetition that the patient will, by and large, vigorously resist (Stolorow et al., 1987; Kohut, 1971, 1978).

In reflecting on our experience of resistance from both the evolutionary perspective and the comparative psychoanalytic viewpoint sketched above, we have made a number of clinical observations on some of the phenomena called "resistance." In the discussion that follows we focus on a way of thinking about some commonly noted manifestations of basic resistance to the analytic process. Viewed as universal expressions of the patient's and analyst's innate expectations and strategies for dealing with conflict and deception in the analytic situation, some of these resistances can be understood as illuminating the evolved deep structural meanings, for both patient and analyst, that are inherent in the analytic process. An understanding of these evolved, universal meanings creates a new context within which we can further evaluate the classical, self psychological, and interpersonal perspectives on resistance.

The Patient's "Innate Skepticism"

The evolutionary view suggests that an inherent knowledge about the inevitable disparities between one's interests and those of others exists as a fundamental feature of the psyche. This is accompanied by an innate sense of the potential for deception (and perhaps self-deception) in all interactions, and the existence of complex dynamic strategies for probing these hidden, relational realities. Thus, along with the innate capacity for transference-based renegotiation, an innate, intricately structured *skepticism* is universally built into the human mind. As we discussed in Chapter Eight, this "innate skepticism" is partly fueled and maintained by a system of self-serving and self-protective endogenous motivations, the capacity to store such motivations out of awareness for future use (repression) and the ability to retrieve those repressed aspects of the self under altered relational conditions (regression) and to use them to form a new compromise between one's self-interests and the relational world (re-negotiation).

The kind of skepticism we are referring to may have evolved as an indispensable accompaniment to the huge evolutionary increase in our capacity to learn, that is, as a way of safely employing our relatively "open program" (Mayr, 1974; see the box on p. 70). Since learning is a process of being influenced by the environment and the relational environment will inevitably be biased toward its own interests, the ability to learn brought with it a particular vulnerability. A variety of counterbalancing dynamics became necessary features of our overall psychological system. Clinically, we'll simply refer to some combination of these inner features as our "innate skepticism." Such skepticism is especially ready to be activated in any relational situation that offers itself as a context for regressive renegotiation, and it is a perfectly natural aspect of human experience, more pervasive than Freud realized when he remarked that a *"benign* skepticism" is the most propitious, attitude for a prospective analysand.

Most analytic therapists, however, tend to experience their patients' innate skepticism as calling into question the validity of their own professional efforts. In some sense, of course, such skepticism does call any effort at influence into question. The classical paradigm has tended to represent the patient's skeptical core largely in terms of its meaning for the analyst's subjective world (and interests), calling it a "resistance" in the sense of distortions and defenses that impede the patient's capacity to become immersed in the world of reanimated longings, fears, and other intense feelings that the analyst knows the patient is going to have to experience. Yet in terms of our evolved guidance system, arguably the most compelling (and at least initially, the most trustworthy) guideline for the patient's approach to the analyst, this form of "resistance" probably provides a highly necessary brake on even adaptive transference regression and a functional inhibition of the curative fantasy. In so doing, the patient's inherent skepticism helps to motivate aspects of the patient's initial "investigation" of both the analytic situation and the relationship with the particular therapist—the investigation that, sooner or later, all patients must conduct in order for a meaningful engagement in treatment to take place.

Kin versus Non-Kin Influence

The main feature of "innate skepticism" is the adaptive capacity to evaluate the particular texture of the overlap and divergence of interests in a given relationship, to detect deception in the definition of the situation, and to assess the probabilities of direct and indirect reciprocity. One of the evolved shorthand methods for accomplishing this in our way of approaching relational realities consists of an "intrapsychic algorithm" that is built around the universal distinction in our object relations between unrelated (non-kin) and related others (and between degrees of relatedness among the latter). As we discussed in Chapter Four, all things being equal, the universal fact of highly overlapping interests among close kin means we can expect that, though inevitably biasing to some degree, kin are more likely than unrelated others to influence us in accord with our own inclusive self-interest. The fact that we have evolved ways of internalizing these early kin influences (in the form of basic selfobject schemas, the basic sense of self or identity) is simply an expression of the overall, adaptive utility of an evolved design whereby we tend to allow kin influences to have such significant shaping power.[11]

However, patients in analysis must ultimately allow themselves to alter, in some sense be "weaned" from, particular kin relationships as well as from some of the principles of reciprocity and investment (interpersonal exchange) that organize our experience around kin relationships. As some object relations and interpersonalist theorists emphasize, these principles are often generalized and used inappropriately and repetitively as models for negotiating non-kin relationships (Fairbairn, 1952; Levenson, 1983; Mitchell, 1988). The versions of self and interaction schemes that were internalized from early kin ties, although perhaps costly or destructive in some respects, were formed in the context of relationships with deep, overlapping self-interests with concomitantly deep emotional attachments. On the whole, it made extraordinarily good, functional sense for humans to have evolved a strong

---

[11]The central and universal meaning of the kin–non-kin distinction has long been recognized in anthropology but has never found more than a peripheral place in psychoanalytic models. Modern Western societies have perhaps lived with the illusion that, at least among the mobile middle class, kin relations are not innately privileged in our emotional life (Parsons et al., 1955; see fuller discussion in Chapter Four).

proclivity to develop our most basic picture of ourselves and the world in the context of the unparalleled investment and overlapping interests in the family setting. And, because, throughout our evolutionary past, familial ties could generally be counted on to remain a significant part of most human lives, it made equally good sense for our inner system of introjects and identifications to retain the stamp of close kin identity even in face of some of the psychological drawbacks this identity may entail.

For most patients, adaptive, functional growth requires that kin relationships and the versions of self and other that are adaptations to them be replaced, in part, by potentially more rewarding, non-kin, reciprocal relationships along with revised internal organizing structures for experiencing them anew, not as replications of kin ties. Yet, in these new object ties our patients will never find the degree of love (investment) they found in their family simply for being. They will almost always have to reciprocate to a greater degree in relationships outside the family.

The "Unnaturalness" of the Analytic Situation

Of course, our psyches must, and do, allow unrelated individuals psychological entry into the processes of defining the self. The universal operation of transference as a relational probe leading to the regressive reorganization of self schemas represents the ongoing operation of a process by which, through complex, intersubjective negotiations, we ultimately allow non-kin to have a significant degree of influence over the basic organization of our mind and emotional life. Yet, regardless of which theoretical tradition we use to define its prime therapeutic action, analytic treatment is designed to be a process by which an unrelated other is given an extraordinarily privileged access to very deep layers of the mind. Since it is only *related* others with whom, a priori, we most closely share our core individual genetic identity, the influencing power we grant to unrelated individuals toward whom we deeply open ourselves emotionally and allow ourselves to become identified must normally be commensurate with the tangible experience of a *large degree of long-term, stable reciprocity* (e.g., marriage, intimate enduring friendships, and, occasionally, some mentor, collegial, and business relationships).

Yet, the analyst is an unrelated other who asks the patient to pay (sometimes dearly) for what is always experienced (by even the most grateful patients) as a relatively small investment (in terms of cost to the analyst), with little tangible reciprocal sharing or exchange and, for long periods of time, ambiguous and subtle long-term results. The analyst asks the patient to give free reign to his or her own natural tendency to generate intense transferences about the analyst and the therapeutic relationship. But transferential negotiations, as we have described them, were designed over eons of phylogenetic history to be activated in relationships where there are likely to be long-standing, reliable, mutual reciprocation. Transferences of great intensity and duration are only fully unleashed and lived out in the most intimate interpersonal situations or well-circumscribed social or religious ritual and drama.[12] The evolved ability to generate, live, and use the regressive remobilization in transferential experience is a vital, adaptive capacity that must be carefully monitored lest the element of illusion and symbolic enactment lead too far away from a realistic evaluation of one's interests (i.e., to excessive self-deception and vulnerability to deception by others). In the relational world to which our psyches are adapted, the signals for a safe evocation of this capacity within an intimate relationship are reliable reciprocal exchanges that are frequently evidenced by relatively equal risks and investments over considerable periods of time.

Thus, there is a general process operating. From the patient's point of view, he or she is presented with powerful and inherently dangerous role expectations:

> Though I as analyst give you nothing tangible in return—and, in fact, insist that you pay me—I expect you to trust me and open yourself up to my influence and give free reign to powerful fantasies and wishes. I expect you to allow the full activation of the regression-based change process that was designed over eons of phylogenetic history to be activated in relationships where there are long-standing, reliable, reciprocal interests that are evidenced by mutual, relatively equal risks and investments over considerable periods of time. Further, as your analyst, I imply, that the negotiating power that the activation of these innate, evolved developmental forces confer on me under these relational conditions—as a relatively closed, non-reciprocating, protected, professional participant in this process—will ultimately reorganize you in ways that are more aligned with your real interests than you can even imagine, and that, at this point, either of us can know.

In short, the "setup" in the analytic situation should make the

---

[12]And in some cases, for example, religious cults, it may be that rituals are designed, precisely, to elicit such experiences "inappropriately," that is, to "trick" the individual into disregarding their innate skepticism. Frequently the long-term reciprocity is missing and most religious conversion experiences, while quite dramatic, are short-lived (see Kriegman, 1980; Kriegman & Solomon, 1985a, 1985b).

patient suspicious. From the point of view of the adaptive design of the patient's psyche, consider the meaning of the analyst's ultimate strategic plan to elicit an intense transference attachment and regression. The analyst is trying to create a setting designed to revive aspects of early experiences in part in order to mobilize the process of regressive retrieval, to maximize transference precisely in order to essentially *mimic* the unique, evolved, emotional influence of kin ties or very special, reliable relationships that entail significant long-term investments. The analyst asks that the patient fully yield to the creation of a "potential space" (Winnicott, 1971), if you will, a transitional way of relating, in which the sense of real and unreal, past and present, self and other blend together. Yet, analysts certainly do not tend to define their core selves in a fashion that includes the self (and self-interest) of a specific patient. Consider the relative experience of the analyst and patient if for some reason a working relationship is prematurely lost. Nor does the analytic situation carry the signs of non-kin, reciprocal reality—of highly overlapping, shared interests. From an evolutionary–adaptive perspective, it is hard to imagine any situation more likely to elicit a vigilant readiness to detect signs of deception.

From the therapist's point of view, there are also many intrinsic unnatural dimensions to the analytic situation. Although the therapist is paid and generally finds a professional identity in the work, he or she is also expected to tolerate extraordinary violations of very deep, evolved norms of reciprocity and self-expression—norms that generally push us to express our personal identity and exercise interpersonal influence in human relationships. Whether it consists of the striving for "neutrality" in the classical mode, the "decentering" from one's own subjectivity in self psychology, or the adroit juggling of the paradoxical role of "participant/observer" in the modern interpersonalist tradition, we believe analysts are called on to contravene certain fundamental, evolved inner signals in the perception of and response to another's experience. In many respects, analysts are called on to sustain an exquisite alertness to self-deception and avoidance of deception that, given the realities of normal parent–offspring conflict, actually far exceeds what any healthy, good-enough parent could or should provide.

Finally, as many experienced therapists know (Eigen, 1991; Pizer, 1992) and as we discuss at greater length in the next chapter, *it is a rare therapeutic relationship that does not, in fact, require the analyst to reopen and revise his or her own self-structure in ways that ultimately reciprocate the alterations in the patient's psychic reality.* If our view of analytic structural change as a *reciprocal negotiating process* holds, the analyst must ultimately be willing to eventually alter some of his or her fundamental ways of organizing experience, to alter his or her own self. Thus, we believe that both patients as well as analysts are, by evolved design, prone to enter the treatment relationship with a powerful, entirely natural skepticism about whether such an apparently nonreciprocal relationship with an unrelated other can possibly result in a revision and enhancement of their genuine inclusive self-interest. A universal, evolved relational context powerfully characterizes the experience of intersubjective meanings, interpersonal exchange, and the process of influence in the analytic situation. We believe that none of the models in the analytic tradition provide us with a way of thinking about the analytic situation that gives adequate credence and weight to these broader, universal determinants of the negotiation that takes place between analyst and patient.

Some Typical Clinical Expressions of the Evolved Dimensions of Resistance

Clinical examples abound in every analytic clinician's practice attesting to the fact that patients are deeply and profoundly affected by the universal meanings of the fact that the therapist is not of their flesh and does not appear to be bound by some of the elemental kinds of social and material reciprocity that might begin to establish a sense of clearly reliable, overlapping interests. Yet, in the entire range of analytic literature on the "analytic situation" there has been no way of looking at this universal dimension of the human relationship between patient and therapist that allows for a candid acknowledgment of the meanings of non-kin relatedness and its inherent implications.

For example, Eigen (1991) describes a desperate patient who made one simple request, the agreement to which she set as the precondition for engaging in therapy. "What I need is simple," she said, "it's simply what you'd give a child of your own. The simplest thing in the world." Clearly sensitive to both the crucial relevance of her request as well as the sheer impossibility of fulfilling it, the therapist is aware that at some level he understands what she wants: "I feel," he writes, "that simple thing at the center of my being. It shows in my eyes, my tone, my skin. I live from and through it. It is my home." Indeed, in a poetic lamentation on the dilemma this request necessarily poses for a compassionate, empathic clinician, Eigen shows very clearly that he (like most clinicians) ultimately realizes this "simplest thing in the world" cannot be part of the analytic relationship. But he does not really know why he knows it, or, more important, why it is as relevant as both he and his patient know it to be.

In the absence of an understanding of the evolutionary dimension of the parent–child relationship—of the absolutely unique meaning to the human psyche of having intrinsically shared, overlapping interests with another—the most empathic clinician (however exquisitely immersed in the subjective reality of the patient's world) will not have any clear way of thinking about why he or she cannot have a relationship with a patient that primarily and consistently taps the innate wellsprings of love and caring that come with the most intimate, genetic (and/or the most powerfully reciprocal) relationships. A purely ontogenetic perspective on object relations lacks a view of the way in which we are, in effect, millions of generations older, with greater innate wisdom, than the particular experiences our lifetimes could ever confer. Such knowledge is, indeed, as Eigen intuitively knows, "at the center of [our] being." And it is equally at the center of our patients' being whether or not they, like the desperate woman whom Eigen describes, happen to be endowed (or cursed) with "knowing that they know" how central it is.

Without an evolutionary perspective on relatedness and its innate, evolved meanings, the intuitive knowledge of some of these meanings by our patients tends to be interpreted as resistance in a far narrower, pathologized, or overly individualized sense. Many responsive analysts simply have no way of realizing that, for most patients, the analytic situation elicits profound inner signals that derive from an ancient, adaptive legacy—a legacy that informs and interweaves tightly with personal history but, at root, does *not* have individual, ontogenetic origins. In a primary, clinically relevant way, the resistance prompted by the patient's innate skepticism is organized around something universal rather than individual. Consider the following therapeutic interaction:

> Nancy, a very troubled patient who had been frequently diagnosed as "borderline," became agitated and depressed in response to hearing that her therapist just had a child. The therapist responded to this by trying to articulate what he thought was a crucial element of distortion in her perception of the meaning of the situation. He said, "You feel that there is a limit to the amount of love and concern available in the world and that if I now begin to give to my son I will have less available for you." The implication was that the patient's experience was, in some way, a pathological carryover from childhood (e.g., a reexperiencing in the transference of painful sibling or Oedipal rivalry). Following this, Nancy regressed further and became suicidal!
>
> Later, the therapist confronted a degree of self-deception in his own response. He acknowledged the inherent, current-day *truth* in the patient's feelings, namely, that, of course, the therapist's life energies would be significantly absorbed by a child of his own flesh. Perhaps this would

leave somewhat less of a capacity to be thoroughly invested in Nancy in the way she desired and was, in part, accustomed to.

This essentially universal aspect of the therapeutic relationship was already innately known by the patient (and, after reflection, by the analyst). When it was acknowledged and explained in a way that the patient could experience as genuine, she reported feeling less afraid. She then commented on the fact that maybe her therapist would now learn something about nurturing that might be of use to her, and she rapidly recompensated.

We are reminded of Freud's (1918) cryptic and misunderstood comments in the case of the Wolf Man:

> a child catches hold of . . . phylogenetic experience where his own experience fails him. He fills in the gaps in individual truth with prehistoric truth; he replaces occurrences in his own life by occurrences in the life of his ancestors. (p. 97)[13]

As we discussed in Chapter Two, Freud's rough approximation of an evolutionary perspective tended to portray universal, adaptive schemas, such as Nancy's innate knowledge of the impact of her therapist's inevitable investment in his new child, as somehow outranking and, in an either/or fashion, superseding what is learned from personal experience. "Whenever experiences fail to fit the hereditary schema they become remodeled in the imagination" (Freud, 1918, p. 119).

This tremendous oversimplification of actual experience may often

---

[13]Freud's dilemma with the Wolf Man concerned what Freud believed to be the patient's apparent knowledge of the primal scene in the absence of a memory of it having actually been observed. Did it literally have to have been directly observed in order for the meanings of parental intercourse to be known? If it had not been observed, where do the fantasies that seem to be built around it come from? Freud's "phylogenetic solution" to the problem became a way of avoiding a reduction of the answer to these questions to the common pitfall of an exclusively ontogenetic perspective, that is, the reduction of a patient's experience to a "psychic reality" that somehow has less than full validity because it cannot be attributed to an "objective" individual, historical experience. To this we might add another pitfall: the creation of a completely subjectivist, relativistic view of reality in order to render completely moot the question of the truth status of the patient's subjective experience. Or, for that matter, the assumption that, barring any evidence, a supposedly objective historical event must have occurred as a literal and concrete explanation for the fantasies. Crude and oversimplified as his evolutionary thinking tended to be, Freud clearly grasped the fact that it was impossible to account for a wide range of what he called "the residual phenomena of analysis" (1939) if we assume that one's drives and one's personal history of social interactions must, somehow, fully account for what one knows, fears, intuits, or sets out to discover about the world.

reactively lead to a facile dismissal of the evolutionary dimension. Such an automatic and closed system would *not* represent a sound adaptive way for evolved, innate aspects of meaning or knowledge to work.

One "catches hold of phylogenetic experience" (or, really, the expectations selected for by phylogenetic experience) in the context of an intricate set of expectations learned from individual and cultural experience. Consider the following interaction that, in its essence, we believe, most analytic therapists repeatedly encounter.

> Tom was caught in a regressive spiral of anger and withdrawal regarding his analyst's impending vacation. To compound matters, he felt very ashamed and defective at having reacted so strongly, with so much jealousy and sense of abandonment. His thoughts and fantasies returned incessantly to scenes of his analyst playing with her husband and children. The analytic relationship was, by comparison, not real; it was just an illusion, and a painful one at that.
>
> The analyst tried to convey that she could understand his pain at her departure, and, particularly, the meaning of the pain in envisioning these family scenes. Not only did it evoke a painful longing for what Tom did *not* experience with his own mother and family but, she added, "because we are not family, and we are not married, it is all the more important for you to affirm that what needs to exist for you in our relationship is really here. You *must* assure yourself of the presence here of the important features that can and do come from being related to someone: the reliability and concern for your real interests. I wonder if your fantasies about this with me are an important way of telling yourself that you must make sure you can count on enough of this here, especially because ours is *not* a true familial tie."
>
> What the analyst addressed were the *universal*—yet very deeply, personally felt—meanings and truths that were intertwined with the patient's individual concerns. The analyst's own acute awareness of the very real differences between her ties to her own family and those to the patient prompted what was essentially an acknowledgment of the underlying validity of the patient's concerns and the adaptive core within the patient's resistances—the reasonable apprehensions on which they were based. This consequently sensitized the analyst to the patient's transferential quest, as it were, to investigate the possibility of the creation within the analytic relationship of some functional equivalent of the kin investment, as well as for insight into the meaning of the legitimate anxieties that the patient experienced and needed the analyst to acknowledge. The analyst clearly acknowledged and delineated the limits of the kind of investment she could make in the patient. She affirmed his crucial need to establish ways of knowing that his true interests could come to be known and enhanced by the relationship with the analyst. He felt less need to vigilantly guard against the potential in the analytic situation for a degree of self deception (that is, the immersion in intense transference fantasies)

because the *adaptive value* of those fantasies—their relation to his own interests—had become manifest. Subsequently, Tom brought forth many new personal associations from his own developmental history. His associations, of course, expanded and deepened the intricate web of meanings involved in his powerful emotional response to the analyst's vacation.

The capacity of the therapist to recognize these essentially universal meanings and universal features of the psyche can be immensely reassuring to the patient who is in search of putting a mixture of these universal and idiosyncratic, personal meanings—aspects of which may well have needed to be repressed—into a new context that brings out their underlying adaptive significance. Addressing the universal (essentially phylogenetic and innate) aspect of the resistance to the transference brings ancient, ancestral meanings into the intersubjective context in a manner that can make personal, individual meanings more accessible to empathic understanding and analysis.

In repeated encounters like these, the analyst articulates a fundamental recognition of the universal inherent dimensions of the patient's transferential quest and the need to thoroughly investigate the validity of its pursuit in the analytic situation. He or she conveys his sense of the depth and meaning of the patient's experience and of the natural striving and probing within the transference for signs of a powerfully committed investment within the analytic relationship. This approach goes substantially beyond simply acknowledging the patient's longings for an adequate self–selfobject relationship, the provision of a holding environment and the recognition of the patient's personal dread of repeating experienced familial failures. The universal awareness of the meanings of the analyst's non-kin status and strange (hard-to-recognize and fully appreciate) reciprocal tie to the patient are typically overlooked. Despite their own personal (training) analyses, analysts are likely to ignore the patient's realistic concerns about the kind of investment and reciprocity in the analytic relationship. The patient's experience in these realms is nonetheless analytically significant for its universality. We have found that sometimes the relative transcendence of the idiosyncracies of individual, ontogenetic experience by these universal realms of experience almost necessitates the elucidation of the universal in order for the idiosyncratic meanings discovered in the transference to have the "ring of truth" as opposed to the appearance of a "proper, correct" interpretation which is accepted in an intellectualized, identificatory, and ultimately defensive manner. The patient's innate awareness of the fact that his or her motives diverge from those of the analyst, as well as the limits of the analyst's investment, is an issue with which the analytic method must contend at its most basic level.

Though many of the larger clinical implications of the complicated process of regressive renegotiation of identity are far from fully understood, a recurrent theme throughout the next chapter consists of examining several paradoxical aspects of the analytic situation. We shall try to understand why the empathic stance has various adaptive meanings to both patient and analyst that are not fully acknowledged and accounted for by the self psychological, intersubjective, classical, or interpersonal models—meanings that are partially rooted in the expectations and innate relational strategies encoded in our evolved deep structure.

PSYCHOPATHOLOGY AS A FALL FROM EVOLUTIONARY GRACE

The recent work by Glantz and Pearce (1989) is another example of the evolutionary psychology that is relatively behavioral in focus (see box in Chapter Three, page 65). Their discussion of psychopathology in an evolutionary context heavily stresses the importance of confusion and anxiety over the breakdown of innate expectations concerning reciprocity in the genesis of psychopathology. They base their important contribution on a recognition of our evolved need and capacity to negotiate social exchanges in kin and reciprocal settings where the costs and benefits to self and others can be known or intuited in some reliable fashion. In this emphasis, Glantz and Pearce's work supports our focus on the way in which dealing with universal expectations about reciprocity is, in some sense, central to the whole therapeutic enterprise. However, in several other respects, Glantz and Pearce's approach differs radically from our own:

1. They assume that much of the behavior of their patients, outside and apparently within the treatment relationship, reflects the simple fact that the human psyche no longer exists within the environment for which it was adapted (the hunter–gatherer band) and will "go haywire" (p. 76), become "confused," and generate "inaccurate models of the world." In other words, individual sources of pathology pale before the massive, general uprooting from the "environment of evolutionary adaptedness" that we all experience.

2. Despite their attunement to universal issues of reciprocity, Glantz and Pearce largely overlook the highly complex ways in which expectations about reciprocity, conflict, and mutuality become major (potentially productive) parts of the therapeutic medium—the therapeutic relationship itself. They view transference as simply a recreation of old (kin) ties and quickly dismiss it as maladaptive. We, of course, focus on the evolved, adaptive role of transference in the complex process of change.

Moreover, we do not believe that the hunter–gatherer band (the ancestral relational matrix) represents the largely harmonious cohesive setting they envision it to be. We believe that they pay insufficient attention to the major complications and paradoxes inherent in human prolonged childhood (within the loving but biased, self-interested family). They are thereby forced to overemphasize the *psy-*

chological significance of the fact that we no longer live as nomadic hunter gatherers.

As Glantz and Pearce note, much of what we are designed to manage and cope with (e.g., parent–offspring conflict, reciprocal altruism, prolonged dependency on biased others, adolescence, etc.) now takes radically different *manifest* form from the hunter–gatherer band. And these manifest social differences may create innumerable forms of stress on our evolved psychological capacities. There has been a shift from a life that is lived amongst kin and other direct face to face reciprocal relationships to one in which large impersonal institutions and formal systems of law mediate many of life's crucial transitions and exchanges. In our view, this necessitates far greater tolerance of aloneness, the need for a more individually sustained, internalized identity, and a capacity for indirect symbolic mediation of our relationships than is optimal for our evolved dynamics to handle.

At some level, a significant aspect of psychopathology may be understood in these terms. But once we move beneath this behavioral level with an individual in treatment, we find ourselves dealing, once again, with the conflicts, paradoxes and needs that have characterized the challenges of our species relational world since before the Pleistocene. Beneath the striking socio-ecological differences between Pleistocene hunter-gatherers and ourselves, the same underlying psychological challenges that characterized our ancestral environment continue to be primary features of our modern relational world. We believe that there is little reason to assume that our psyche is especially less well adapted to meeting these challenges now than it was in our distant evolutionary past.

CHAPTER ELEVEN

The Ambiguities of Empathy and the Creation of an Alliance with the Patient's Inclusive Self-Interest

> The first distortion of truth in "the myth of the analytic situation" is that it is an interaction between a sick person and a healthy one.
> —H. RACKER, *Transference and Counter Transference*

It becomes increasingly clear that much of what has come to emerge as the evolutionary perspective on the analytic process revolves around certain fundamental paradoxes—paradoxes that originate in what we have been calling the ancestral, relational dilemmas that have shaped many of the inherent tensions within our psyche. In the clinical context, there is the paradoxical fact that the relationship between patient and analyst is, like that between any two individuals, inevitably fraught with pervasive elements of conflict and potential deception; yet, significant change in the structure of the self is only likely to occur—to be "permitted" to occur by the evolved core of the patient's psyche—in a relationship in which patients experience a profound alliance with their interests and a basic alignment of the therapeutic process with their intuitive sense of their core (true) selves.

Equally, the human mind was designed to be most profoundly shaped in its basic identity by the influences operating in the environment of close kin relationships—a setting in which a certain natural "alliance of interests" could, throughout our evolutionary history, be assumed to operate over and above the conflicting aims within the family. Yet, it is precisely some failure in this expectable, familial, alliance of interests—a failure often covered by layers of deception—that accounts for the patient's disturbed sense of self and loss of inner alignment with a working sense of his or her inclusive self-interest.

How can the analytic situation ever hope to mobilize and exercise the power to influence structures that have been warped by the very context in which the patient was designed to be most safely molded? Can the analytic process influence structures that were perhaps never designed to be deeply influenced by unrelated, largely nonreciprocating others? Do these questions illuminate some of the "impossibility" of the profession?

Intersubjective Biases, Deceptions, and the Creation of an Alliance of Interests

Perhaps the most crucial aspect of the analytic situation lies in the process by which patient and analyst come to establish the vital sense that the analyst is sufficiently allied with the patient's core self and inclusive self-interest. While such an experience is crucial if the innate capacity for transferential experience and relational influence is to be activated, such an alliance can, in fact, never be fully achieved. In a fundamental manner, despite deception and self-deception, we believe this is innately known by both patient and analyst.

Patient and therapist begin as two unrelated human beings with an immense divergence between their aims and desires. Their individual, self-interested perspectives are primary (on the whole, they would each be in a far more vulnerable position if, conceivably, this were not the case). We expect that in the process of attempting to understand each other, like the parent and child and every other human intimacy, their self-definitions will inevitably clash and the intersubjective endeavor will be rife with elements of deception and self-deception. Of course, this view of the thoroughly good-enough therapeutic relationship stands in sharp contrast to the view that, as highly invested professionals, most analytic therapists are consciously prone to take. Most theorists operating within the relational model subscribe to a markedly different underlying premise that Atwood and Stolorow (1984) explicitly recognize to be at the philosophical core of an intersubjective approach to self psychology. They state, without further question or elaboration that

> it is...assumed that [the] understanding [between patient and therapist] can develop in a collaborative endeavor posing no intrinsic threat to the self definitions of the persons involved. (p. 29)

Although the classical paradigm does entail a far more profound recognition of the inevitability of conflict between individuals (indeed, as we have argued at several points this side of human motivation is greatly overstressed in the Freudian and Kleinian canon), when it comes to the question of the encounter between the analyst and the patient, the classical/ego psychological model essentially joins ranks with the relational tradition. Both models tend to attribute the conflicts in the therapeutic relationship primarily to failures to adopt the situationally called for intimacy and mutuality by one of the participants. Case accounts in the classical/ego psychological tradition locate this failure largely in the patient's pathology; assuming a sufficiently "well analyzed," neutral therapist.[1] The intersubjectivists and self psychologists typically attribute the conflict to the therapist's failure to sufficiently "decenter" from their own subjectivity in order to sustain an adequately empathic self–selfobject bond. The interpersonalists tend to see much conflict and misunderstanding in the analytic relationship as a thoroughly inevitable and ultimately productive way of communicating within the relational field (Levenson, 1983). Yet, they tend to revert back to views that are essentially similar to—perhaps, indirectly derivative of—classical notions of how the patient is engaged in a compulsive process of re-creating archaic distortions and repetitions within the interpersonal fabric of the therapeutic relationship.

It is an extraordinarily complicated and, we must admit, *unnatural* process by which one human being comes to know and become deeply allied with the interests of another to the degree that is sometimes achieved in the analytic relationship. When the natural, evolved barriers to this alliance with the patient's interests are recognized (leaving aside for a moment the supplementary barriers rooted in therapist's and patient's idiosyncratic distortions and "pathological" defensive structures) we arrive at a view of the creation of a therapeutic alliance as, necessarily, a two-way, dyadically negotiated process. Both patient and analyst must deal with powerful, universal signals that lead each of them to sustain—and fully expect the other to sustain—a somewhat *biased* subjective perspective on things and a strategic system for concealing this bias from themselves and the other.

The notion that the alliance must be a two-sided negotiation pro-

---

[1]Note that a view of the inevitability and universality of relational clashes and deceptions that is rooted, as is the Freudian tradition, in proximal mechanisms such as *drives* can readily lead us to lose sight of the inherent conflict in human *relationships*. The result is the tendency to view so-called pathological processes as entirely responsible for the conflict. We believe the evolutionary vision of conflict between *whole* individuals—prompted perhaps by *elements* of their inner, proximal design—may be one in which we are less likely to lose sight of this critical dimension of human nature.

cess that deals with these evolved barriers is different in emphasis from classically inspired views such as Greenson's (1965) notion of a "working alliance" rooted primarily in the patient's reasonable acceptance of the analyst's way of working, and the "therapeutic alliance" (Zetzel, 1956) rooted in the patient's positive, pre-Oedipal transference. The creation of an experienced alliance with the patient's *interests* also differs from the intersubjectivist notion of alliance based on the consistent application of a "sustained empathic inquiry" (Stolorow et al., 1987), which, unlike classical views, strongly emphasizes an immersion in the perceptual truth of the patient's subjective world. By considering some of the differences between the evolutionary view and the intersubjectivist/self psychological perspective, we try to clarify some of the ways in which the evolutionary focus on the management and regulation of competing "interests" leads to an overall clinical vision that is distinct from the other major analytic traditions.

The Evolved Signal Functions of Empathic Communication

Prior to the mutual achievement of the analyst's alliance with the patient's interests (and then, during the continual process of its breakdown and mutual repair) there is one kind of interpersonal signal that lets patients know that a genuine alliance with their interests is even a real possibility: the signal that the therapist has a more or less reliable interest and emotional capacity to decenter from his or her own personally adaptive bias and join in viewing things from the subjective point of view of the patient. As we view it, this in good part is what is signaled by the adoption of an "empathic stance" (Kohut, 1959, 1982; Schwaber, 1979; Ornstein & Ornstein, 1980), "sustained empathic inquiry" (Stolorow et al., 1987), the emergence of an empathy-based "selfobject transference" (Kohut, 1971, 1977, 1984), and the establishment of a "holding environment" (Winnicott, 1963; Modell, 1976).

Let us define empathic communication very broadly. It entails more than simply a "mirroring" of the predominant, conscious affective states and meanings experienced by the patient. This broader definition revolves around an "observational stance" within the patient's subjective universe (Schwaber, 1979; Ornstein & Ornstein, 1980). And it can also include an attunement to less conscious, but vital, adaptive strivings for self-realization and underlying subjective truths within symptoms and transferences. Empathic communication can be seen as providing a very complex set of signals that convey the message that the achievement or reestablishment of the analyst's alliance with the patient's interests is a viable, mutual goal.[2]

Before further discussion of the complex topic of empathy, we must make some basic observations clear. On the whole, our clinical experience compels us to conclude that, over and above any assumed need to compulsively repeat or defensively distort experience that we might attribute to a patient's pathology, it is enormously crucial for patients to experience their therapist as consistently capable of allying with their own subjectivity and attuned to the existence of an overarching, adaptive, developmentally progressive process inherently operating within even their most tortured and confused psychological states. This experience in the therapeutic setting—the creation and maintenance of an effective self-selfobject relationship, if you will—not only enhances self-cohesion and self-esteem, but also serves to expand awareness and significantly diminish defensiveness.

Yet as Mitchell (1988) notes, the reasons why we regularly observe these enormously facilitative effects of sustained empathic communication are likely to be far more complicated than is apparent in the theories that advocate its central importance. Likewise, from an evolutionary perspective, the meaning of such communication, and thus the reasons why it often operates so effectively, go beyond the considerations stressed in the self psychological literature.

An immersion in the patient's subjective experience, a closely sustained attunement to personal meanings and grasp of developmental strivings, may replicate, in very heightened form, certain signals con-

---

[2]It has been noted that empathy is a value-neutral concept (Kohut, 1971, 1977, 1980, 1984). Thomas Kohut (1985) emphasized that empathy and altruism or compassion are different concepts. Empathy and empathic communication can be used for evil; to injure as well as to help. Here we are speaking of empathic communication as an attempt to join with the patient in seeing things from the patient's subjective viewpoint.

Moreover, psychoanalysis as a clinical intervention is not simply based on empathy as a mode of observation, as a scientific tool for information gathering about the patient's subjective world. Rather, the analytic use of empathy lies in its communication to the patient. What is seen as "curative" in the self psychological tradition is the empathy based selfobject function played by the analyst (Kohut, 1984). If the analyst were to "scientifically" communicate (i.e., objectively and accurately but without feeling or personal involvement) an understanding of the patient's experience, much of the selfobject function would be lost. It is the willing sharing of the patient's experience and the desire to communicate this to the patient that is often an integral part of the patient's experiencing the analyst as a selfobject. An essential aspect of the selfobject function is this empathic *union* in which the analyst intentionally joins in viewing things from the patient's subjective viewpoint. (See note on p. 94.)

tained to varying degrees within more natural human interactions—signals that are, indeed, present in normal families and may not have been sufficiently communicated to the patient given his or her individual needs. Sustained empathic communication signals the message that—over and above the intuitively known inherent divergence of interests between the two individuals—the other person is more likely than most to be willing and capable of becoming an ally in the realization of one's own aims. The analyst's consistent decentering and abandoning of his or her own (personally more adaptive) subjective view, even though only temporary and never fully achieved, is a sign of genuine investment that might, at moments, be experienced outside of the analytic setting with a true friend or be implicitly conveyed in a variety of ways among close kin.

On the whole, however, the message signaled by empathic communication in the treatment setting is only one part of the exchange of communications between parents and children. From the evolutionary perspective, sustained empathic communication in the treatment setting cannot be understood primarily as developmentally corrective in the Kohutian or Winnicottian sense of making up for a particular set of deficiencies in obtaining the type of experiences that predominantly characterize the good-enough family environment yet were missing in the patient's development. While degrees of empathic attunement are likely to be conveyed in a variety of ways by good-enough parents because of the enormous overlap between their interests and those of the child, the evolved realities of parent–offspring conflict and deception suggest that such empathic attunement is likely to be regularly mixed within a far more diverse brew of messages of all qualities and types. No friend, intimate other, or parent could have the capacity to sustain an empathic attunement with the patient to the degree that is possible for a well-paid, carefully trained, unentangled,[3] compassionate analyst.

The view that the operation of empathy is particularly crucial in regard to the treatment of developmental arrests in the formation of self-structures that emerge *prior to "structuralized conflicts"* (Stolorow &

---

[3]Unentangled in the sense that the patient's particular life, success, failure, beliefs, actions, biases, etc., are far *less* important to the analyst than they are to, for example, a parent, lover, or a spouse. It is much easier to maintain an empathic stance when what you are hearing does not place the majority of your own interests in jeopardy. As we know, genuinely empathic clinicians frequently have enormous trouble maintaining their empathy with their spouse and children. It is not primarily the emotional intensity stirred by these intimate situations, but rather the potential implications for one's interests (which the emotional intensity is designed to signal) that causes the "entanglement."

Lachmann, 1980) leaves out a critical dimension of development as understood from the evolutionary perspective.[4] Not only is conflict present and psychologically relevant from the beginning of life due to intrinsic, inherent differences in individual aims and the uniqueness of the child's true self, but the very power of the message conveyed in sustained empathic communications, that is, that the "other" can be counted on to share one's vantage point, is always relative to the intuitively known existence of competing interests in all relationships. Empathic communication always takes its meaning against the "background of conflict" inherent in all relationships. All significant relationships represent some degree of threat to the self.

As in the infant research that shows the newborn to be pre-equipped to organize perceptions, form attachments, etc. (Stern, 1983; Emde, 1980), we believe there are no structures that emerge *prior to conflicts*. Much psychological structure, existing in rudimentary form even at birth, is designed, in part, to anticipate and manage interpersonal conflict. In fact, as we suggest in Appendix C (p. 298), the explosive evolutionary growth of the human brain appears to be a direct result of the selective pressure emanating from the need to manage complex, conflict-ridden human relationships.

The Inherent Ambiguities of the Empathic Stance

Empathic attunement and communication are apt to be vital dimensions of all human communication—dimensions that locate the relationship along some continuum of conflict and mutuality and competing and shared interests. We have evolved to read sustained empathic attunement as a signal that a given relationship is a particularly good context for the activation of the innate process of self-revision: the experimental reopening of the fullest range of repressed versions of the self (regression) and experimental probing of the relational context for the possibilities of new types of reciprocal exchange and self-revision. Yet, even mature empathic responsiveness, not sympathy or over-identification, is by no means a simple and unambiguous signal. First of all, the consistent maintenance of an empathic stance runs counter to the

---

[4]From the evolutionary perspective, even the treatment of "borderline" or "psychotic" patients can be seen as operating following the principles of the regressive renegotiation of identity in which the goal is to "undo repression" and achieve a greater integration of the "true self" into the patient's core sense of being (see Appendix F, "Regressive Renegotiation in the Treatment of the 'regressed' patient," p. 308).

basic design of a (thoroughly good-enough) analyst's own mind. The analyst, we believe, is designed and, if reasonably healthy, lives within a self-structure that is biased toward his or her own inclusive self-interest and is buttressed by a fair degree of self-deception and deception.

A recognition of this crucial aspect of the analyst's essential humanness by no means minimizes the importance of (1) the analyst's personal capacity to "decenter" (Stolorow et al., 1987) from his or her own subjectivity in entering the patient's subjective world; or (2) the analyst's relative freedom from those aspects of the classical tradition that have reified the analyst's (and the parent's) reality while drastically underestimating the essential truth (Kohut, 1984) in the patient's perceptual reality (and that of the child). However, it puts into a somewhat different perspective the tendency among some intersubjectivist theorists to attribute virtually all empathic (or selfobject) failures to unanalyzed countertransferences on the part of the analyst, or to theoretical biases that objectify the "truth" in the analyst's viewpoint (Stolorow et al., 1987).[5] Well beyond its being a normal, human empathic response, empathic attunement—as it is clinically defined and understood as the central analytic stance—is, to a great extent a cultivated, learned way of interacting: one that seems far more natural to some therapists than others, and runs counter, to some degree, to the way we have evolved to treat the mind of another.[6]

---

[5] Although most intersubjectivist and self psychological theorists never take the absolute position that all empathic or selfobject failures can be understood in terms of technical failures or countertransference impediments in the analyst, virtually every case report of such therapeutic "impasses" is, in practice, viewed in this way.

[6] It seems to us that a good deal of the tension that exists between the self psychologists and the contemporary interpersonal school revolves precisely around (1) the interpersonalists' accurate emphasis on the multidimensionality of the analytic relationship and the recognition that "the empathic stance," especially when viewed as technical prescription, can itself become, in our terms, a means of deception and self-deception (it can become a way of attempting to convey a connectedness with the patient without acknowledging to oneself and the patient the necessity of undergoing a full and genuine process of being influenced and changed by each other); and (2) the persisting difficulty the interpersonal school has in appreciating the immense *interpersonal* meaning of the therapist's capacity to observe and respond consistently from the empathic stance: the fact that empathic attunement serves as an invaluable means of communicating to the patient the therapist's ability and willingness to "de-center" (Atwood & Stolorow, 1984) from his or her own subjectivity; as such, "de-centering" provides a crucial interpersonal sign that the analyst is genuinely capable of becoming influenced by the patient in ways that will ultimately make more likely the negotiation of a true alliance with the patient's interests. As we noted in Chapter Ten, the interpersonalist narratives are often still infused with the view of the patient (child) as a distorter of experience whose aims in the therapeutic relationship lie predominantly in a compulsive repetition.

An appreciation of the universal, evolved nature of bias and deception in the analyst—and the equally universal, evolved intuitive knowledge of the existence of this bias on the part of the patient—actually heightens our appreciation of the meaning of experiencing empathy from another. A recognition of these universals serves also to emphasize those personal factors and theoretical stances that enable a therapist to eventually experience his or her patient as a "subject" (Schwaber, 1979; Ornstein & Ornstein, 1980). However, the interpersonal meanings of empathic communication become more complicated.

For one thing, it is quite possible for empathy to be practiced with a fair degree of verisimilitude, as a technique, rather than as the genuine intimate act and sign of mutuality that is so profoundly, intrinsically valued. Indeed, patients know, or come to know, that another human being whose only substantial utterances take the form of validating affirmations of the patient's own subjective world and developmental strivings are likely, themselves, to be engaged in one or another form of self-deception and deception. Such a sustained, subjective convergence is simply unparalleled in any other realm of human experience, and it was undoubtedly as rare in the ancestral environment in which the dynamics of our intuitive knowledge of others were formed.

Moreover, the analyst is well paid to engage in genuine efforts at empathic relating and has major self-interested motivations (not to mention other necessary professional gratifications) in assuming, within the drama of the treatment, the empathic "role." These realities of the investment by the analyst simply affirm the fact that the motives for the analyst's immersion in the patient's subjective experience are quite complicated. Thus, a highly consistent attempt to communicate to the patient exclusively from a vantage point within the patient's subjective world will tend to be only cautiously accepted by some patients, and for others will often become quite naturally suspect as a strategy for hiding the therapist's biased self.

Let us consider, for example, the way consistent empathic communication can sometimes engender terrified and hostile reactions in some patients with core problems in self-structure (Teicholz, 1988). In our experience, when so-called negative therapeutic reactions of this kind occur in the context of relatively consistent mirroring on the part of the therapist, there are often several ways to understand what is taking place. Not infrequently, what may be experienced and labeled as "empathic" by the therapist (and the supervisor or consultant) may well be efforts at kindness, sympathy, or compassion, or simply an avoidance of confrontation. What is taken for empathy may even include accurate understandings of *aspects* of the patient's experience that for his or her own subjective reasons the analyst is prone to focus on, aspects that are, nevertheless, quite removed from the patient's salient subjective reality and the central adaptive strivings that the patient is currently trying to formulate and express. In such situations, over and above the therapist's expression of such emotional sensitivity, concern, and partially accurate understanding, what is called for is an exploration and elucidation of the divergent meanings that are held by therapist and patient and a better appreciation by the therapist of the simple core meaning of empathy as an observational stance (Schwaber, 1979) that entails a real effort at "decentered" perception.

Nonetheless, some patients do experience enormous terror at the possible repetition of past traumas evoked by such relatively consistent, empathic communication. In these cases, the sustained empathy invites a major mobilization of precisely the hopes and longings that have been thoroughly unresolvable and brought the most pain to the patient's life (Kriegman & Slavin, 1989; Brandchaft, 1983). And the defenses against a feared retraumatization lead to a spiral of terror and rage. Teicholz (1988) interprets these reactions as indicating the need for the analyst to introduce more of the reality of his or her own personality into the work. Similar to the emphasis in the work of Fromm-Reichmann (1950), Teicholz argues that the patient needs to make use of the basic coherence of the analyst's own self-organization, whether in the form of limits, a clear commitment to the work, hope for progress, etc. An immersion in the patient's subjectivity at these moments simply elicits a heightened terror of fragmentation.

From an evolutionary perspective, the meaning of the interpersonal transactions that take place at these moments is quite complex. It is clear that the therapist must often shift into a mode of communication that serves to introduce meanings from the therapist's frame of reference even when these meanings clash with those of the patient. Even Stolorow et al. (1987) cite the case of Jane in which the analyst literally "stopped her from talking and insisted that for once she was to listen to what he had to say" (pp. 153–154). What he demanded she listen to was a view of her situation that differed fundamentally from her subjective frame of meaning. Though the authors claimed, quite possibly accurately, that the analyst's move demonstrated "his understanding of her deepest longing and needs," they do not comment on his striking shift away from a stance that focuses on validating the patient's perceptual reality. In contrast, Teicholz draws our attention to a need for a very different kind of communication that is not encompassed by the notion of empathic attunement to the patient's subjective reality. Ultimately, Teicholz, like the intersubjectivists, resolves the problem by broadening the definition of empathy to include communications that "provide needed selfobject functions,"—regardless of whether the communications are in conjunction or dysjunction with the patient's frame of meaning.

We believe we can ultimately achieve greater consistency and clarity if we reconceptualize the interpersonal exchange represented by these effective therapeutic transactions. Broadening the meaning of empathy to include selfobject needs is, in effect, a way of equating empathic communication with an alliance with our patients' *interests*—their genuine underlying interests—not just their currently dominant conscious representation of them. Once we have essentially defined an alliance with the patient's underlying interests, it may be more useful to view empathic communication (defined in terms of a consistent recognition of the patient's subjective frame of meaning) as a *means* toward the end, or goal of communicating the analyst's capacity to ally with the patient's interests. From this point of view, empathic observation and communication are crucial because we have encoded within us an evolved responsiveness to them as *signals* regarding the likelihood of the establishment of a reliable alliance with the superordinate human goal of continually defining and enhancing inclusive self-interest. Yet, there are other means toward this goal that are also crucial aspects of a therapeutic relationship. *And*, there are other meanings of empathy in some contexts.

From an evolutionary perspective, the fact that we inevitably must respond to patients in terms of our own realities—revealing our own knowledge (as well as biases) and at times self-deceptions and deceptions—illuminates still another facet of those negative therapeutic reactions that appear to occur in the context of a relatively effective empathic responsiveness. The attempt to remain exclusively attuned to what appear to the therapist to be the dominant themes and meanings in the patient's subjective world is, in fact, sensed by many patients as a self-protective strategy on the part of the therapist. The fact that, as Brandchaft & Stolorow (1990) recognize, an empathic stance can be used to hide some of the therapist's feelings such as anger or fear (or to reify the patient's belief in the actual possibilities for the fulfillment of archaic needs thus bolstering the therapist's grandiosity), reflects the partial and incomplete nature of the empathic stance as a human mode of communication. Over and above any particular individual defensiveness that we may attribute to the therapist, the overly consistent use of the empathic mode will, for some patients, be sensed as the therapist's hiding some aspect of him- or herself, or pursuit of his or her own interests—interests that, as the patient well knows but therapists are loath to face, indeed, diverge in some significant ways from those of the patient. We must, thus, clearly face the fact that an immersion in the patient's subjective world, as Modell (1984) has described, must be complemented, at times, by what is, in effect, the open expression of the analyst's reality.[7] If we interpret, to paraphrase Winnicott, it is to show the patient our subjective biases and our distinct self-interest (so that he or she can decide what to make of them).

> Consider the very troubled patient, Tanya, who was experiencing extreme distress at "having to pay to be cared about" by her therapist. What seemed like very careful attempts to empathically clarify the personal meanings of this in terms of either the patient's immediate experience—or prior disappointment—were usually met with disdainful rage. At some point the therapist explicitly stated that the monetary dimension of their relationship clearly connoted the way in which their interests do, in fact, diverge. In charging her, he said that she knew he was clearly pursuing his own interests that, in this respect, were quite different from and in conflict with hers. This acknowledgement of their divergent interests was coupled with the therapist's explicit recognition of Tanya's crucial need to carefully evaluate and verify all of those other aspects of their relationship in which the therapist's direct alliance with her interests was central, palpable, and could be trusted.

Because Tanya, like all of us, was designed to be wary about interpersonal deception, only a clear delineation of the real limits of the alliance could be heard as an acknowledgement of the conflicting aims, biases, and at least potential deceptions in the analyst's viewpoint. Tanya engaged in a transference probing of the therapist's comprehension of and capacity to be aware of and candid about the ways in which the empathic stance, in which the therapist consistently takes the patient's subjective point of view, exists only as a safe "island of mutuality" within the larger divergence of interests that exists even within the analytic setting.[8] What took place was a moment in a mutually created corrective emotional experience, in which both parties are

---

[7] While Modell's description of the shift from the holding environment (cocoon) is cast in more Winnicottian language and in terms of a more limited group of patients, it emphasizes the crucial role of empathic resonance as a means toward a relationship in which there is *also* a significant role for the therapist's expression of perspectives that diverge significantly from the subjective framework of the patient.

[8] This is a complex, poorly understood but crucial issue that merits a slight digression. As we discussed in Chapter Four, there is a unique, unparalleled, expectable overlap of interests between parent and child in the good-enough environment. Yet, from the outset, the subjectivities of parent and child must diverge enormously in the service of their distinct, if overlapping, interests. A clinically useful fiction with which all but the most classical analysts operate, is that our patients have had developmental experiences that deviate substantially from what would have obtained in an environment that did not

(continued)

significantly influenced. More "neurotic" (i.e., healthier) patients are far more forgiving at these moments. They are far less demanding, more subtle teachers concerning the subjective biases in their therapists. But it is crucial for them too.

The Need to Know the Analyst and the Reciprocity of Change

From an evolutionary perspective, the communication of an alignment with the patient's subjective world provides only one of the multiple crucial signals necessary to allow patients to believe that they are fully engaged in the construction of a relationship in which the analyst is allied with their interests. *Ultimately one of the major ways in which therapists fail their patients revolves around the therapist's use of self-deceptive strategies for protecting or enhancing his or her interests (identity) in a fashion that is cast in terms of the interests of the patient.*

In our view such a confusion of interests—and the resultant loss of the capacity to define, know, and promote one's interests—is the common denominator in a huge range of less than good-enough, or pathogenic environments. And it is not uncommonly replicated in many therapies. In a sense the interpersonal goal of the ideal of "analytic neutrality" in the classical model can be seen in more interpersonal terms as an effort to guard against precisely this confusion of interests, just as the concentration on empathic attunement can be seen as an effort to focus on the side of the therapeutic relationship in which inherent clashes of interest do not exist. The concept of neutrality is rooted in a perspective that tends to assume that for a well-analyzed

---

necessitate painful inner compromises and elaborate self-protections in the course of growing up. In other words, the alienation they experience has been engendered by parental inadequacies, failures, and familial pathology. It is incontestable that virtually all of our patients have experienced the familial environment as failing to fully provide for their needs, and that even many highly functioning, analytic patients have, indeed, experienced types of trauma and deprivation that served as the relational data, so to speak, around which they have developed highly self-protective strategies for organizing their experience of themselves and the world. And the elucidation of the individual, idiosyncratic, experiential paths by which these inner structures were formed is, of course, useful in creating a context for revising these strategies. But as the backdrop for such historical reconstruction, must we assume that there are "normal" environments free of the enactment of substantial parental and familial bias toward interests other than those of the patient? Is it possible to keep in mind the extraordinary degree of both the natural alliance of interests in our patient's families as well as to recognize that the inherent conflicts and deceptions they encountered were a version of the human condition, that is, their particular experience of the lack of recognition of the true self of which no childhood can be free?

therapist significant elements of bias and deception will not be an intrinsic part of the therapeutic relationship. Because "drives" can presumably be neutralized or sublimated and these are what, at bottom, account for the clashes between individual subjectivities in the classical narrative, it becomes conceivable that the personal reality of the analyst will not as a whole inevitably clash with the personal reality of the patient. From the less reductive evolutionary perspective, in which endogenous motives are inherently designed, precisely, to bias the individual, any such conflict-free interpersonal space is unlikely to ever exist.[9]

---

[9]In fact, this points out a need for a *type* of analytic neutrality, or rather the type of relative passivity that characterizes the analyst's interactions more than therapists' actions in other psychotherapeutic modalities. As Winnicott (1969) put it, "The principle is that it is the patient and only the patient who has the answers. We may or may not enable him or her to encompass what is known or become aware of it with acceptance" (p. 102). From the evolutionary perspective we would predict that most inclinations to speak or act (on the part of the analyst) will naturally be guided by motivations designed to enhance the analyst's self interest. As an evolved organism, we could expect nothing else from any therapist if he or she acts on their natural impulses. Every analyst, in order to function in the empathic manner we have been describing, is going to have to actively *suffer* considerable, unnatural self denial. Powerful forces that can never be made completely conscious make it inevitable that, at times, the analyst's needs will be actively pursued *to the patient's detriment*. Because of this—despite genuine altruistic motives, ethical and moral guidelines, and reciprocal exchanges that characterize good treatment—*part* of the treatment relationship will inevitably entail a *battle* or struggle of conflicting interests. At times, *spontaneous* action on the part of the analyst is essential for the patient to have an opportunity to "interpret the analyst" (Hoffman, 1983) and to experience the analyst's response as genuine. However, careful listening and self scrutiny prior to most action (a relative passivity and abstinence) is essential to the exceedingly complicated, *unnatural* task of trying to adopt an empathic point of view based upon experiencing the patient as subject not object.

In criticizing the "monadic school," i.e., the view of psychic structure as deriving from the vicissitudes of endogenous motives rather than from relationships with others, Mitchell (1988) notes and then rejects the very real danger to the analytic process that had been encoded in classical drive theory:

> If one views the analysand's experience of the analysis and the analyst as fundamentally monadic, the analyst is outside that experience and should *stay* outside it.... Efforts to relate the material to himself would make the analyst an intruder, a contaminant. (p. 300)

In the view we are presenting, we certainly do not believe the analyst can take an objective stance outside of the analytic interaction. However, we believe that both evolutionary theory and clinical practice converge: inevitably, when the analyst enters into a collaborative effort to mobilize the patient's program for regressive renegotiation, the analyst will intrude and contaminate the patient's efforts to find his or her own meaning. It is precisely the fact that with hard work a solution to this problem can be negotiated, i.e., worked through in the treatment relationship, that makes it a real experience that the patient can use as a model for discovering her/himself in other relationships.

The implications of these views move us away from the idea that the analyst can ever assume a single technical stance, or even a formulated set of shifting stances, that induce the process of change. While recognizing the absolute centrality for the patient of the experience of having his or her perceptions recognized and validated, we believe that many of the times when this does not occur we are not dealing with empathic (or selfobject) failures in the sense of technical failures or breaks in some process that could or should be sustained. Many of these moments represent normal, continuing reversions away from the highly difficult task of maintaining a focus on the patient's subjective viewpoint toward the more natural tendency to express the analyst's own individual bias, manifested often by the tell-tale signs of self-deception and deception that protect it. It is in these ongoing "countertransferential self revelations" (Russell, 1990) that the empathic clinician gives the patient the opportunity to know and correct, as it were, for the inherent deceptions entailed in the good-enough analyst's sustained, empathic inquiry. Hoffman (1983) describes this process beautifully in what he calls "the patient as interpreter of the analyst" (see also Szasz, 1963). What is interpreted, as we see it, are often the alternating manifestations of shared and divergent human subjectivities. Aspects of what Searles (1975), Tower (1956), Levenson (1983), and others have described as the patient's need to know the analyst may only come to light when such differences of interests are seen and acknowledged. Indeed, this may be a broader way to understand why, as Kohut recognized, that what is learned around empathic breaks is of central value to the patient and to the analyst.[10]

More broadly, what we are saying is that *the evolutionary perspective leads us to a view of the analytic process as a profoundly reciprocal one—a process of mutual influence*—in which the patient's probing of the analyst's subjectivity, the analyst's transference, is just as crucial as the child's probing efforts (described in Chapters Five and Seven) to ferret out and find the parent's true identity, in order to correct for deception and to better define his or her own identity. The patient interprets the analyst by means of transferential probes which serve to induce certain reactions as well as intuitive readings of the analyst's behavior in other less transferentially charged ways. The engagement of patient and analyst is thus an intimate, mutual exchange of information about each other. In this sense, as Russell (1990) notes, "countertransference" dis-

---

[10]Kohut's notion of optimal frustration in which the patient is nontraumatically disappointed leading to transmuting internalization of the selfobject function played by the analyst may, thus, operate in a parallel way.

closures go on all the time. Throughout this exchange, the patient continually monitors the therapist's readiness to open and alter his or her own subjective world, allowing it to be influenced by questions and views of what life is about from the patient's unique perspective. The patient's overarching, ongoing goal is to establish a sense that the analyst can form and remain in an alliance with the patient's interests. This evolved, adaptive process is what can lead the patient to negotiate an enhanced intuitive grasp of his or her own "inclusive self-interest."

From this perspective, the patient does not "have" a corrective emotional experience; nor, certainly, does the therapist provide one in any deliberately enacted way. Rather, as Modell (1990) notes "it is the patient who introduces the corrective emotional experience" (p. 74). In our view this takes place through a complicated set of negotiations with the therapist. The patient needs to teach the therapist how to play his or her part in this process in which certain of the patient's needs are met while others are relinquished through analytic interpretation and developmental mourning. *Unfulfilled needs can only be relinquished and mourned once their validity is recognized, for only then can the vital information that they carry be incorporated into a new organization of self-experience.* Only after the vital information carried has been substantially recognized is relinquishing the direct fulfillment of the wish no linger linked with the fear of having a vital aspect of the self lost forever. In outline, this is the heart of the "therapeutic action" of interpretation from the evolutionary perspective.

Further Comparative Considerations

In his "theory of the crunch" Russell (unpublished) depicts a process of gradually mounting divergence and conflict between the subjective worlds of patient and therapist that eventuates in a "crunch," that is, a clash in which the differences must emerge and, if the work is to continue, find some mutually negotiated solution.[11] There must always be a complex relationship between two evolved, actively strategizing psyches in analyst and patient, only one aspect of which consists of the experience of alignment or conjunction in their subjective worlds. The other part consists of their interaction as two separate individuals in the course of which the divergence in their interests comes to light and will be continuously monitored for self-deception and deception in both

---

[11]Though originally cast in terms of borderline patients, Russell (personal communication) now views this process far more broadly as a universal therapeutic phenomenon.

parties. As we discussed in Chapter Ten, the mutually interacting transferences evoked by this relationship will—indeed, are designed to—ensure that some replication of important role relationships will be enacted in their relationship. The resulting mixture of reality and fantasy, longing and loss will be continuously negotiated, as analyst and patient create an intersubjective reality. This intersubjective reality includes the meanings inherent in the differences between this peculiar, unnatural analytic relationship and that of parent and child, lovers, spouses, friends, or other forms of directly reciprocating human connection.

The patient's history is a mutual discovery. In some sense, so is what we mean by his or her inclusive self-interest. While a history and identity exist in some form from the outset, what is sought in an analytic treatment can only be fully discovered (constructed) through negotiation in the context of an authentic relationship—a relationship that demands changes in both the patient and the analyst. As Goldberg put it:

> An authentic analysis is one that enables patient and analyst to participate in the patient's remaking; and no analyst can emerge from this sort of experience unscathed. (Goldberg, 1990, p. 148)

Some General Observations

It would be hard to overstate our growing conviction in the absolute centrality of the need to achieve a capacity to listen as much as possible from the perspective of our patients' own experience. We have become increasingly aware of the inherent, universal divergence (sometimes conflict) between the interests of therapist and patient in the thoroughly good-enough treatment relationship, and we have become respectful of and, we believe, productively candid about what we have come to view as our patients' apparently innate knowledge of these realities. After all is said and done about matters of "technique," we find ourselves giving immensely greater scrutiny to what we feel is the central challenge of being a therapist: finding a way to sustain a close attunement to the reality and truth in our patients' subjective worlds while acknowledging and maintaining our often conflicting identity, role, and self-interest in the treatment relationship.

We have found ourselves fundamentally shifting away from an emphasis on our patients' "pathology" as a predominant focus: away from an understanding of transference and resistance as "distortions" of reality, as opposed to the ways in which these operations of the psyche function as tactical efforts to communicate crucial adaptive information. To some extent this modified understanding of transference and resistance is rooted in a growing belief in the existence of an overarching adaptive process that exists as a basic feature of the human psyche—a process that is designed to use a range of psychic means to continuously, adaptively revise (or renegotiate) "inclusive self-interest" or identity.[12] At present we have come to view the treatment relationship as fundamentally based on an alliance with our patients' innate capacity to maximally mobilize and sustain this renegotiation process.

In our view, aspects of the self psychological/intersubjectivist stance are clinically vital and indispensable to effective analytic practice. However, our emphasis on discovering and articulating the adaptive validity of meanings uniquely known and expressed by the patient derives from the (evolved) link between subjectivity and self-interest. It is not based on the global adoption of a radically subjectivist or exclusively hermeneutic epistemology. Our view respects the intersubjective nature of how we and our patients come to mutually construct a narrative about their lives. Yet, we approach this task with the belief that, beyond our subjective constructions, there is a "reality" about which both analyst and patient will inevitably engage in a negotiation process fraught with deceptions and self-deceptions.

The overriding implication of the evolutionary perspective is, thus, the emphasis on the essential humanity of both analyst and patient. By this we mean a certain overriding sensibility in which both the patient and analyst are, first and foremost, evolved creatures who share a basic design that is, in most instances, far more important than the ways in which their assigned (technical) roles and presumed differences in psychological organization (pathology) would have them differ. In essence, a recognition of this essential symmetry in the dynamics of influence in the psychoanalytic process parallels Erikson's (1963) insight into the negotiation process in the family:

> Babies control and bring up their families as much as they are controlled by them; in fact, we may say that the family brings up a baby by being brought up by him. (p. 69)

As soon as one begins to "abstract out" technical notions or ideas about therapeutic stances, one immediately risks losing sight of the essential reciprocity in the analytic relationship; in addition, the particulars of the specific relationship begin to recede as does the awareness

---

[12]In a sense that modifies and extends the meaning of the term in Erikson's work.

of complex embeddedness of *any* clinical stance in the universal realities of the relational matrix that are innately and inherently a part of all human interactions, that is, conflict and deception/self-deception. No theoretical perspective substantially reduces the lived uncertainties of the therapeutic encounter. At best, the evolutionary perspective becomes a platform that is sufficiently broad and sufficiently apart from any particular analytic model to provide a place to stand while we allow the major analytic perspectives to "deconstruct and dereify" (Hoffman, 1987) each other. What is different is that from the vantage point of this platform we remain attuned to the ways that, in the peculiarly modern, analytic encounter, we may be able, as Cosmides (1989) put it, to "respect the natural contours of the mind." In the treatment relationship, there are two ancient, evolved relational entities—the psyche of patient and of analyst—enacting an intensified version of a complicated intersubjective dance we have been engaged in for hundreds of thousands of generations, since the early origins of our species.

PART V

Toward an Evolutionary Foundation for Psychoanalysis

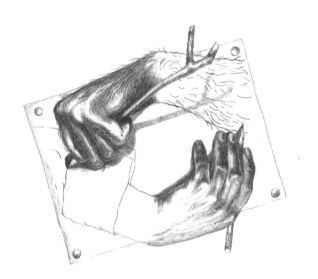

CHAPTER TWELVE

The Evolved Design of the Psyche and the Classical–Relational Dialectic: Toward a Rapprochement of Competing Psychoanalytic Narratives

> [T]here is the possibility that these two apparently irreconcilable contexts... will at some later date be brought into a new synthesis... On the other hand, it is also possible that the existence of insoluble paradoxes may reflect an intrinsic quality of the human mind.
> —A. MODELL, *Psychoanalysis in a New Context*

> If I am not for myself, then who will be?
> If I am for myself only, then what am I?
> If not now, when?
> —HILLEL 'THE ELDER'

Dialectical Tension: Thesis and Antithesis

Like the proverbial blind men examining an elephant, psychoanalysis has repeatedly focused on parts, or fragments, of human nature. Each tradition, classical and relational, has grasped an important *aspect* of human nature, elaborated it conceptually and clinically and then extrapolated its "part" into a vision of the whole. Yet, as Mitchell (1988) notes, psychoanalytic traditions are in no sense simply "blind." They may be blind to aspects of the whole, but they come equipped with complex preconceptions that serve to *construct* how the particular part of "the elephant" they happen to touch actually feels within their grasp. To a significant degree, overall visions *create* (or color) specific perceptions.[1]

In this book, we have viewed the major psychoanalytic traditions both as biased ways of constructing (or narrating) human nature and the psyche and as ways of attempting to discover and describe aspects of a reality that does, in fact, have some larger essential existence within the natural world. The whole elephant, as it were, must exist in order for a so-called blind man to have any conceivable reason (or capacity for that matter) to elaborate his partial experience into a conception of a whole creature.

So it is, in our view, with the limited, biased, but hardly blind psychoanalytic narrative traditions. They have constructed images of the psyche that, in some respects, relate to each other as the sides of a coin. As a coin flips through the air, at one moment we see one face; an instant later we see the other side. Because of the structure of the coin (the psyche) itself, we never see both sides at once. Yet, we know that each side must somehow imply the existence of the other. No one argues over whether "coin nature" is essentially "headed" or "tailed." Yet, in psychoanalytic debates the heads and tails of the human psyche are commonly *separated* from the essential, overarching structure of the

---

[1]Mitchell (1988) also uses the metaphor of the blind men and the elephant but reaches a strikingly different and telling conclusion. Because theories shape perceptions—a view that we wholeheartedly accept—he concludes that the blind men are not, in fact, all examining an elephant: some may be examining a giraffe! We agree that we powerfully influence the world we construct in the very act of perceiving it. Yet, in the evolutionary view, there are realities beyond any (adaptively) biased individual and collective experience.

It is illustrative to compare our view of the different psychoanalytic positions as capturing essential truths regarding a larger elusive and complex reality with Mitchell's point of view. He sees each position as a construction that can be so dependent on the perceiver's viewpoint that fundamentally different realities can be created, realities that may be so different that we have no reason to expect them to have any inherent compatibility. This fits nicely with the other metaphor he uses in presenting his version of the interpersonalist position: the Escher drawing of two hands drawing one another. In this view of the psychoanalytic relationship, indeed of all relationships, the participants construct one another through their interaction with no independent reality pre-existing or underlying the mutual construction. In our evolutionary view, there are powerful pre-existing forces that shape us prior to our interactions and enable us to be constructed by (and to construct) those very interactions. The circularity of the two contemporary Escher hands is broken: the ancestral hand constructs the modern which, in turn, contructs the view of the ancestor (see p. 261).

psyche as a whole. Speaking from a contemporary social-constructivist perspective, Hoffman (1991) notes that:

> In our zeal to correct the overemphasis in classical psychoanalytic theory on the individual dimension, it is important that we not swing to an overemphasis on the relational dimension, thereby isolating each from the other ... [A solution] requires a synthesis of the two perspectives with appropriate redefinitions of each in the light of their interdependence. (p. 103)

Our application of the evolutionary approach to psychoanalysis is an attempt to find and apply such an overarching perspective—a special observational vantage point—that will, in some sense, illuminate the whole. The evolutionary perspective, of course, brings its own narrative structure to the task of understanding human nature. And, as such, the evolutionary perspective represents one step in the continuing process of constructing, deconstructing, and reconstructing what is essential and universal in the human condition.

In our view, the tension between the classical and relational psychoanalytic narratives is best understood as "dialectical" in nature. By this we mean that it is useful to think of the existing psychoanalytic paradigms as representing the historical tendency for depth psychological thinking to polarize around two, internally coherent sets of assumptions and observations—a classical "thesis" and a relational "antithesis"—about human nature and the human condition. These opposing sets of assumptions concern the basic nature of the child's mind; the reason that significant conflict is observed in the interaction between parent and child; the meaning of the child's tendency to divert (repress, split off, disavow) awareness of aspects of this conflict; and the implications of all this for the process by which the child negotiates the transition to adulthood and the adult negotiates and renegotiates crucial elements of the self throughout the life span. Calling the distinct narrative structures by which we commonly define and view these processes a "thesis" and an "antithesis" implies mainly that (1) neither narrative is complete; (2) while each narrative, in some sense, actively "calls for" the truths embodied in the other, they are, in their *currently* elaborated theoretical forms, rooted in incompatible premises; and finally (3) that each will eventually contribute its essential message and then essentially recede as a new synthesis, a more embracing narrative structure takes form.

Although such dialectical change may resemble attempts to "integrate" the two traditions, it differs substantially from integrative approaches that deal with psychoanalytic models and clinical perspectives at the level of derivative theoretical and clinical concepts divorced from their underlying narrative structures. When one attempts to utilize concepts that have been derived from fundamentally different visions—visions that, to some degree, color and determine the very perceptions that comprise the data one used to build theory—the process is like mixing apples and oranges (Greenberg & Mitchell, 1983).

The dialectical synthesis we propose as an alternative to such an integration, is the result of a process of deconstruction and reconstruction utilizing a vantage point (evolutionary theory) that lies beyond traditional analytic positions and debates. The resulting "evolutionary narrative," of which psychoanalytic traditions currently represent parts, amounts to what is arguably a major philosophical shift in Western thought.

Dialectical Tension: Synthesis

The evolutionary biology of kinship theory (Hamilton, 1964) and reciprocal altruism (Trivers, 1971) presented in Chapter Four represent a reconceptualization of the boundaries of the individual psyche as well as the very meaning of self-interests. This philosophical shift is the basis for the evolutionary narrative of the purpose and structure of human aims and desires, that is, our basic representational and motivational system. Rooted in a solution to fundamental theoretical dilemmas in biological theory, this altered philosophical position expresses the intrinsically social, relational essence of the psyche without omitting those critical features that remain inherently, competitively individual in their aims. In this fashion, the evolutionary narrative differs markedly from both the philosophical traditions that underlie the polarization in psychoanalysis:

1. The individualistic (atomist) tradition of eighteenth-century British philosophy (Locke and Hobbes, and its nineteenth-century utilitarian derivatives, e.g., Bentham, Mill) in which the individual is the primary unit of analysis and individual interests are assumed to diverge substantially; and

2. The social (collectivist) tradition of Continental thought (e.g., Rousseau, Hegel, Marx, Durkheim) in which the interactive field or collectivity is primary and individual interests are viewed as convergent, potentially harmonious parts of the relational whole.

As we discussed in Chapter One, we believe that this philosophical clash is considerably more complex than Greenberg and Mitchell's (1983) view of the "deeper divergence" between psychoanalytic models. Ultimately, the evolutionary narrative can create a substantially revised philosophical foundation for psychoanalysis at its most fundamental level.

As we have suggested throughout this volume, *the individualist–collectivist dichotomy is structured into the human psyche itself*. The tensions that have been expressed as polarized philosophical positions and as visions of human nature exist within the complex "semisocial" design of the psyche. These tensions include (1) the ambiguous boundaries that separate self and other and facilitate the movement of the locus of the sense of self back and forth from the skin-bounded individual to massive identification with (kin and reciprocally) related others; and (2) the innate, inescapable, and continuing tension between those essential affects (or drives) that promote the definition and actualization of the narrower, skin-bounded self and those (equally innate) affects that promote a deep connection with the interests and subjective experience of related others.

We are saying, in essence, that deconstructed into their essential meanings and reconstructed into the evolutionary narrative, the classical and relational psychoanalytic narratives ultimately dissolve. Their constituent metaphors contribute to the creation of a new whole, a new psychoanalytic paradigm. To envision this dialectical abstraction in some detail and substance, let us return to the scheme we developed in Chapter One, which revolved around the four dimensions of the vision of the human condition on which classical and relational narratives clash.

The Classical and Relational Narratives Revisited

The Basic Unit of Analysis Revisited: The Extended (Inclusive) Self

Let us begin by reviewing the first dimension of the classical–relational dichotomy (Table 1, p. 268): the *basic unit of analysis* has been historically constructed as either the *individual* or the *interpersonal field*. The evolutionary perspective contains a framework in which our customary notions of individuality and relatedness dissolve. To paraphrase and transform Winnicott's metaphor, evolutionarily we expect that there is, simultaneously, a unique, genetically distinct infant, as well as a nursing couple with its shared overlapping realities, aims and genotype. Psychological deep structures that demarcate self and other are likely to have evolved in a way that profoundly respects the realities of separate *and* overlapping interests. Thus, innate, self-serving individuality and innate, connectedness (including other-directed altruism) are equally primary.

TABLE 1. Classical Versus Relational Narratives

Four dichotomies	The classical narrative	The relational narrative
1. Basic unit of analysis	**The individual psyche:** emphasis on intrapsychic dynamics	**The interpersonal field:** the individual can only be understood within an interactive context
2. Basic source of patterning (or structure) in the psyche	**Vicissitudes of endogenous drives:** relational ties are derivative	**Vicissitudes of interpersonal interactions**
3. Relationship between basic aims of self and other	**Inherent clash of normal individual aims:** selfishness, rivalry, competition are motivationally primary (corollary: the normal [tripartite] psyche is "divided" in a way that reflects the inevitable conflicts between the individual and others)	**Emphasis on mutual, reciprocal, convergent aims:** significant interpersonal clashes are due to pathology or environmental failure (corollary: the psyche [the self] is an essentially "holistic" entity that reflects the possibility of relative harmony between the individual and others)
4. Role of deception and self-deception in subjective experience	**Defense and transference as distortions of reality:** resistance serves to conceal truth (corollary: repetition is a means of not knowing; people defensively repeat the acting out of unconscious motivations to maintain repression, to preserve ties, or to maintain instinctual fixations)	**Defense and transference as inherently valid expressions of personal reality:** resistance serves to elicit recognition of individual truth for further growth (corollary: repetition is an artifact created as people repeatedly try to elicit a needed response from the relational world to reinitiate thwarted growth and revise the self; people repeat in the process of preserving and protecting a vulnerable self—e.g., by maintaining needed relational ties—so that revision can be tried again in the future)

In Figure 13, you see a very schematic contrast between the classical (Freudian and ego psychological) "individualistic" bias toward envisioning the psyche as a separate, freestanding, tension-reducing entity and the equally biased, "mutualistic" picture of self and other in relational theories in which it has been implicitly assumed that relatedness is somehow more basic than separateness and individuality and that the overlap of interests (the shared mutualistic aims) in the normal environment is far greater than, in fact, it could ever be. Below these, we represent in visual metaphor the evolutionary perspective. As we suggested in Chapters Four through Seven, the evolutionary perspective generates the view that the psyche has been selected to operate with a basic ground plan, an evolved deep structure, in which self-interest is represented and promoted in narrowly individual *and* social forms. Our ways of experiencing reality, our basic motives, and relationships with others are guided, in part, by an adaptive system that has been shaped by a relational matrix that inevitably contains powerful, inherently competing, individualistic *and* inherently mutualistic, overlapping aims.

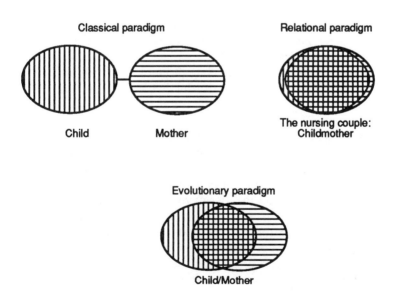

FIGURE 13. Classical, relational, and evolutionary assumptions about the rational world. Adapted from Slavin and Kriegman (1990).

Beyond the Endogenous versus Experiential (Intrapsychic vs. Interpersonal) Dichotomy

Recall now the second dimension of the clash between the classical and relational narratives that we outlined in Chapter One: the dichotomy between *endogenous inner forces* and *relational experiences* as the primary way in which *structuralization of the psyche* takes place. Is the psyche a response to, and a construct of environmental interactions, or is it essentially an internally constructed entity pushed by innate, endogenous forces? From the evolutionary perspective this dichotomy takes on a completely different meaning. The deep structure of the psyche (the basic system and program that underlie and make possible all individual psychic structure) is seen as an "evolved adaptation" that has been shaped by relational conflict over vast evolutionary time. In a sense, the psyche functions as a "fitness optimizing organ" operating through mechanisms designed to regulate the perception of and response to the conflicting pressures inherent in life within the relational world. Significant, universal, endogenous aspects of our psyche are, thus, seen as ultimately having been patterned by an "interpersonal field," as Greenberg and Mitchell (1983) have characterized the determinant of psychic structure in the relational perspective. However, at the level of *deep structure*, this "field" is not equated with the developmental experience of each individual. Rather, it embraces the relational world of countless life cycles, the world in which hundreds of thousands of generations of our ancestors (and their unsuccessful compatriots) strove to maximize their inclusive fitness.

Thus, the psychological precipitate of life within such a relational world, including those parts of the psyche that are traditionally viewed as nonrelational, such as endogenous drives themselves, are seen as having been patterned by the ancestral relational field. As we shift our focus from *mechanism* (repression, drives, the true self, mirroring, empathic or unempathic response, etc.) to *function*, we begin to think in terms that move us away from the inner versus outer dichotomy. Function always essentially entails an *inner psyche* operating *within a relational world*. Recurrent outer realities have become represented in inner, intrapsychic adaptations—adaptations that, in turn, exercise their own "endogenous" power through innate affects, expectations, and strategies. Blind mechanisms are not blind, but their vision (what they "see") often differs from what appears through our other ways of knowing the world. Instinctual drives, repression, and the true self, for example, are acutely sensitive means for assessing (for over many millennia they have been painstakingly sculpted by) the inherently biased relational world. Within that world, they confer a fundamental degree of flexibility in the pursuit of one's self-interest.

The ultimate paradox is that *in order to construct an adaptive functional self or individual identity in a relational fashion—out of inferences based on intersubjective experience within a biased relational surround—that "self" cannot simply be a relational product.* It must be shaped by universal, ineluctable forces and schemas that emanate from the intrapsychic workings—the deep structure—of the evolved, individual psyche. Ultimately all such inner forces and schemas are, themselves, ancient products of the paradoxical tensions in human social relations. This view is far from the more common assumption that the self and its dynamics are, somehow, created anew by each of us out of the sum of our experiences in our individual lifetimes.

Inherent Conflict versus Inherent Mutuality in the Relational World

Recall now, from Chapter One, the third crucial dimension on which the classical and relational narratives clash: the assumption of *inherent conflict* versus *inherent mutuality* as primary in the *interpersonal world*. As we tried to show in Chapters Four through Seven, the evolutionary view reframes the issue. Conflict is expected to be intrinsic to the very nature of human relationships themselves, quite apart from the whole question of drives or environmental failure. This means that the classical drive theory tradition was, indeed, attempting to explain something—endogenous conflict—that needed explaining. Yet the explanation that Freud created to account for the conflict—bodily based, somatic, *intrapsychic drives* that cause *interpersonal* conflict—put the cart before the horse. Drives do not define the purposes of the organism, the aims of the psyche. They are one form of inner authority, one vehicle, or affective prompter, which makes it more likely that these aims will be realized. In confusing this issue, equating mechanism with purpose, drive theory seriously distorts our understanding of human ends.

While conflict is inherent in the relational world, a significant degree of mutuality and cooperative action must also exist in an equally fundamental way. This means that the basic inner design of our motives and our entire inner adaptive system will be as fundamentally, innately organized around the inclusion of the interests of *others* in what we most deeply value and desire. Over vast evolutionary time, the complex, contradictory realities of this "universal relational matrix" represent the chief selection pressures that shaped important aspects of the psychodynamic "deep structure" of the human psyche.

Freud's vision of the relational world in *Totem and Taboo* (1912–1913) and *The Phylogenetic Phantasy* (1915d) almost exclusively emphasized the conflictual aspects of the ancestral human environment, especially the conflicts of interest existing between generations. Even if he had a better grasp of the workings of natural selection, of social selection pressures, Freud still might not have been able to modify some of the basic flaws in drive theory. The classical metapsychology approximates a viable model of the psyche of a creature adapted to *one* major facet of relational reality. In skirting the whole issue of Freud's basic assumptions about the (ancestral) relational world, focusing simply on nonsocial adaptive capacities, Hartmann could not make use of the evolutionary perspective any more critically than did Freud.

Conversely, the "evolutionary" work of Bowlby and Erikson (and, in less explicit ways the range of other relational theorists) radically altered the picture of the social realities, the actual evolutionary relational world to which our psyche is adapted. By essentially replacing the "narrative of conflict and deception" in the Freudian and Kleinian views with a view of the normal relational world as consisting of individuals whose motives are either neutral or largely converge and mutually harmonize, these theorists created the rationale for the opposite, equally biased narrative tradition—the relational model.

Defensive Deception versus Valid Self Protection and Communication

Finally, consider the fourth dimension of the clash between the classical and relational narratives: do the highly subjective versions of experience that can be called "defenses" represent *deceptive distortions* of reality? Or do such organizations of experience and self-presentation more fundamentally convey *valid personal truths*? In the classical narrative, defenses such as repression are primarily seen as ways of concealing and distorting reality. In the relational narrative, particularly in the work of Winnicott and Kohut, defenses are more apt to be interpreted as ways of protecting, maintaining, and ultimately expressing valid, vital aspects of the self and as safeguarding and promoting the process of development. Such self-enhancing and self-protecting views of reality may be particularly important in reinitiating thwarted development.

As we discussed at length in Chapter Ten, the evolutionary view suggests that there are strong adaptive reasons why the subjective alterations of experience in a process such as repression must intrinsically entail *both* features. First, the child's motivations have inherent conflicts within them. At times, when acting selfishly in a manner that is at odds with *the child's* altruistic motives, or when acting altruistically in a manner that is at odds with *the child's* selfish motives, the child may need to self-deceptively hide the motive that is discrepant with the action. One cannot be convincingly enraged, for example, if one's concern and need for the other is conscious—thus "showing through" to others and/or inhibiting full expression of the rage—for this would diminish the other's belief in the danger posed by the rage. Likewise, love cannot be fully and spontaneously expressed if one's more selfish motives are simultaneously conscious, thus interfering with a sense of genuineness in the emotional expression. Therefore, individuals must hide from themselves aspects of their motivations. The pervasiveness of this link between self-deception and deception was one of Trivers's (1976a) great insights into the workings of nature from the cellular level to that of the human psyche.

We must hide from others certain of our motives for our expressed motives to be believable. While this applies to *both* altruistic[2] and selfish motives, it must be remembered that, in some respects, we all ultimately engage in a completely natural, yet undeniably self-centered quest for advantage even in relation to our closest kin. This is the aspect of human action, including inevitable attempts to hide and distort reality in order to deceive and gain advantage over others, that has been signified by classical models of the id and instinctual drive motivation. Yet, in these models, this aspect has been essentially projected into the body in the form of imperious, beastlike, somatic tensions that are, in effect, blamed for the existence of conflicts that are actually far more intrinsic to the self-promoting design of the psyche as a whole. Similar to the classical model of defense, repression frequently functions to conceal and distort motives and views that are distinctly biased toward the individual's own interest. And patients, as well as analysts, have been designed to regularly engage in such "distortions."

Simultaneously, though, the evolutionary model incorporates the relational view: individuals have been designed to repress aspects of their affects, aims, and true identity, in order to sequester them safely, to protect them from the inevitable side of the relational surround that

---

[2]The degree to which our love, concern, and need for others must be hidden when it is in our interest to appear angry, hurt, and indifferent has never been fully captured in the classical model. Yet, every couples therapist has observed how difficult and dangerous it frequently feels for partners to express their love and concern after they have been hurt or betrayed. In the complex negotiations that comprise a marital relationship, it often appears necessary to "convince" the other that only their changing will enable the relationship to remain alive. This gambit can only work if the individual can successfully (self-deceptively) hide his or her love, concern, and need for the other (Kriegman, in preparation).

is geared more closely to its own interests than to those of the individual. The biases and self-interested motives of the normal caretaking environment and, to some degree, all subsequent relational settings make it necessary for individuals to have a way of developing subjective versions of their needs that remain somewhat independent of the environment. The evolutionary perspective suggests the crucial adaptive need for an innate, subjective sense of one's "true" (Winnicott, 1965), "authentic" (Fromm, 1941), or "nuclear" (Kohut, 1984) self. Moreover, the evolutionary narrative suggests that we have been designed to insure that such vital configurations of impulses, aims, and fantasies are effectively held in reserve in order to be retrieved and reanimated under altered relational conditions. The process of retrieval, however, will also entail the quest to actualize more selfish aspects of self-promotion that are inherently mixed with the innate striving for self-repair and overcoming structural deficits. Thus, while there is support for the intrinsically deceptive, distorting aspects of defense, the evolutionary perspective is clearly consistent with the view of Winnicott and Kohut that developmental strivings toward the resumption of thwarted growth must play a central, perhaps superordinate, place in the organization of the psyche.

In the evolutionary view of the function of repression, the process of conflict regulation is, simultaneously, both intrapsychic and interpersonal in character. The evolutionary perspective embraces elements of what are now often referred to as one-person (intrapsychic) and two-person (interpersonal) psychologies (Modell, 1984). Both of these perspectives are needed to fully describe the evolutionary conception of human adaptation to the mixture of conflict and mutuality in the relational world. Countless generations of recurrent, ancestral, intrafamilial conflict have equipped the child with a deep structural capacity to apprehend and manage a range of expectable, but ambiguous and deceptive, tensions between its own self-interest and those of closely related others. The particular form, intensity, and, to some extent, content of these conflicts, as well as the individual's ways of coping with them, will, of course, be shaped to a great extent by one's particular interpersonal experience, one's individual history in a particular environmental context. (See Table 2 for a summary.)

Summary and Conclusions: Toward an Evolutionary Psychoanalysis

We do not, in any sense, advocate a shift to thinking about the psyche in directly biological terms. Many aspects of the meaning and com-

TABLE 2. Classical Versus Relational Narratives, Revisited

Four dichotomies	The classical narrative	The relational narrative	The evolutionary synthesis
Basic unit of analysis	**The individual psyche:** emphasis on intrapsychic dynamics	**The interpersonal field:** the individual can only be understood within an interactive context	**The genetically based self—the "gene's-eye" view:** individual boundaries are partially innate and include aspects of related others; the self is intrinsically *semisocial*
Basic source of patterning (or structure) in the psyche	**Vicissitudes of endogenous drives:** relational ties are derivative	**Vicissitudes of interpersonal interactions**	**The psyche as an evolved deep structural adaptation:** universal deep structure allows both drive and relational patterning of the psyche to operate; each individual psyche carries within it the evolutionary history of the interpersonal
Relationship between basic aims of self and other	**Inherent clash of normal individual aims:** selfishness, rivalry, competition are motivationally primary (corollary: the normal [tripartite] psyche is "divided" in a way that reflects the inevitable conflicts between the individual and others)	**Emphasis on mutual, reciprocal, convergent aims:** significant interpersonal clashes are due to pathology or environmental failure (corollary: the psyche [the self] is an essentially "holistic" entity that reflects the possibility of relative harmony between the individual and others)	**Inherent conflict and inherent mutuality in the good-enough environment:** the self-interests of genetically distinct yet related individuals necessarily conflict and overlap in the evolved relational world (corollary: the normal psyche [the self of inclusive fitness] is designed to operate as an overarching, holistic entity that manages the ongoing negotiation of the inherent tension/division between selfish and mutualistic aims)
Role of deception and self-deception in subjective experience	**Defense and transference as distortions of reality:** resistance serves to conceal truth (corollary: repetition is a means of not knowing, we repeat by acting out of unconscious motivations as a defense for maintaining repression, to preserve ties, or to maintain instinctual fixations)	**Defense and transference as inherently valid expressions of personal reality:** resistance serves to elicit recognition of individual truth for further growth (corollary: repetition is an artifact created as we repeatedly try to elicit a needed response from the relational world to reinitiate thwarted growth and revise the self; we repeat in the process of preserving and protecting a vulnerable self—e.g., by maintaining needed relational ties—so that revision can be tried again in the future)	**Defense and transference as adaptive deception (and self-deception) in the promotion of vital individual truths:** resistance serves to conceal truths while ascertaining whether valid, individual truths can be recognized and utilized for further growth (corollary: creative repetition; we search for patterns and match past situations in the process of regressive renegotiation; we probe new relational contexts, seeking *reasons for maintaining or relaxing repression* and to explore old compromise solutions in the attempt to negotiate a more advantageous fit between our inclusive self interest and the relational world)

plexity in human experience have been captured within the diversity of psychoanalytic models and epistemologies, from the relatively experience-distant, reductionist–physicalistic classical emphasis on individual endogenous drives and presumably "blind" mechanisms to the experience-near, phenomenological–hermeneutic study of intersubjectivity at the forefront of the modern relational tradition. In our view, all psychoanalytic models, implicitly or explicitly, intentionally or disavowedly, are *necessarily* rooted in complex biological assumptions and are thus potentially part of a larger, overarching, and unifying, biological reality. Contemporary evolutionary theory thus provides a needed platform on which to stand outside of the psychoanalytic traditions in order to view the field as a whole, to evaluate and potentially revise its parts. Without this essentially experience-distant "platform" as a reference point, hermeneutic approaches to psychoanalysis as a purely clinical discipline, or a clinical study of individual systems of subjective meanings, risk becoming "hermetic," that is, sealed off and overly derivative of those idiosyncratic subjective biases held in common by members of a particular school of thought. Yet, to the extent that such an experience-distant platform—the evolutionary as well as any other—cannot effectively capture and embrace the structures of human subjectivity, it becomes a kind of "scientism," that is, a ritualistic pursuit of an illusory "objective" truth based on methods and concepts from physics and chemistry that define away the most vital data of human lived psychological life. The test of the validity and usefulness of such an outside platform as a vantage point ultimately lies in its helping us embrace crucial *experiential* data: the way in which such an outside vantage point may enhance our empathic understanding of the subjective structures of inner meaning elicited, in part, within the intersubjective field of the psychoanalytic clinical dialogue.

We believe the evolutionary biological perspective, as we have developed it, enables us to navigate between the methodological Scylla of hermeneutics and constructivism that is only partially aware of its own basic assumptions about what is universal (essential) in the human condition, and the Charybdis of a ritualistic scientism that is as likely biased by its own assumptions about the human condition while naively assuming that its methods—suited to the study of inanimate particles and forces—will generate an "objective" picture of the psyche. Substantively, the evolutionary narrative permits us to embrace elements of both the classical and relational psychoanalytic traditions after we have deconstructed the underlying meanings of the metaphors and narratives that have traditionally defined these models. Both existing analytic traditions can be understood as contributing vital parts, indeed complementary, reciprocal parts to a larger, broader picture of human psychodynamic adaptation. This book has been an attempt to demonstrate how contemporary evolutionary biology may point the way toward a dialectical movement into a new, more embracing narrative structure—a new paradigm that incorporates elements of each existing narrative tradition yet interprets and translates these features into dimensions of an overarching whole.

The Psyche as an Evolved Adaptation that Has Been Structured by Relational Conflict

The basic patterning of the human psyche is ultimately explainable as an evolved "deep structural adaptation" that has been shaped over vast evolutionary time to optimally represent inclusive self-interest while regulating the inevitable conflicting pressures inherent to life within the human relational world.

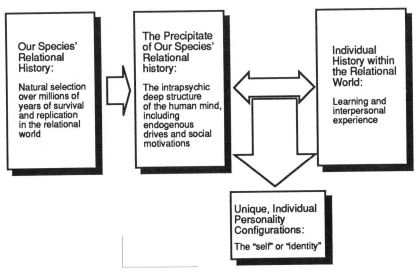

FIGURE 14. Evolutionary psychoanalysis.

Intrinsic Relational Conflict in the Expectable, Good-Enough Environment

The evolutionary perspective depicts an average, expectable, good-enough environment that consists of distinct, unique individuals whose interests necessarily diverge and, to some degree, inevitably compete.

This evolutionarily based conception of conflict is at marked variance with the tendency in existing relational narratives to focus primarily on those dimensions of conflict that can be attributed to inadequacy, pathology, or abnormal lack of attunement in the caregiving environment. In other words, the evolutionary perspective suggests that conflict and its accompanying strategic deceptions and self-deceptions are intrinsic features of all object relational ties and interactions. In contrast to the "narratives of environmental failure" found in the writings of Kohut, Sullivan, Guntrip, and Fairbairn, a powerfully conflictual dimension is understood to be absolutely normative in the fully good-enough environment.

This evolutionary narrative yields a view of inner, intrapsychic struggles characterized by *primary relational conflict*, as distinct from the drive-based conflict depicted in the classical model, the environmentally induced conflict of the relational perspective, or the unspecified "inherent difficulties in relating" and complexities that lead to relational conflict in certain contemporary interpersonalist views (e.g., Mitchell, 1988). In other words, the conflict that exists in human relations is found in the interactive dynamics of the relations themselves as opposed to being something that is, in a sense, *imposed* upon human relations by drive demands, environmental pathology, or ubiquitous, unspecified difficulties in maintaining attachments. To be sure, relational dynamics are influenced by the adaptively relevant information provided by endogenous drives, as one source of input on the more selfishly individual side. On the other side there are the equally primary "prosocial" motivations. The patterning of the whole is profoundly shaped by the character and reliability of the particular relational world into which we are born as well as the form of each individual's innate character. Our "selves" are superordinate structures that include the continuous synthesis of an overarching, relatively unified identity. However, this relative unity is built on a profound dividedness at the core of the human psyche—a "dividedness" that is embodied in the fact that our primary motivational structure appears to be broadly organized around two, simultaneous, yet inherently competing sets of aims:

1. A relatively altruistic set of "social" aims, related affects, and self and object representations that are adaptations to the realities of our shared, overlapping interests. Such motives (i.e., genuine object love, altruism, and the attachment-based affects) are as real, primary, and innate as the aggressive and libidinal instinctual aims that the classical model views as motivational bedrock. As we suggested in Chapter Four, in the classical view, such "social" motives have been mistakenly characterized as a complex, yet ultimately defensive and reactive, overlay on more primary aggressively self-interested aims;

2. A complementary and inevitably conflicting set of more aggressively self-interested, "driven" aims that represent more directly "selfish," asocial, uniquely individual needs. Such aims are, in part, sustained and guaranteed to operate in the interests of the individual by virtue of their making use of endogenous drives—or nonrelationally based motives—as vital, imperative signals of adaptively relevant information. Such peremptory "pushes" from within will effectively counter the powerful, inherently biased forces (of others' self-interests) in the normal relational environment.

Intrapsychic conflict thus represents an "archaic heritage," as Freud (1923) put it, by which is meant an ancient, inner structural legacy that is our inherited capacity to organize and cope with the incredibly complex realities of divergent interests in the evolved relational world. Such a model echoes Bakan's (1966) conception of the inherent opposition between "agentic" (individualistic, selfish) aims and "communal" (other-directed, or group-oriented) aims in human motivation. Its emphasis is similar to Klein's (1976) effort to reformulate the universal basis for inner conflict in terms of a clash between inherently opposed, equally primary, human aims in contrast to the ego psychological notion of an inherent clash between reality-based ego and a "biological" id (Eagle, 1984). In its reinterpretation of instinctual drives as a means subordinate to and operating in the service of broader, relational ends, this resembles Fairbairn's (1952) attempt to recast the libido theory in an object relational context. Yet, in clear distinction to all such previous attempts at reformulating the nature of inner conflict, the evolutionary model is rooted in a general theory of the inherently conflictual-mutualistic relationship between individuals and their relational environment; and it is supported by a broad conception of how the overall functional architecture of the psyche was shaped by the selective pressures in our evolutionary history.

As Figure 15 shows, there is an evolved, inherent tension in the human psyche. This tension was formed in the context of intimate conflictual interactions in the relational world through which a basic tension and division in our motives was shaped: directly self-interested, asocial inner forces were structured into the psyche in dynamic opposition to equally primary, more mutualistic, prosocial inner aims. This stands in contrast to both the classical tripartite model of intrapsychic conflict derived from endogenous drives and the relational model of conflict derived from environmental failure. All of this is, in the deepest sense, endogenous because its *deep structure is intrapsychic*. Even though it is derived from what was interpersonal and social in our evolutionary history, each individual psyche now carries within it the evolutionary history of the interpersonal (Teicholz, 1989).

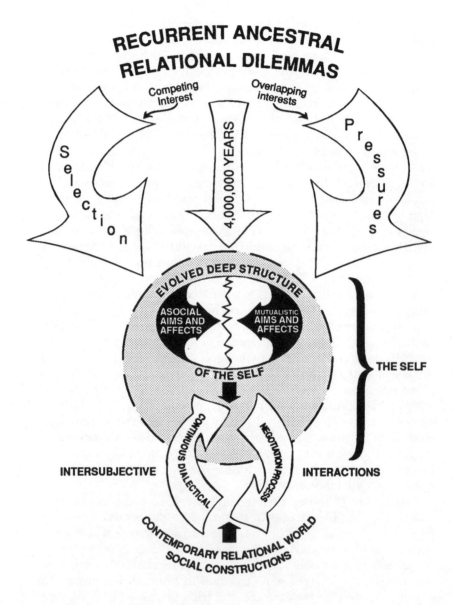

FIGURE 15. Transcending the dichotomy of "the relational" (external/social) versus "the endogenous" (internal/bodily-based): the evolution of the divided (semisocial) self.

An evolutionarily based "metapsychology" thus depicts us as innately individualistic and innately social; as endowed with inherently selfish, aggressively self-promoting aims, as well as an equally primary altruistic disposition toward those whose interests we share. We are, in short, never destined to attain the kind of highly autonomous individuality enshrined in the classical tradition, nor are we the "social animal" of the relational vision. *We are essentially "semisocial" beings whose nature, or self-structure and motivational system, is inherently divided between eternally conflicting aims.*

Overall, we believe our depiction of evolved deep structure argues for a different view of the complex ways in which the self eventually emerges as a socially constructed entity. Although essentially an emergent entity, the product of a lifelong, dialectical negotiation actively pursued by the individual within an intersubjective field—the self is only able to come into being because of the prior existence of a complex set of evolved, adaptive capacities or strategies. These evolved, functional capacities, for example, repression (and adaptive regression), endogenous drives, and the "true" self, are somehow "encoded" in our genes and have become woven into a complex system of evolved adaptations, a system that has been shaped by recurrent, relational mutuality and conflict over vast evolutionary time. These evolved functional capacities guide the intersubjective negotiation process, ensuring that it is likely to be carried out in a fashion that adequately represents, promotes, and actualizes the inclusive fitness of the individual.

In effect, the deep structure of the psyche can be thought of as an enormous fund of accumulated information, strategic knowledge, and wisdom encoded in the form of structures and dynamic processes that are designed to deal with the recurrent relational dilemmas encountered in our ancestral, relational environment. These recurrent dilemmas produced hundreds of thousands of generations of social selection pressures emanating, in good part, from the paradoxes entailed by the need to construct and maintain a viable, unique, individual self in a biased familial environment—an environment universally characterized by an ambiguous mixture of conflicts and deceptions inextricably interwoven with overlapping interests and genuine mutuality.

We are, thus, universally equipped to deal with the following:

1. The highly predictable, invariant aspects of the relational world within which the emergent self must be negotiated, that is, the simultaneously competing and overlapping interests of the "genetic relational matrix," including parent–offspring conflict, the inevitability of deception, and the necessity for reciprocity (and the existence of cheating) within and outside the kin setting; and

2. The predictably novel (predictably unpredictable) dimensions of

the negotiation process, for example, our innate individual dispositions, our intrinsic uniqueness, and the novel and virtually unforeseeable configuration of interests and challenges that exists in interactions with the other unique individuals in our formative, intersubjective worlds.

It is inconceivable that we would be able to solve the staggeringly complex adaptive problems we face without innate, a priori structure. More specifically, without innate capacities rooted in historical (phylogenetic) experience, it would be impossible to anticipate and penetrate the normal deceptions in the relational world to "catch hold," as Freud (1918) put it, of the hidden realities of competing interests in the thoroughly "good-enough" environment.

Within the evolutionary picture of the overarching architecture of the psyche, the most effective way to conceptualize the self is as follows:

1. A provisional and continuously self-dismantling/self-creating structure built up of multiple repressed schemas that are regressively revived and revised through a transferential probing of the object world. This entails a process of negotiation that demands a fluid movement between unconscious and conscious schemas. The overarching aim is to create a working model of the complex tapestry of shared and divergent interests, the validity of which is signaled by innate affective reactions like a sense of "wholeness," integrity, and spontaneity.

2. An inherently divided structure, divided along a fault line between two basic sets of affectively organized anticipations about reality, that is, narratives of reality as an essentially mutualistic world composed of individuals with essentially overlapping aims and goals (literally overlapping self interests) and narratives depicting a conflictual world fraught with divergent interests, hidden meanings, deceptions, and distortions. Thus, the child has an inborn dual vision, as it were, an expectation of and active search for both of the equally ancient, equally integral dimensions of our relational makeup.[3]

---

[3]Some self psychologists (Shane & Shane, 1988) have tried to capture aspects of this inherent internal conflict in their formulations (Kriegman & Slavin, 1990). Lichtenberg (1989) has attempted to delineate separate and distinct motivational systems that sometimes "compete" for dominance. He sees an overarching self-structure that attempts to organize and prioritize these different motivations. While our view of the self as a fitness-optimizing, overarching, mental organ (Kriegman & Slavin, 1990) has some similarities with those other views we see aspects of these "competing dominances" not merely as differing motives all striving for expression but, in certain cases, as motives that exist in inherent opposition to one another. Thus, the evolutionary view presents us with an overarching self-structure with inherent conflict as a design solution for maximizing success in a conflictual relational world; a solution that almost inevitably generates significant developmental problems, some of which inevitably become intertwined with what we term "psychopathology."

The Evolutionary Knife Cuts Both Ways

The evolutionary perspective yields a new, unique paradigm that enables us to steer a conceptual course that avoids the more problematic assumptions of the existing psychoanalytic traditions. To the extent that we "read" the classical analytic agenda as synonymous with drive theory, the structural model, and many of its ego psychological revisions—with its insupportable assumptions about the primacy of a limited set of asocial, "animal" drives in human nature (Eagle, 1984; Holt, 1976; Klein, 1976)—the evolutionary perspective clearly does not support the classical position. The notion that defense mechanisms have "a single purpose, that of assisting the ego in its struggle with its instinctual life" (A. Freud, 1936, p. 51) becomes an untenable view of the human mind. Drives, and the structural model of drive–defense conflict, assume a *subsidiary* role within a larger, relationally designed and configured psyche. But, to the extent that the classical agenda is read as a "narrative of conflict," it alone captures certain major, significant features of the relational world and the inherently "divided" way we are adapted to it. Its metaphors of inner conflict depict the deep divisions and tensions within the self that are indispensable concomitants of an adaptation to the conflictual relational world.

Conversely, to the extent that we read the relational tradition as replacing the "narrative of conflict and deception" with a view of the normal relational world consisting of individuals whose motives essentially converge and mutually harmonize—with its related tendency to downplay the role of endogenously derived inner conflict in the creation of "psychic reality" (Cooper, 1983; Wallerstein, 1983)—the relational position cannot find validation from the evolutionary perspective. Instead, we can read the relational model as, essentially, a vision of human nature in which our psyche is understood as adaptively organized around built-in motives and capacities for conducting vital social relationships in the service of optimal, authentic self-development. In this way the relational model finds a strong, clear echo in evolutionary biological thought.

The evolutionary biological perspective permits us to fully embrace both the classical and relational traditions as valid parts, indeed complementary parts—but only parts—of a larger, broader picture of human psychodynamic adaptation.

Appendices

APPENDIX A

Has There Been Sufficient Time for the Evolutionary Design of Complex Psychological Structures through Random Variation and Natural Selection?

> Though the Life Force supplies us with its own purpose, it has no other brains to work with than those it has painfully and imperfectly evolved in our heads.
> —GEORGE BERNARD SHAW

We have chosen to refer to the adapted products of natural selection using the term "design" in a fashion that is consistent with contemporary evolutionary theory, that is, with no teleological implications of directionality or purpose in the process by which adaptive structures are formed (see Appendix B, "Freud, Lamarckism, Haeckel's Law, and Modern Evolutionary Thought," p. 290). The universal, evolved design of the mind is thus viewed as profoundly shaped by hundreds of thousands of generations of exposure to repeated, historical (phylogenetic) social selection pressures—"recurrent adaptive problems that selected for minds that came pre-equipped with mechanisms tuned to solving these problems" (Tooby & Cosmides, 1990, pp. 25–26). The child is seen as heir to this legacy of ancestral, encoded "wisdom," its "draft blueprint" for personal agency which, through the deep structure of its mind, enables it to use relational experience as a guideline for its own adaptation.

Part of the reason that teleological thinking is so attractive in analyzing complex design is that it is hard to imagine how such intricate well-coordinated designs could arise through random variation and natural selection (see also Dawkins, 1986). The time frame that would have been necessary seems enormous and incomprehensible. Religious creationist critics of evolutionary theory use this to argue that the theory of evolution is false. Other intellectual critics tend to accept the theory of evolution but then argue that the complex design solutions could not possibly have evolved through random variation and natural selection in the time frame available for human evolution, and that, therefore, we are reading into the human psyche too much adaptive structure. So, let's look briefly at the time frame that has been available for natural selection to shape the human psyche.

Our current best estimates indicate that our line diverged from the great apes over 5 million years ago (Trivers, 1985). If we consider the dramatic changes that controlled breeding has attained in domesticated species, it becomes easier to conceptualize how any mental structure or trait that provided even a minuscule selective advantage over other alternatives, would long ago have become the predominant form.

For an example, take a quick look at two versions of a trait that offer almost no visible selective advantage or disadvantage. If those who possessed version A averaged 1.2 successful, healthy offspring, while those with version B averaged 1.2001 successful, healthy offspring, then even a careful observer would be unable to measure a difference in reproductive success given modern methodology. However, in just 99 generations, individuals with version B would average 692,981 more descendants than version A individuals, in a situation where version A has less than 1/100 of one percent selective disadvantage. In situations where the selective pressure is more significant (i.e., measurable using available methodology, as many selection pressures are) the time necessary for evolutionary change is well within that which has been available.

> In humans . . . it takes about 10,000 years for a trait with a 1% selective disadvantage to go from 99% of the population to less than one percent. . . . Yet 10,000 years is but the blink of an eye in evolutionary time. . . . The best evidence suggests that we are at least 5 million years separated from a common ancestor with the chimpanzees. The traits that differentiate us from chimpanzees have passed through at least 200,000 generations of selection. (Trivers, 1985, p. 29)

We have had about 5 million years of selection since the beginning of technology (tool use); 30 million years (3 million generations) of selection since the beginning of adaptation to living in social groups; and more than 50 million years (10 million generations) of adaptation to living in small face-to-face groups of kin, "friends," and dominants (Trivers, 1985).

Thus, the evolution of social patterns has had a 50-million-year history for our ancestral line. Since divergence from the ancestral line we share with our nearest living primate relatives, another 200,000 generations have passed. This is more than enough phylogenetic history for the further shaping and refining of inner processes and structures for regulating interactional patterns—for the shaping of a significant part of human nature and the human psyche—part of which were shaped during a period of history when our line was coincident with other species, and part of which is a result of our unique human history. In fact, there is reason to believe that the rapid increase in brain size may have resulted from the selective pressures that result from the *social* context; specifically from living in an extended family and larger social environment with its enormous complexities. It is currently assumed by most paleontologists that this "social selection" may account for far more of our increase in brain size than tool use, hunting skills, etc. (also see Kriegman, 1988; Trivers, 1971; Appendix C, "The Evolution of the Human Cortex and Its Relation to Civilization and Guilt").

APPENDIX B

Freud, Lamarckism, Haeckel's Law, and Modern Evolutionary Thought

Lamarckism

Freud's well-noted Lamarckism was, in fact, part of the *Zeitgeist* of Freud's time. Indeed, Darwin was a Lamarckian if we define Lamarckism as the belief that acquired characteristics could be transmitted to descendants. Until the elucidation of the genetical theory of natural selection, the "modern synthesis" in the work of Haldane (1932), there was no clear understanding of how natural selection operated exclusively through mutation (genetic variation) and differential survival and reproduction. Darwin himself accepted Lamarck's mistaken notion of the inheritance of acquired characteristics (Mayr, 1983b).

However, there was always a fundamental difference between Lamarck and Darwin. This difference was Darwin's adherence to the nonteleological principle of natural selection, in contrast to another of Lamarck's mistaken beliefs in a teleological view of the process of change consisting of directed variation, that is,

> a need driving the organism from within, resulting in a tracking of the environment by an organism seeking perfection... Darwin's final theory is devoid of this teleological notion of directed change that is crucial to Lamarck's system. (Parisi, 1987, p. 242)

For Darwin, the creative force was variation (never fully understood by him) in interaction with natural selection. It was natural selection on which Darwin focused: the environment determined the success of the organism and thus selected the type of organism that would be "allowed" to reproduce and rear successful offspring. It is this conception of natural selection as the shaping force that has always differentiated Darwin's nonteleological views from Lamarck's teleological views. And it is the more superficial Lamarckism, that is, the inheritance of acquired characteristics, not the profoundly teleologically flawed notion of organism directed change, that Parisi (1987) and others refer to when they claim that Freud somehow really did not hold to a naturalistic, Darwinian view; that he adopted a "teleological view of human beings" (Parisi, p. 243). This is fundamentally untrue.

To understand why, we must first be clear about the distinction between a functional–adaptive (Darwinian) argument and a teleological one. Both attempt to explain the remarkable diversity of life and the amazing "fit" between species and their environments. A functional–adaptive explanation for this fit cites the fact that a phenomenon has certain advantageous consequences as the presumptive basis for explaining how it came to be. Labeling this process teleological is due to the confusion of the *outcome* (design in nature) of a completely nonteleological process (natural selection) with the notion of a *plan or purpose* that intentionally produces that outcome.

In a teleological, directed variation explanation, using the example that appears in Figure 16, one would assume that because certain finches needed longer, more pointed beaks, they were able to generate this variation themselves because, somehow, all life forms seek a preordained, ideal fit between themselves and the environment. Such tele-

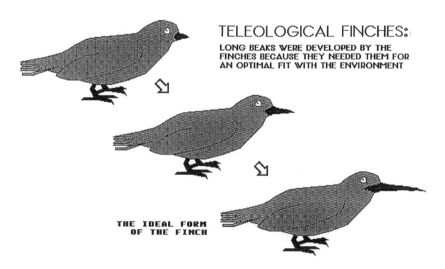

FIGURE 16. Teleological finches: Long beaks were evolved to provide an optimal fit with the environment.

ological explanations can be quite attractive and even appear necessary when one studies the uncanny fit between organisms and their specific environments. In a functional–adaptive explanation, a long-pointed beak (on one of Darwin's finches, for example) that functioned superbly in pulling worms out of heavy bark would be presumed to have arisen due to genetic variation in beak length and shape but to have been favored by natural selection because it served this function better than other forms. Functional–adaptive arguments play an integral role in contemporary Darwinian theory; teleological arguments have no legitimate place in it (Williams, 1966; Mayr, 1983b). Differential survival and reproduction are the ways in which natural selection operates in a nonteleological fashion.

Consider two hypothetical individuals in our evolutionary past who differ in their fundamental capacity to nurture. We can imagine that the two will differ in terms of the number of surviving offspring they leave behind. If the more successfully nurturant individuals average three successful offspring to the less nurturant individuals' two, then after only three generations the more nurturant type will have 27 descendants to the other's 8. Clearly the more nurturant type will come to typify the species. While such dramatic differences in reproductive success were probably rare, in Appendix A, "Has There Been Sufficient Time for the Evolutionary Design of Complex Psychological Structures?" (page 287), we suggested that there has been plenty of time for nonteleological processes to have structured fairly complex designs into existing species. Since this confusion between nonteleological processes and teleological thinking frequently leads to objections to evolutionary explanations, let us look at how the "father of teleology," Aristotle, struggled with just this issue in regard to an evolutionary explanation of the formation of existing species.

Aristotle described four factors or varieties of "causes" to take into account when explaining events in the world. The first factor is the material involved. For example, bronze cannot burn, wood can. The second is the form of the material. For example, wood can be in the form of a tree, a bed, or a house. The third factor is the efficient cause, that is, the force that initiated an event or change. Aristotle defines the fourth or "final cause" as:

> the goal or the end for the sake of which the change takes place. For example, health is the end for the sake of which one takes a walk. Why does he take a walk? We say in order that he may be healthy. And in saying this ... we are showing the cause of his walking.

There is a specific commonality between teleological explanations and functional–adaptive evolutionary ones. In both types of explanations the end result is important. However, in evolutionary explanations, the end had no effect on the events that caused it to come about. Historic events led up to the present actuality rather than the present actuality shaping historic events. To use Aristotle's example, the future state of health does not exist and can therefore have no effect on our walking. However, our existing knowledge that exercise can lead to better health (based on past events) and our existing wish to be healthy can lead to our walking. Thus, we can try to understand how biological organisms pursue adaptive ends without switching into teleological thinking. Aristotle himself struggled with an early form of evolutionary theory, and in the attempt of the world's foremost teleologist to remain true to his teleological vision we can see the clear distinction between teleological and functional–adaptive reasoning.

> [According to] the 'proto-Darwinism' of Empedocles... advantageous mutations were originally due to chance. There were once creatures with parts of their bodies arranged in all sorts of bizarre combinations, and the extinction of all but mankind and the other extant animals had been a matter of survival of the fittest. Such an evolutionary theory ran counter to Aristotle's belief in inherent purpose, and he did in fact believe the species... of animals to be so unvarying as to exclude the possibility of new species arising as the theory of evolution demands. (Guthrie, 1981, pp. 111–112)

> Aristotle referred to the doctrines of Empedocles... [who] held three views directly opposed to Aristotle's. First, [Empedocles] argued that nature is blindly mechanistic. Hence, nature does not change simply to achieve natural goals. Second, Empedocles said more complicated species developed from less complicated species. And third, he said some species have survived and others have died off because the survivors adapted to their environment and the extinct species did not. (Brickhouse, 1990)

Aristotle begins his refutation of Empedocles' views by paraphrasing Empedocles' objection:

> [N]ature does not act for the sake of a goal or because it is better that nature acts as it does. [According to this objection] just as Zeus makes it rain not in order that the corn may grow but of necessity, so if the grain is destroyed by the rain while still on the threshing floor, it did not rain for the purpose of destroying the grain. Why then should it

not be the same with the parts of the body that exist by nature. For example, why should it not be by necessity that the front teeth grow up sharp and suitable for tearing food and the back teeth are broad and flat and suitable for grinding food. And the same can be said about the other parts of the body where they *seem* to exist for a purpose. In all such cases where these characteristics come together just as if they came about for a purpose, those survived that were suitably organized by chance. Those not so suitably organized perished and still perish. (Aristotle, *The Physics*, Book 8)

Aristotle went on to offer a series of arguments to rebut this view of mechanized nature. The first argument he presented is similar to the blind watchmaker argument (Dawkins, 1986) against evolution. Basically it consists of incredulity in the face of the regularity and extreme predictability of very complicated arrangements in the natural world. Reliable, predictable events surely could not be occurring by chance, and thus they must be for a purpose. Second, he compares natural events with those brought about by art. Clearly there is an inherent goal or *design* in both types of phenomena. If an artistic creation was brought about for a purpose, the same must be true of similar events in nature. Finally he says:

That which the lower animals do is neither the result of craftsmanship, nor research, nor deliberation. Some actually raise the question of whether spiders and ants and other animals act from intelligence or something else. Going gradually down the scale, it appears that plants, too, have organs produced for the sake of the end. For example, the leaves are produced for the sake of shading the fruit. Hence, if it is by nature and for a purpose that the swallow makes its nest, and the spider its web, and plants make leaves for the sake of the fruit, and the roots turn not up but down for the sake of nourishment, it is clear that the cause is such as we have described and exists in the things that come about by nature. . . . Anaxagoras asserts that the possession of hands makes man the most intelligent of living creatures. Actually, it is clear that it is because he is the most intelligent that man got hands. Hands are like an instrument. But nature, which is like an intelligent man, always gives the instrument to the animal which can use it just as it is more appropriate to give flutes to the flute player than to give to someone who possesses flutes the skill to play them. If, therefore, nature always does what is best among possible ways of acting, it is not through the possession of hands that man is the most intelligent creature. Instead, he possesses hands because he is the most intelligent creature.

In Aristotle's view there is an unmistakable sense in which nature acts with intention, foresight, planning, and design. In modern evolutionary theory—essentially similar to the views of Empedocles as paraphrased by Aristotle—none of these are necessary to explain evolution (Mayr, 1983b, 1988; Dawkins, 1976, 1986).

There are probably few better examples of how functional–adaptive and teleological thinking were confused historically than in a passage from Freud's *Three Essays on the Theory of Sexuality* (1905) and a related footnote added in 1920. Although Freud was actually trying to make a perfectly acceptable functional–adaptive argument, we can see (in his own wary, second thoughts) how he feared that he was making a Lamarckian (and thus teleological) argument because he used Lamarck's *other* mistaken notion of the inheritance of acquired characteristics. In discussing the nature of infantile oral gratifications (sucking pleasures), Freud hypothesized that infants are innately equipped to feel motivated by the pleasure in sucking because sucking is directly advantageous to an infant's survival. As he put it:

> This satisfaction must have been previously experienced in order to have left behind a need for its repetition; and we may expect that Nature will have made safe provision so that this experience of satisfaction will not have been left to chance. (Freud, 1905, p. 184)

In a footnote in 1920, Freud added:

> In biological discussions it is scarcely possible to avoid a teleological way of thinking, even though one is aware that in any particular instance one is not secure against error. (1905, p. 184)

Freud confused his perfectly acceptable Darwinian functional–adaptive arguments about the selective advantages of innate oral sucking pleasure with what he knew was associated with Lamarckism (inheritance of acquired characteristics) and thus might have to be considered scientifically suspect for he knew that Lamarckism was also associated with teleological thinking (Slavin & Kriegman, 1988).

It is clear that natural selection operates according to certain "rules": What is selected is what "works," is "adapted to," or "fits with" the environment. As can be seen most clearly in convergent evolution (when strikingly similar adaptations evolve totally independently),[1] natural selection hardly operates randomly. Species attributes (e.g., sucking pleasure) have, indeed, as Freud noted, "not been left to chance."

---

[1] Examples of convergent evolution include the independent evolution of strikingly similar mammalian and squid eyes and bat and bird methods of flight. The peculiar pattern of motion called "sidewinding" evolved independently in several species of snakes and the ability to generate electricity evolved independently in at least six animal species.

When he used phrases like "Nature will have made safe provision," Freud clearly implied that pressures have operated to shape the organism in ways that are advantageous to its survival. Unable to conceptualize this thoroughly contemporary, functional logic in terms of the genetics of variation and modification by natural selection, Freud feared he was slipping into Lamarckian/teleological thought when, in fact, he was presenting a perfectly valid, nonteleological, Darwinian argument.

Haeckel's Law

The notion that ontogeny recapitulates phylogeny has always had immense intuitive appeal. As Gould (1977) notes, even today some sophisticated biologists—not to mention most other highly educated people—still believe that, somehow, individual human embryos pass through a series of stages that replicate the adult stages of phylogenetically "lower" organisms. To a limited extent, the embryo does pass through stages that resemble some of the embryonic stages of phylogenetically earlier species. Yet the appeal of the idea that ontogeny recapitulates phylogeny derives some of its metaphorical power from the way it generates a sense of the great continuity of organic life as well as suggesting to some a more hierarchical notion that adult humans represent a cumulative pinnacle in the development of life on Earth (Mayr, 1983b). Perhaps the simplest definition of these terms was put forth by Mayr (1959):

> Ontogeny is the decoding of coded information, phylogeny is the creating of ever new codes of information and the survival of the most successful ones. (p. 181)

In *Totem and Taboo, A Phylogenetic Fantasy,* and *The Wolf Man,* Freud (1912–1913, 1915d, 1918) did link some of his speculations about the emergence of castration anxiety, guilt, the superego, and the predisposition to neurosis in childhood to a presumed sequence of traumatic, evolutionary events in our development as a species (Gubrich-Simitis, 1987). The specifics of these Haeckelian fables have as little validity as did Freud's Lamarckian notions about the inheritance of acquired characteristics as the mechanism of organic change. However, the essential notion that the child inherits some form of deep structural dispositions and innate anticipations and knowledge of the relational world—the world into which the child is "prepared" to be born and function—is an integral aspect of modern evolutionary theory (Mayr, 1974; Tooby & Cosmides, 1990; Alexander, 1990). (See also quote from Tooby & Cosmides on p. 287.)

APPENDIX C

The Evolution of the Human Cortex and Its Relation to Civilization and Guilt

Trivers's presentation of reciprocal altruism suggests that such behavior and its underlying motivational system began to develop in the early history of our species, or even earlier (with our primate predecessors or even with their ancestors). As kin altruism is even more primitive (ancient) there may have developed a tendency to utilize the same cognitive and motivational systems employed for kin altruism for reciprocal altruism. First, among kin an admixture of kin and reciprocal altruism probably developed and this later extended to unrelated others. The following reconstructive evolutionary history of altruism and intelligence (Kriegman, 1988) is based on modern evolutionary theory (Hamilton, 1964; Axelrod & Hamilton, 1981; Trivers, 1971, 1985).

Compassion or sympathy for one in pain (the empathic sharing of the experience of pain and the desire to alleviate the sufferer's anguish) probably first evolved within the context of kin directed behavior. This is so, because in a kin relationship an altruistic act increases the inclusive fitness of the altruist even if the act is not reciprocated. Therefore, the evolution of such behavior within kin relationships did not depend on there first being a tendency for the recipient to reciprocate. The generalization of this behavior to *reciprocal* altruism among kin (and then to unrelated others) became possible when sufficient cognitive abilities were developed so that the altruist could distinguish between those more likely to return the altruistic act (reciprocal altruists) and those unlikely to do so (cheaters). The cognitive abilities necessary for the successful trading of altruistic acts are quite complex (Trivers, 1971). If powerful adaptive advantages were obtained through participation in a reciprocal altruism system, this must have created additional selective pressure for the further development of the increased cognitive abilities needed to successfully engage in trading altruistic acts without being cheated. Trivers suggested that reciprocal altruism evolved alongside increased cortical capacity and was a major source of the selective pressure driving and shaping the development of such.

Traditional conflict psychology clearly sees altruistic behavior as a recent development brought into being *after* increased brain size began to lead to the formation of civilization: civilization, made possible by increased cognitive capacities and the related ability to delay gratification, led to pressure to restrict instinctual behavior, resulting in altruism. In Trivers's analysis, the adaptive advantage of reciprocal altruism existed at the earliest stages in the development of human intelligence. This forced rapid enhancement of cognitive capabilities and the shaping of "civilizing" (social) tendencies in the individual. These adaptive tendencies enabled the individual to participate in a developing social order and, thus, to garner the optimal advantages of the successful trading of altruistic acts. As altruistic tendencies began to spread, the adaptive advantage of altruism increased. The following paragraph describes this snowballing effect which is consistent with Axelrod and Hamilton's (1981) model showing how reciprocal strategies could arise in a population and ultimately outcompete purely selfish strategies which would then be unable to reinvade the population.

The adaptive advantages of reciprocal altruism may have been a major source of the selective pressure that brought about intellectual advancement and as a result made social systems possible and advantageous. As increased cooperation became possible, those early humans who were able use guilt to control their sexual and aggressive behavior were able to attract altruistic actions toward themselves. Potential altruists were developing the cognitive ability to discern reciprocal altruists from cheaters. The cheaters could be identified, in part, by their selfish/instinctual behavior and their apparent lack of guilt, shame, or compassion. There was then an increased benefit from being able to successfully participate in a reciprocal altruism *system* that began to develop (an organized social grouping with rules to govern human relationships). This led to additional selective pressure for the increased cognitive abilities that would allow the participant to comprehend the subtleties involved in engaging in the trading of altruistic acts without being cheated. As this developed, it further enhanced the benefits of an organized social system and guilt became even more important for it insured the individual's inclusion in the social group. Once individuals could be identified who displayed guilt and could be trusted to use it to temper their selfish behavior, the ability to remain in and participate in a reciprocal system—which because of this new guilt related stability could function at a higher level of efficiency—became even more important.

Thus, the capacity for guilt may be an important component of the reciprocal altruism system. It may have provided an important counterbalance to the pressures that selected for successful undetected cheating. Although it may have played a role in the further development of and maintenance of such a complex reciprocal exchange system, it was not the *cause* of it or the force that enabled it to arise in our evolutionary past. Guilt[1] developed along with other sources of motivation toward trading altruistic acts and functioned as a signal to others that "here stands a reciprocal altruist." The selective pressure shaping the development of guilt was the adaptive increase in one's inclusive fitness due to a fuller inclusion in the reciprocal altruism system as opposed to exclusion based on being identified as a cheater. Although some altruistic behavior may be influenced by guilt, the reciprocal altruism system (society) was not brought into play by the early development of guilt. Rather than guilt being thought of as the "cause" of altruism, in this analysis, it may be more accurate to think of altruism as the cause of guilt!

These aspects of the emotional motivational system of the human social animal have not been fully accounted for in the classical psychoanalytic paradigm. Modern evolutionary theory suggests that the selective pressures making it likely that people would want to affiliate and cooperate (the empathic tendency, compassion, and the need to utilize others as selfobjects) were probably early developments *prior to guilt*. Guilt may have aided the development of highly complex, stable, and intimate social structures (which allowed for the frequent trading of altruistic acts without rampant acting out of sexual and aggressive drives), but the selective pressure that brought about the development of guilt was the ability to gain full inclusion in, and the fullest benefits of, the reciprocal altruism system. This system, possibly in rudimentary form but probably already including the need for "human" community existed as a precursor to the development of guilt.

This suggests that the traditional individualistic (drive/structure) psychoanalytic notion of guilt-tempered selfish drives cannot fully explain most altruistic, cooperative, and communal social behavior. The tendency toward empathic union with others and altruistic behavior may not be a recent cortically based overlay in conflict with a more primitive biological core of self-serving sexual and aggressive drives, but rather may itself be based on a very primitive biological core. When

---

[1]Though an important distinction can be made between shame and guilt (Morrison, 1989), this discussion applies to both phenomena.

we understand what shapes reciprocal altruism in nature (the prerequisites on page 102), we see that these conditions are present in greater proportion for our species than any other known species. We may be one of the most reciprocally altruistic—if not the most reciprocally altruistic—species in the animal kingdom.

Note that this view is clearly consistent with conflict. In fact, as we have already suggested, the most psychoanalytically significant aspects of human psychological design have probably evolved to cope with the inevitable, profound conflict inherent in human relations. However, contrary to the classical perspective, we now see conflict within a larger social network of kin and reciprocal altruism. It is now widely believed that human intelligence may have evolved primarily for social uses—to negotiate one's way through kin relationships in which one's self-interest is related to the self-interest of others in a quite complicated manner, and to allow for the successful trading of altruistic acts (friendships, mutual parasite control [grooming], protective and/or aggressive alliances, economic [business] arrangements, etc.) without being cheated—as opposed to having arisen for the manipulation of the non-human environment.

APPENDIX D

The Confusion of Proximal and Distal Causes

As we have seen, evolutionary theory presents us with a psychobiological picture of the human organism that includes a primary relational orientation that is essential to the self-interests of this semisocial organism. This orientation ensures that each individual will strive to be included in the protective envelope of the resource-laden human community; even if this means sacrificing some driven pleasure-seeking desires.

This need for inclusion in what the self psychologists call the self-selfobject milieu is not reducible to tension-reducing drive gratifications, though pleasure seeking and pain avoidance surely play a role in forming and maintaining important aspects of human ties. The longing for inclusion in the self–selfobject milieu, which can contribute to instinctual repression through the need to be perceived as a reciprocal altruist, and the fear of losing such a valuation, can be phenomenologically observed as an irreducible sense of well-being and/or joy when grounded in a well-functioning self–selfobject milieu, and irreducible disintegration anxiety in the absence of such (Kriegman, 1990).

Freud's Confusion of Proximal and Distal Causes

This relational need is related to the "fear of loss of love" (Freud, 1926), and there is clearly a sound adaptive basis for such a fear in evolutionary theory. Freud (1930) was correct in seeing that what is feared is somehow connected to lost opportunities for drive gratifications: that what can be lost are physical, material things (including people), and that these essential material resources/objects are why one must compromise one's desires in order to participate in civilization. What Freud failed to see is that this was the *distal* cause for the fear of loss of love. It is a description of the selective pressure present throughout the *history* of our species: those who feared loss of love, and thus were willing to forego some of their more selfish pleasures, were seen as more trustworthy reciprocal altruists and obtained the powerful adaptive advantage of being more fully included in reciprocal exchanges. Those who did not develop this fear were less successful in leaving viable copies of their genetic material behind, for they tended to be excluded from reciprocal exchanges, for example, friendships, economic exchanges, alliances.

However, this is the distal cause or the selective pressure that shaped relational needs. It is not a description of the psychological mechanism that was shaped by that pressure. The *proximal* cause for a psychological event or a behavior is the shaped structure or mechanism operating in the present. The child does not necessarily have the anaclitically derived experience (either consciously or unconsciously) of, "Oh, no! Now the 'other one' doesn't love me and I will not be able to gratify my tension-based needs and thus I will die or be overwhelmed by instinctual anxiety." Rather the mechanism controlling the child's experience may well be the fear of being disconnected, unloved, and abandoned, even though the former description, of loss of opportunities for drive gratification, may be essentially more accurate in depicting the historical selective pressures (distal causes) that shaped the current individual's proximal need for a stable self–selfobject milieu and the disintegration anxiety that results from its absence or inadequacy.

Freud appears to have confused the *distal* and *proximal* causes. Often he accurately identified the distal cause and then "projected" it into the currently operating human psyche and referred to it as a proximal cause; thus, giving rise to teleological reasoning. By confusing these two levels of analysis, Freud tries to proximally reduce one into the other. Thus, in Freud's notion of anaclitically derived attachments to others, the human need for relatedness (an irreducible proximal mechanism) is seen as being reducible into a proximal need for the drive gratifications obtained through human relationships. The latter may be more accurately conceived of as a distal cause for the development of the proximal need for human relatedness (Kriegman, 1990).

This is not to imply that self-serving drive gratifications play a relatively unimportant role in human relationships. Obviously they do not. The evolutionary perspective simply suggests that powerful *social* needs are biologically ancient and thus have an enormous selective history shaping them. Therefore, they are extremely unlikely to be recent evolutionary developments that are reducible into a more primary *a*social drivenness.

Modern applications of evolutionary theory to psychoanalysis must avoid repeating Freud's conceptual error. For example, we believe that Nesse (1990) confuses distal and ultimate causes when he uses evolutionary theory to maintain a more classical position. He refers to altruism as a "defense" that is used to deceive others. Altruistic motives have no independent existence, and are seen as a transformation of more selfish motives. Nesse's argument is that those who were able to hide these selfish motives benefited by being perceived to be reciprocal altruists, and thus obtained the benefits of trading reciprocal acts. Thus short-term gain is exchanged for long-term gain, or for a reciprocation to come in the future. As in Freud's reasoning, Nesse has identified the most likely *ultimate* explanation for this type of altruism, that is, he has accurately identified the selective pressure that shaped the trait: those who forfeited short-term gains for greater long-term gains were probably more successful and thus altruism was shaped.

The misplacing of ultimate causes into the presently functioning organism, as if ultimate causes are currently operating as proximal causes, enables both Nesse and Freud to see the ultimate selfish, gene-promoting utility of social attachments, including compassion and altruism. However, this ultimate utility is the distal cause, not the proximal product. The proximal product is often a genuine wish to remain connected to another, to empathically share another's experience, and to compassionately aid others in need. Only by imputing the ultimate cause into the current day as a proximal cause (operating in the individual's unconscious) can the bias of the classical paradigm be retained.

APPENDIX E

Specific, Functional, Affective, Motivational Systems from an Evolutionary Perspective

Many theorists have translated drives into affects as primary shapers of human experience. We suggest that there is indeed a need for specific, functional, affective, motivational systems that are subsidiary parts of a larger functional whole. Such subsidiary parts function within the larger functional adaptive design of the psyche as elements in a larger architecture. When viewed from the evolutionary perspective, affect theory takes on meanings that may not otherwise be apparent. For example, let us consider shame as it appears in Sylvan Tomkins' (1987) well-elaborated affect theory that has been utilized by several psychoanalytic theorists.

For Tomkins (1987), shame is the innate affect triggered when excitement or interest is rapidly inhibited. Yet, such a conception does not focus on the *interpersonal function* of shame and the interpersonal contexts that may have been innately programmed to bring out the shame response. For example, the tendency to engage in sexual activity in private—to experience shame if discovered—is a nearly universal cultural phenomenon. It is not difficult to conceive of the possible adaptive advantages of such a tendency, for example, safety, or to avoid arousing competitors. Although we cannot be sure about the selective pressure that shaped this pattern of behavior, it seems likely that such a universal phenomenon is not an accidental occurrence. Thus, sexual shame may not result primarily from the experience of rapid inhibition of excitement, though this may be a part of many children's early experience of the adult response to their sexual interest and curiosity and would be likely to increase a sense of shame.

Also, consider stranger anxiety. This aversion to strangers occurs at an age when children first become effectively mobile (are less likely to be carried) and thus this anxiety insures that they will stay close to their parents when they are moving around other people. The greatest danger to children at this age is not from predators but from other people. As we now know, sexual abuse in childhood is fairly common and it may be as important to children that they not be alone with strangers as that they maintain their connection to their life-supporting parents. This experience of anxiety that fairly reliably appears within a certain period of development suggests that there are specific affects that are tuned into specific relational contexts without learned experience bringing them into existence.

There may be many such limited programs in the human motivational–behavioral repertoire. It is our view that they must be considered in the context of the larger, functional psychological architecture for their full meaning to appear. That is, that they should be considered subsidiary components in a larger functional, dynamic psychological architecture. As elements in such a larger system they do yield some clinical implications. Consider, for example, Mitchell's (1988) incisive argument against the "developmental tilt"—the tendency to interpret adult experiences and actions as a revival of childhood (as opposed to seeing needs and feelings that represent themes that exist in *both* childhood and later in life, changing only in form):

> What the analysand experiences as neediness is often more usefully understood . . . not as reflecting infantile fixations or developmentally arrested needs, but as a complex mixture of perfectly appropriate adult desires interfused with intense anxiety. These analysands often come from families where depending on other people for anything was regarded as weak or babyish (often as "bad"). . . . These analysands probably *were* thwarted as infants. . . . [W]hat they experience as neediness in their adult lives has . . . more to do with the ideas and feelings the individuals developed, through these early experiences, regarding desire in general. When such a person experiences any intense wish or longing . . . he or she tends to become flooded with anxiety. The desire is felt as weakening them, as making them vulnerable, demanding, bad. (p. 164)

Although we feel that his point regarding the developmental tilt is valid, consider the possibility that Mitchell has, in effect, come up with a developmental tilt of his own. He is attributing what may well be an inborn, universal tendency—the experience of intense desire and neediness as shameful in adulthood—to childhood experience. It is a commonplace observation that the developing child begins to alternate between self-reliance and neediness as his or her capacity to function independently develops. Children frequently experience a fierce, adaptive sense of independence and a great deal of shame and distress when they fail at a task and need to be rescued. This sense of personal failure and embarrassment may have little to do with having experienced shaming responses from others. A significant part of the analysand's reluctance to reveal weakness and intense neediness may be linked to a natural shame over weakness that, because of variability in the level at which specific innate programs operate in different individuals, may in certain analysands have little to do with lived experience.

> Let thy discontents be thy Secrets;—if the world knows them, 'twill despise *thee* and increase *them*. (Benjamin Franklin, *Poor Richard's Almanac*)

There may simply exist an innate knowledge that people tend to value vigorously healthy and capable others more than they value those who are overwhelmed by needy feelings and less able to function as effective reciprocal altruists. Thus, there may be significant innate shame at full analytic self-revelation experienced by some patients, such as those that Mitchell describes, that may be only partially related to lived (childhood) experience. Mitchell operates without any notion of innate propensities and meanings brought to interpersonal experience from within the individual (see Chapter Six), and so may be in danger of reintroducing a version of the very developmental tilt he criticizes.

APPENDIX F

Regressive Renegotiation in the Treatment of the "Regressed" Patient

In treating "primitive" patients, it is sometimes suggested that they are too regressed; that we are dealing with a failure of repression. However, from the evolutionary perspective, the notion of regressive renegotiation may hold even with such patients. Consider the case of Ms. B on page 170. She went through life never feeling "real" or that the terror of nonexistence was more than a hair's breadth away. While she would alternately fall into terrible despair and/or (often simultaneously) uncontrolled rages, these "regressive" experiences were mostly unavailable in between her "psychotic episodes." During these "healthier" times she felt engaged in a chronic, exhausting struggle against terror and disappearance. In addition to her therapist needing to respect her rage as a problematic part of her (see discussion on page 172), it became necessary to see the rage (and its converse, her hopeless despair) as the most *real* parts of her!

Eventually this led to understanding why she felt so unreal almost all of the time. In all of her interactions, she felt she had to pretend to be someone she wasn't. She felt a need to almost totally deny the existence of her true self. The despairing, hopeless infant, (which took the form of misdiagnosed "autism" in childhood and later was labelled as "schizophrenia") as well as the enraged, destructive infant (which was referred to as "psychotic rage" and "the need to destroy her treatment") were, in fact, the only vital representations of her true self that she ever experienced. In her words, "I can't win." This meant that if she tried to follow the dictates of others and stifle all her rage, (resulting in despairing, motionless inactivity) those around her actually became more enraged with her than when she exploded in rage—even though in the latter case they would struggle with her and place her in four-point restraints and seclusion rooms (where she spent *years* of her life).

If she tried to communicate her despair, no one wanted to hear it (in treatment it did often seem endless, hopeless, and inevitable that it would lead to suicide).

Even with such a "primitive" patient, it becomes clear that the therapeutic work entails creating an analytic relationship in which the repressed true self can emerge and be cautiously accepted as vital parts of her. With this patient, the repeated reworking, acceptance, and interpretation of the regressive reemergence of seemingly infinite despair and frighteningly destructive rage—both of which directly conflicted with the needs and interests of the caregiving surround that sustained her from the age of 3 onward—has led to the only sustained periods of hospital free functioning in almost 30 years.

References

Alexander, R. D. (1979). *Darwinism and Human Affairs.* Seattle: University of Washington Press.
Alexander, R. D. (1990). Epigenetic rules and Darwinian algorithms: The adaptive study of learning and development. *Ethology and Sociobiology, 11,* 241–303.
Axelrod, R., & Hamilton, W. D. (1981). The evolution of cooperation. *Science, 211,* 1390–1396.
Atwood, G., & Stolorow, R. (1984). *Structures of Subjectivity: Explorations in Psychoanalytic Phenomenology.* Hillsdale, NJ: Analytic Press.
Badcock, C. (1986). *The Problem of Altruism: Freudian-Darwinian Solutions.* Oxford: Basil Blackwell.
Badcock, C. (1988). *Essential Freud.* Oxford: Basil Blackwell.
Badcock, C. (1990). *Oedipus in Evolution.* London: Basil Blackwell.
Bakan, D. (1966). *The Duality of Human Existence: An Essay on Psychology and Religion.* Chicago: Rand-McNally.
Balint, M. (1965). *Primary Love and Psycho-analytic Technique.* New York: Liverwright.
Balint, M. (1968). *The Basic Fault.* London: Tavistock.
Barash, D. (1979). *The Whisperings Within.* London: Penguin.
Basch, M. F. (1984). The selfobject theory of motivation and the history of psychoanalysis. In A. Goldberg & P. E. Stepansky (Eds.), *Kohut's Legacy: Contributions to Self Psychology* (pp. 3–17). Hillsdale, NJ: Analytic Press.
Basch, M. F. (1988). *Understanding Psychotherapy: The Science Behind the Art.* NY: Basic Books.
Bateson, G. (1972). *Steps to an ecology of mind.* New York: Ballantine.
Beebe, B., & Lachmann, F. (1988). Mother-Infant Mutual Influence and Precursors of Psychic Structure. In A. Goldberg (Ed.), *Frontiers in Self Psychology* (pp. 3–25).
Bibring, E. (1943). The conception of the repetition compulsion. *Psychoanalytic Quarterly, 12,* 486–519.
Bickenton, D. (1990). *Language and Species.* Chicago: University of Chicago Press.
Bion, W. R. (1962a). *Learning From Experience.* New York: Basic Books.
Bion, W. R. (1962b). A theory of thinking. In *Second Thoughts.* New York: Jason Aronson (1967).

Bird, B. (1971). Notes on transference: Universal phenomenon and the hardest part of analysis. *Journal of the American Psychoanalytic Association, 20,* 267–301.

Blos, P. (1967). The second individuation process of adolescence. *Psychoanalytic Study of the Child, 22,* 162–186.

Blos, P. (1972). The epigenesis of the adult neurosis. *Psychoanalytic Study of the Child, 27,* 106–135.

Blos, P. (1979). *The Adolescent Passage.* New York: International Universities Press.

Blurton Jones, N. G., & da Costa, E. (1987). A suggested adaptive value of toddler night waking: Delaying the birth of the next sibling. *Ethology and Sociobiology, 8,* 135–142.

Bollas, C. (1989). *The Forces of Destiny: Psychoanalysis and Human Idiom.* London: Free Association Books.

Bower, B. (1991). Oedipus wrecked. *Science News, 140,* 248.

Bowlby, J. (1969). *Attachment and Loss, Vol. I. Attachment.* London: Hogarth Press.

Bowlby, J. (1973). *Attachment and Loss, Vol. II. Separation: Anxiety and Anger.* New York: Basic Books.

Bowlby, J. (1980). *Attachment and Loss, Vol. III. Loss: Sadness and Depression.* New York: Basic Books.

Brandchaft, B. (1983). The negativism of the negative therapeutic reaction and the psychology of the self. In A. Goldberg (Ed.), *The Future of Psychoanalysis,* New York: International Universities Press.

Brandchaft, B., & Stolorow, R. (1990). Varieties of therapeutic alliance. *Annual of Psychoanalysis, 18,* 99–114.

Brazelton, T. B. (1969). *Infants and Mothers: Differences in Development.* New York: Delacorte Press.

Brickhouse, T. C. (1990). *Aristotle* (audiotape). Nashville: Carmichael & Carmichael.

Bromberg, P. (1989). Interpersonal psychoanalysis and self psychology: A clinical comparison. In D. W. Dedrick & S. P. Dedrick (Eds.), *Self Psychology: Comparisons and Contrasts* (pp. 275–291).

Browne, M. W. (1992). Biologists tally generosity's rewards. *New York Times,* April 14, C1–C8.

Bruner, J. S., Jolly, A., & Sylva, K. (1976). *Play: Its Role in Development and Evolution.* New York: Penguin.

Cassirer, E. (1962). *An Essay on Man.* New Haven: Yale University Press.

Chomsky, N. (1972). *Language and Mind.* San Diego: Harcourt Brace.

Cooper, A. (1983). The place of self psychology in the history of depth psychology. In A. Goldberg (Ed.), *The Future of Psychoanalysis* (pp. 3–18). New York: International Universities Press.

Cooper, S. (1991). Recent contributions to the theory of defense mechanisms. *Journal of the American Psychoanalytic Association, 37*(4), 865–891.

Cosmides, L. (1989). The logic of social exchange: Has natural selection shaped how humans reason. *Cognition, 31,* 187–276.

Cosmides, L., & Tooby, J. (1987). From evolution to behavior: Psychology as the missing link. In J. Dupre (Ed.), *The Latest on the Best: Essays on Evolution and Optimality*. Cambridge, UK: Cambridge University Press.

Daly, M. (1989). Parent-offspring conflict and violence in evolutionary perspective. In R. W. Bell & N. J. Bell (Eds.), *Sociobiology and the Social Sciences*. Texas Tech University Press.

Daly, M. & Wilson, M. (1988). *Homicide*. Hawthorne, NY: Aldine de Gruyter.

Daly, M. & Wilson, M. (1990). Is parent-offspring conflict sex linked? Freudian and Darwinian models. *Journal of Personality, 58*(1), 163–187.

Darwin, C. (1858). *The Origin of Species*. London: Murray.

Darwin, C. (1871). *The Descent of Man and Selection in Relation to Sex*. New York: Random House.

Dawkins, R. (1976). *The Selfish Gene*. New York: Oxford University Press.

Dawkins, R. (1986). *The Blind Watchmaker*. New York: W. W. Norton.

DeVore, B. I. (1971). The evolution of human society. In J. F. Eisenberg & W. S. Dillon (Eds.), *Man and Beast: Comparative Social Behavior* (pp. 297–311). Washington, DC: Smithsonian Institution Press.

de Waal, F. B. M. (1982). *Chimpanzee Politics: Power and Sex Among Apes*. New York: Harper & Row.

de Waal, F. (1989). *Peacemaking among the Primates*. Cambridge, MA: Harvard University Press.

Durkheim, E. (1897). *Suicide: A Study in Sociology* (J. Spaulding & G. Simpson, Trans.). NY: Free Press (1966).

Eagle, M. N. (1984). *Recent Developments in Psychoanalysis*. New York: McGraw Hill.

Ehrenberg, D. B. (1975). The quest for intimate relatedness. *Contemporary Psychoanalysis, 11*, 320–331.

Eigen, M. (1991). Boa and Flowers. *Psychoanalytic Dialogues, 1*(1), 106–118.

Edelman, G. M. (1987). *Neural Darwinism*. New York: Basic Books.

Emde, R. N. (1980). Levels of meaning for infant emotions: A bisocial view. In W. A. Collins (Ed.), *Development of Cognition, Affects, & Social Relations*. Hillsdale, NJ: Ehrlbaum.

Erikson, E. (1956). The problem of ego identity. *Journal of American Psychoanalytic Association, 4*, 56–121.

Erikson, E. (1963). *Childhood and Society* (2nd ed.). New York: W. W. Norton.

Erikson, E. (1964). *Insight and Responsibility*. New York: W. W. Norton.

Erikson, E. (1968). *Identity Youth and Crisis*. New York: Norton.

Fairbairn, W. R. D. (1952). *An Object-Relations Theory of the Personality*. New York: Basic Books.

Fajardo, B. (1988). Constitution in infancy: Implications for early development and psychoanalysis. In A. Goldberg (Ed.), *Learning from Kohut: Progress in Self Psychology* (Vol. 4, pp. 91–100). Hillsdale, NJ: Analytic Press.

Farber, S. (1981). *Identical Twins Reared Apart: A Reanalysis*. New York: Basic Books.

Fenichel, O. (1941). *Problems of Psychoanalytic Technique*. New York: Psychoanalytic Quarterly, Inc.

Fenichel, O. (1945). *The Psychoanalytic Theory of Neurosis*. New York: W. W. Norton.

Fodor, G. (1983). *The Modularity of the Mind*. Cambridge, MA: MIT Press.
Fox, R. (1989). *The Search for Society: Quest for a Biosocial Morality*. New Brunswick, NJ: Rutgers University Press.
Freud, A. (1936). *The Ego and the Mechanisms of Defense*. New York: International Universities Press.
Freud, A. (1958). Adolescence. *The Psychoanalytic Study of the Child*, 13, 255–278.
Freud, S. (1878). Uber Spinalganlien und Ruckenmark des Petromyzon. *Sitzungsberichte der kaiserlichen Akademie der Wissenschaften*. Mathematisch-Naturwissenschaftliche Classe, 78, III. Abtheilung: 81–167.
Freud, S. (1895). Project for a scientific psychology. *Standard Edition*, 1, 283–398.
Freud, S. (1900). Interpretation of dreams. *Standard Edition*. 4 & 5.
Freud, S. (1905). Three Essays on the Theory of Sexuality. *Standard Edition*, 7, 125–243.
Freud, S. (1912). Dynamics of transference. *Standard Edition*, 12, 97–108.
Freud, S. (1912–1913). Totem and taboo. *Standard Edition*, 13, 1–161.
Freud, S. (1913). On beginning the treatment. *Standard Edition*, 12, 121–144.
Freud, S. (1914a). On narcissism: an introduction. *Standard Edition*, 14, 67–102.
Freud, S. (1914b). Remembering, repeating and working through. *Standard Edition*, 12, 145–156.
Freud, S. (1915a). Instincts and their vicissitudes. *Standard Edition*, 14, 117–140.
Freud, S. (1915b). Repression. *Standard Edition*, 14, 141–158.
Freud, S. (1915c). On transference love. *Standard Edition*, 12, 157–171.
Freud, S. (1915d). An overview of the transference neurosis. In I. Grubrich-Simtis (Ed.), *A Phylogenetic Fantasy: Overview of the Transference Neuroses*. Cambridge, MA: Harvard University Press (1987).
Freud, S. (1915e). Thoughts on war and death. *Standard Edition*, 14, 275–300.
Freud, S. (1916). Introductory lectures on psycho-analysis. *Standard Edition*, 15 & 16.
Freud, S. (1917). A difficulty in the path of psycho-analysis. *Standard Edition*, 17, 136–144.
Freud, S. (1918). From the history of an infantile neurosis. *Standard Edition*, 17, 3–122.
Freud, S. (1920). Beyond the pleasure principle. *Standard Edition*, 19, 3–64.
Freud, S. (1921). Group psychology and the analysis of the ego. *Standard Edition*, 18, 65–143.
Freud, S. (1923). The ego and the id. *Standard Edition*, 19, 1–66.
Freud, S. (1925). An autobiographical study. *Standard Edition*, 20, 7–74.
Freud, S. (1926). Inhibitions, symptoms and anxiety. *Standard Edition*, 20, 75–175.
Freud, S. (1930). Civilization and its Discontents. *Standard Edition*, 21, 59–145.
Freud, S. (1933). New introductory lectures. *Standard Edition*, 22, 5–182.
Freud, S. (1937). Analysis terminable and interminable. *Standard Edition*, 23, 216–254.
Freud, S. (1939). Moses and monotheism. *Standard Edition*, 23, 3–137.
Freud, S. (1940). An outline of psychoanalysis. *Standard Edition*, 23, 144–207.
Fromm, E. (1941). *Escape from Freedom*. New York: Avon.
Fromm-Reichmann, F. (1950). *Principles of Psychotherapy*. Chicago: University of Chicago Press.

Fuss, D. (1989). *Essentially Speaking*. New York: Routledge.
Gardner, H. (1983). *Frames of Mind*. New York: Basic Books.
Gedo, J., & Goldberg, A. (1973). *Models of The Mind*. Chicago: University of Chicago Press.
Gill, M. M. (1984). Psychoanalysis and psychotherapy: A revision. *International Journal of Psychoanalysis, 11*, 162–179.
Gill, M. M., & Hoffman, I. Z. (1982). *Analysis of Transference II: Studies of Nine Audio-recorded Psychoanalytic Sessions*. New York: International Universities Press.
Glantz, K. & Pearce, J. (1989). *Exiles from Eden*. New York: W.W. Norton.
Glover, E. (1955). *The Technique of Psychoanalysis*. New York: International Universities Press.
Glover, E. (1956). *On the Early Development of Mind*. New York: International Universities Press.
Goldberg, A. (1986). Reply to discussion of P. Bromberg of "The wishy-washy personality." *Contemporary Psychoanalysis, 22*, 387–388.
Goldberg, A. (1988). *A Fresh Look at Psychoanalysis*. Hillsdale, NJ: Analytic Press.
Goldberg, A. (1990). *The Prisonhouse of Psychoanalysis*. Hillsdale, NJ: The Analytic Press.
Goleman, D. (1985). *Vital Lies, Simple Truths: The Psychology of Self Deception*. New York: Simon & Schuster.
Gould, S. J. (1977). Biological potentiality vs. biological determinism. In S. J. Gould (Ed.), *Ever Since Darwin*. (pp. 251–259) New York: W. W. Norton.
Gould, S. J. (1983). Nature holds no moral message. *Bostonia, 57*, 3, 36–41.
Gould, S. J. (1987). Freud's phylogenetic fantasy. *Natural History, 12/87*, 10–19.
Gould, S. J., & Lewontin, R. C. (1979). The spandrels of San Marco and the panglossian paradigm. A critique of the adaptationist programme. *Proceedings of the Royal Society of London, B 205*, 581–598. Reprinted in E. Sober (Ed.). (1984). *Conceptual Issues in Evolutionary Biology*. Cambridge, MA: MIT Press.
Green, J., Bax, M., & Tsitsikas, H. (1989). A longitudinal study of the first six months of life. *American Journal of Orthopsychiatry, 59*, 82–93.
Greenberg, J. R. (1986). Theoretical models and the analyst's neutrality. *Contemporary Psychoanalysis, 22*, 87–106.
Greenberg, J. R., & Mitchell, S. A. (1983). *Object Relations in Psychoanalytic Theory*. Cambridge, MA: Harvard University Press.
Greenson, R. (1965). The working alliance and the transference. In *Explorations in Psychoanalysis*. New York: International Universities Press.
Groddeck, G. (1949). *The Book of the Id*. New York: Vantage Books. (Original German edition published in 1923)
Grotstein, J. (1985). A proposed revision for the psychoanalytic concept of the death instinct. *Yearbook of Psychoanalysis and Psychotherapy* (Vol. 1, pp. 299–326). Hillsdale, NY: New Concept Press.
Grubrich-Simitis, I. (1987). Metapsychology and metabiology. In I. Grubrich-Simitis (Ed.), *A Phylogenetic Fantasy: Overview of The Transference Neuroses*. Cambridge, MA: Harvard University Press.
Guntrip, H. (1971). *Psychoanalytic Theory, Therapy, and The Self*. New York: Basic Books.
Guthrie, W. K. C. (1981). *A History of Greek Philosophy* (Vol. 6). Cambridge: Cambridge University Press.

Haldane, J. B. S. (1932). *The Causes of Evolution*. New York: Longmans, Green.
Haldane, J. B. S. (1955). Population genetics. *New Biology, 18,* 34–51.
Hamilton, W. D. (1964). The genetical evolution of social behavior. *Journal of Theoretical Biology, 7,* 1–52.
Hamilton, W. D. (1969). Selection of selfish and altruistic behavior in some extreme models. In J. F. Eisenberg & W. S. Dillon (Eds.), *Man and Beast: Comparative Social Behavior,* (pp. 59–91). Washington, DC: Smithsonian Press.
Hartmann, H. (1958). *Ego Psychology and The Problem of Adaptation*. New York: International Universities Press. (Original work published 1939)
Hartmann, H., Kris, E., & Lowenstein, R. (1951). Some psychoanalytic comments on "culture and personality." *Papers on Psychoanalytic Psychology. Psychological Issues* (Monograph 14). New York: International Universities Press, 1964.
Heidegger M. (1962). *Being and Time*. New York: Harper & Rowe. (Original work published 1927)
Hoffer, W. (1956). Transference and transference neurosis. *International Journal of Psychoanalysis, 37,* 377–379.
Hoffman, I. (1983). The patient as interpreter of the analyst's experience. *Contemporary Psychoanalysis, 19*(3) 389–442.
Hoffman, I. (1987). The value of uncertainty in psychoanalytic practice. *Contemporary Psychoanalysis, 23,* 205–215.
Hoffman, I. (1991). Toward a social constructivist view of the psychoanalytic situation. *Psychoanalytic Dialogues, 1*(1) 74–105.
Holmes, K. R. (1983). Freud, evolution, and the tragedy of man. *Journal of the American Psychoanalytic Association, 31,* 187–210.
Holt, R. (1976). Drive or wish? A reconsideration of the psychoanalytic theory of motivation. In M. Gill & P. Holzman (Eds.), *Psychology vs. metapsychology, Essays in memory of George S. Klein* (Psychological Issues, Vol. 9, No. 4, Monograph 36). New York: International Universities Press.
Holt, R. (1989). *Freud Reappraised*. New York: Guilford.
Jacobson, E. (1964). *The Self and the Object World*. New York: International Universities Press.
Jones, E. (1953 to 1957). *The Life and Work of Sigmund Freud* (Vols. 1, 2, 3). New York: Basic Books.
Kaplan, H. I., & Sadock, B. J. (1988). *Synopsis of Psychiatry*. Baltimore: Williams and Wilkins.
Katan, A. (1937). The role of displacement in agoraphobia. *International Journal of Psychoanalysis, 32* (1951), 41–50.
Kernberg, O. (1976). *Object Relations Theory and Clinical Psychoanalysis*. New York: Jason Aronson.
Kernberg, O. (1980). *Internal World and External Reality*. New York: Jason Aronson.
King, D. (1945). The meaning of normal. *Yale Journal of Biological Medicine, 17,* 493–501.
Kitcher, P. (1985). *Vaulting Ambition: Sociobiology and the Quest for Human Nature*. Cambridge, MA: MIT Press.

Kitcher, P. (1987). Confessions of a curmudgeon. *Behavioral and Brain Sciences, 10*, 89–99.

Klein, G. S. (1976). *Psychoanalytic Theory: An Exploration of Essentials.* New York: International Universities Press.

Klein, M. (1946). Notes on some schizoid mechanisms. In *Envy and Gratitude and Other Works.* New York: Delacorte (1975).

Kohut, H. (1959). Introspection, empathy, and psychoanalysis. *Journal of the American Psychoanalytic Association, 7*, 459–483.

Kohut, H. (1971). *The Analysis of The Self.* New York: International Universities Press.

Kohut, H. (1972). Thoughts on narcissism and narcissistic rage. *The Psychoanalytic Study of the Child, 27*, 360–400.

Kohut, H. (1977). *The Restoration of the Self.* New York: International Universities Press.

Kohut, H. (1978). Narcissism as a resistance and as a driving force in psychoanalysis. In Paul Ornstein (Ed.), (pp. 517–561). *The Search for the Self.* New York: International Universities Press.

Kohut, H. (1980). Reflections. In A. Goldberg (Ed.), *Advances in Self Psychology,* New York: International Universities Press.

Kohut, H. (1982). Introspection, empathy, and the semi-circle of mental health. *International Journal of Psychoanalysis, 63*, 395–407.

Kohut, H. (1983). Selected problems of self psychological theory. In J. D. Lichtenberg & S. Kaplan (Eds.), *Reflections on Self Psychology* (pp. 387–416). Hillsdale, NJ: Analytic Press.

Kohut, H. (1984). *How Does Analysis Cure?* Chicago: University of Chicago Press.

Kohut, H., & Wolf, E. S. (1978). The disorders of the self and their treatment: An outline. *The International Journal of Psycho-analysis, 59*(4), 413–425.

Kohut, T. (1985, October 5). Discussion of an early version of Kriegman (1988). Paper presented at the Eighth Annual Conference on the Psychology of the Self, New York.

Konner, M. (1982). *The Tangled Wing: Biological Constraints on The Human Spirit.* New York: Harper & Row.

Kriegman, D. (1980). A psycho-social study of religious cults from the perspective of self psychology. Unpublished doctoral dissertation.

Kriegman, D. (1988). Self psychology from the perspective of evolutionary biology: Toward a biological foundation for self psychology. In A. Goldberg (Ed.), *Progress in Self Psychology* (Vol. 3, pp. 253–274). Hillsdale, NJ: Analytic Press.

Kriegman, D. (1990). Compassion and altruism in psychoanalytic theory: An evolutionary analysis of self psychology. *Journal of the American Academy of Psychoanalysis, 18*(2), 342–367.

Kriegman, D. (In preparation). Man and Woman: An evolutionary psychoanalytic approach to couples therapy.

Kriegman, D., & Knight, C. (1988). Social evolution, psychoanalysis, and human nature. *Social Policy, 19*(2) 49–55.

Kriegman, D., & Slavin, M. O. (1989). The myth of the repetition compulsion and the negative therapeutic reaction: An evolutionary biological analysis. In A. Goldberg (Ed.), *Progress in Self Psychology* (Vol. 5, pp. 209–253). Hillsdale, NJ: Analytic Press.

Kriegman, D., & Slavin, M. O. (1990). On the resistance to self psychology: clues from evolutionary biology. *Progress in Self Psychology* (Vol. 6, pp. 217–250). Hillsdale, NJ: The Analytic Press.

Kriegman, D., & Solomon, L. (1985a). Cult groups and the narcissistic personality: the offer to heal defects in the self. *International Journal of Group Psychotherapy, 35,* 2, 239–261.

Kriegman, D., & Solomon, L. (1985b). Psychotherapy and the "new religions": are they the same? *Cultic Studies Journal, 2,* 1, 2–16.

Kris, E. (1952). *Psychoanalytic Explorations in Art.* New York: International Universities Press.

Kris, E. (1956). On the vicissitudes of insight in psychoanalysis. *International Journal of Psychoanalysis, 37,* 1–11.

Kuhn, T. (1962). *The Structure of Scientific Revolutions* (2nd ed.). Chicago: University of Chicago Press.

LaBarre, W. (1954). *The Human Animal.* Chicago: University of Chicago Press.

Lacan, J. (1977). *Écrits.* (A. Sheridan, Trans). New York: W. W. Norton.

Langer, S. (1942). *Philosophy in a New Key.* New York: Penguin Books.

Langs, R. (1978). The adaptational-interactional dimension of countertransference. *Contemporary Psychoanalysis, 14,* 502–533.

Lapierre, D. (1985). *The City of Joy.* New York: Warner.

Lawner, P. (1985). Sincerity, authenticity, and the waning of the Oedipus complex. Paper presented to the Manhattan Institute for Psychoanalysis.

Locke, J. (1690). *Essay Concerning Human Understanding.*

Leak, G. K., & Christopher, S. B. (1982). Freudian psychoanalysis and sociobiology. *American Psychologist, 37,* 313–332.

Levenson, E. (1983). *The Ambiguity of Change.* New York: Basic Books.

Levi, P. (1986). *Moments of Reprieve.* New York: Simon & Schuster.

Lichtenberg, J. (1983). *Psychoanalysis and Infant Research.* Hillsdale, NJ: Analytic Press.

Lichtenberg, J. (1989). *Psychoanalysis and Motivation.* Hillsdale, NJ: The Analytic Press.

Lidz, T. (1963). *The Family and Human Adaptation.* New York: International Universities Press.

Lloyd, A. T. (1990). Implications of an evolutionary metapsychology for clinical psychoanalysis. *The Journal of the American Academy of Psychoanalysis, 18*(2), 286–306.

Loewald, H. (1965). Some considerations on repetition and repetition compulsion. In *Papers on Psychoanalysis.* New Haven: Yale University Press (1980).

Loewald, H. (1969). Freud's conception of the negative therapeutic reaction, with comments on instinct theory. In *Papers on Psychoanalysis.* New Haven: Yale University Press (1980).

Loewald, H. W. (1980). *Papers on Psychoanalysis.* New Haven: Yale University Press.

Lorenz, K. (1955). *Man Meets Dog.* London: Methuen.
McAlister, M. K. & Roitberg, B. D. (1987). Adaptive suicidal behavior in pea aphids. *Nature, 328,* 797–799.
Mahler, M., Pine, F., & Bergman, A. (1975). *The Psychological Birth of the Human Infant: Symbiosis and Individuation.* New York: Basic Books.
Masson, J. M. (1984). *The Assault on Truth.* New York: Farrar, Strauss, & Giroux.
May, R. (1978). Oedipal grief. *International Journal of Psychoanalytic Psychotherapy, 7,* 385–404.
Mayr, E. (1959). Agassiz, Darwin, and evolution. *Harvard Library Bulletin, 13,* 163–194.
Mayr, E. (1974). Behavior programs and evolutionary strategies. *American Scientist, 62,* 650–659.
Mayr, E. (1983a). How to carry out the adaptationist program. *American Naturalist, 121,* 324–334.
Mayr, E. (1983b). *The Growth of Biological Thought.* Cambridge, MA: Belknap Press, Harvard.
Mayr, E. (1988). *Toward a New Philosophy of Biology: Observations of an Evolutionist.* Cambridge, MA: Harvard University Press.
Meade, G. H. (1934). *Mind, Self, and Society.* Chicago: University of Chicago Press.
Miller, A. (1981). *The Drama of the Gifted Child.* New York: Basic Books.
Minsky, M. (1985). *The Society of Mind.* New York: Simon and Schuster.
Mitchell, R. (1985). *Deception: Perspectives on Human and Non-Human Deceit.* New York: State University of New York Press.
Mitchell, S. (1984). Object relations theories and the developmental tilt. *Contemporary Psychoanalysis,* 20(4), 473–499.
Mitchell, S. (1988). *Relational Concepts in Psychoanalysis: An Integration.* Cambridge, MA: Harvard University Press.
Mitchell, S. (1991). Contemporary perspectives on self: Toward an integration. *Psychoanalytic Dialogues,* 1(2), 121–147.
Modell, A. H. (1976). The "holding environment" and the therapeutic action of psychoanalysis. *Journal of the American Psychoanalytic Association, 24,* 285–308.
Modell, A. (1984). *Psychoanalysis in a New Context.* New York: International Universities Press.
Modell, A. (1989). Discussion of Slavin and Kriegman, "Beyond the classical-relational dialectic in psychoanalysis: a new paradigm from contemporary evolutionary biology." Paper presented at the Spring Meeting of the Division of Psychoanalysis (APA), Boston.
Modell, A. (1990). *Other Times, Other Realities: Toward a Theory of Psychoanalytic Treatment.* Cambridge, MA: Harvard University Press.
Morrison, A. (1989). *Shame: The Underside of Narcissism.* Hillsdale, NJ: The Analytic Press.
Munder-Ross, J. (1986). The darker side of fatherhood: Clinical and developmental ramifications of the Laius motif. *International Journal of Psychoanalytic Psychotherapy, 11,* 117–144.

Myerson, P. (1981). The nature of the transactions that enhance the progressive phases of psychoanalysis. *International Journal of Psychoanalysis, 62*, 91–103.

Nesse, R. (1990). The evolutionary functions of repression and the ego defenses. *Journal of the American Academy of Psychoanalysis, 18*(2), 260–285.

Neubauer, P. B., & Neubauer, A. (1990). *Nature's Thumbprint: The New Genetics of Personality.* Reading, MA: Addison-Wesley.

Ogden, T. (1990). *The Matrix of the Mind* (rev. ed.). New York: Jason Aronson.

Ornstein, A. (1974). The dread to repeat and the new beginning: A contribution to the psychoanalysis of the narcissistic personality disorders. *The Annual of Psychoanalysis* (Vol. 2, pp. 231–248). New York: International Universities Press.

Ornstein, A. (1984). Psychoanalytic psychotherapy: a contemporary perspective. In P. E. Stepansky & A. Goldberg (Eds.), *Kohut's Legacy: Contributions to Self Psychology.* Hillsdale, NJ: Analytic Press.

Ornstein, P. (1979). Remarks on the central position of empathy in psychoanalysis. *Bulletin of the Association for Psychoanalytic Medicine, 18*, 95–108.

Ornstein, P. (1991). The clinical impact of the psychotherapist's view of human nature. Paper presented in the "Distinguished Psychiatrist Lecture Series" at the Annual Meeting of the American Psychiatric Association, New Orleans, May 14.

Ornstein, P., & Ornstein, A. (1980). Formulating interpretations in clinical psychoanalysis. *International Journal of Psychoanalysis, 61*, 203–211.

Parisi, T. (1987). Why Freud failed. *American Psychologist, 42*, 235–245.

Parsons, T. (1954). *Essays in Sociological Theory.* New York: Free Press of Glencoe.

Parsons, T., Bales, R. F., Olds, J., Zelditch, M., & Slater, P. (1955). *Family, Socialization, and Interaction Process.* New York: The Free Press of Glencoe.

Peterfreund, E. (1972). *Information Systems and Psychoanalysis, an Evolutionary Biological Approach to Psychoanalytic Theory* (Psychological Issues, Vol. 7, Nos. 1/2, Monograph 25/26). New York: International Universities Press.

Peterfreund, E. (1983). *The Process of Psychoanalytic Therapy.* Hillsdale, NJ: Analytic Press.

Phillips, A. (1988). *Winnicott.* Cambridge, MA: Harvard University Press.

Pine, F. (1985). *Developmental Theory and Clinical Process.* New Haven: Yale University Press.

Pizer, S. (1992). The negotiation of paradox in analytic process. *Psychoanalytic Dialogues, 2*(2), 215–240.

Racker, H. (1968). *Transference and Countertransference.* New York: International Universities Press.

Rappaport, D. (1960). *The Structure of Psychoanalytic Theory* (Psychological Issues, Vol. II, No. 2, Monograph 6). New York: International Universities Press.

Reiser, M. (1984). *Mind, Brain, Body.* New York: Basic Books.

Ritvo, L. (1990). *Darwin's Influence on Freud.* New Haven: Yale University Press.

Ruse, M. (1983). *Taking Darwin Seriously.* London: Basil Blackwell.

Russel, P. (unpublished). The crises of emotional growth.

Russel, P. (1990). Discussion of paper "Countertransference disclosures and the therapeutic action of psychoanalysis" by W. Burke and M. Tansey. Massachusetts Association for Psychoanalytic Psychology scientific meeting, May 2.
Samuels, A. (1990). Original morality in a depressed culture. *Psychoanalysis and Contemporary Thought, 13*, 23–51.
Sandler, J. (1960). The background of safety. *International Journal of Psychoanalysis, 50*, 79–90.
Sandler, J. (1962). The concept of the representational world. *Psychoanalytic Study of The Child, 17*, 128–145.
Sandler, J. (1981). Unconscious wishes and human relationships. *Contemporary Psychoanalysis, 17*, 180–196.
Sandler, J. (1985). *The Analysis of Defense. The Ego and Mechanisms of Defense Revisited* (with Anna Freud). New York: International Universities Press.
Sartre, J. P. (1947). *No Exit*. New York: Knopf.
Sartre, J. P. (1966). *Being and Nothingness*. New York: Washington Square Press. (Original work published 1943)
Schafer, R. (1974). Problems in Freud's psychology of women. *Journal of the American Psychoanalytic Association, 22*(3), 459–485.
Schafer, R. (1976). *A New Language for Psychoanalysis*. New Haven: Yale University Press.
Schafer, R. (1983). *The Analytic Attitude*. New York: Basic Books.
Schmideberg, M. (1970). Psychotherapy with failures of psychoanalysis. *British Journal of Psychiatry, 116*, 195–200.
Schwaber, E. (1979). On the 'self' within the matrix of analytic theory—some clinical reflections and reconsiderations. *International Journal of Psychoanalysis, 60*, 467–479.
Searles, H. F. (1975). The patient as therapist to his analyst. In P. Giovacchini (Ed.), *Tactics and Techniques in Psychoanalytic Theory. Volume II: Countertransference* (pp. 95–151). New York: Jason Aronson.
Segal, H. (1964). *Introduction To The Work Of Melanie Klein*. New York: Basic Books.
Shane, E. (1988). The clinical value of considering constitutional factors. In A. Goldberg (Ed.), *Learning from Kohut: Progress in Self Psychology* (Vol. 4, pp. 104–112). Hillsdale, NJ: Analytic Press.
Shane, M., & Shane, E. (1988). Pathways to integration: adding to the self psychology model. In A. Goldberg (Ed.), *Learning from Kohut: Progress in Self Psychology* (Vol. 4, pp. 71–78). Hillsdale, NJ: Analytic Press.
Shapiro, R. L. (1969). Adolescent ego autonomy and the family. In G. Caplan & S. Lebovki (Eds.), *Adolescence: Psychosocial Perspectives*, New York: Basic Books.
Simon, B. (1991). Is the Oedipus complex still the cornerstone of psychoanalysis? Three obstacles to answering the question. *Journal of the American Psychoanalytic Association, 39*(3), 641–668.
Slavin, J. (1990). On making rules: Towards a reformulation of the dynamics of transference on psychoanalytic treatment. Unpublished manuscript.

Slavin, J. (In press). Do psychoanalysts make good therapists? *Contemporary Psychoanalysis.*

Slavin, M. O. (1974). An evolutionary perspective on the mechanism of repression and the function of guilt. Presented to the Graduate Seminar on Social Behavior, Department of Biology, Harvard University, Cambridge, MA.

Slavin, M. O. (1985). The origins of psychic conflict and the adaptive function of repression: An evolutionary biological view. *Psychoanalysis and Contemporary Thought, 8,* 407–440.

Slavin, M. O. (1985a). A developmental disorder of late adolescence. *International Journal of Psychoanalytic Psychotherapy, 11,* 219–233.

Slavin, M. O. (1986). The quest for a neutral metapsychology: Drive vs. relational theories from the perspective of evolutionary biology. Presented to the Mid-Winter Meeting, Division of Psychoanalysis of the American Psychological Association, Ixtapa, Mexico.

Slavin, M. O. (1988). Opening Ferenczi's trunk: a reexamination of Freud's evolutionary thinking on the occasion of his newly discovered "phylogenetic phantasy." Presented to the Massachusetts Association of Psychoanalytic Psychology, May.

Slavin, M. O. (1990). The biology of parent-offspring conflict and the dual meaning of repression in psychoanalysis. *Journal American Academy of Psychoanalysis, 18*(2), 307–341.

Slavin, M. O. (In preparation). Adolescence and The Problem of Human Adaptation: The Work of Anna Freud, Blos, and Erikson from the Perspective of Contemporary Evolutionary Biology.

Slavin, M. O., & Kriegman, D. (1988). Freud, biology, and sociobiology. *American Psychologist, 43,* 658–661.

Slavin, M. O., & Kriegman, D. (1990). Toward a new paradigm for psychoanalysis: An evolutionary biological perspective on the classical-relational dialectic. *Psychoanalytic Psychology, 7*(Suppl.), 5–31.

Slavin, M. O., & Kriegman, D. (1992). Psychoanalysis as a Darwinian depth psychology: Evolutionary biology and the classical-relational dialectic in psychoanalytic theory. In J. Barron, M. Eagle, & D. Wolitzky (Eds.), *Psychoanalysis and Psychology,* Washington, DC: American Psychological Association.

Slavin, M., & Slavin, J. (1976). Two patterns of adaptation in late adolescent borderline personalities. *Psychiatry, 39,* 41–50.

Solms, M., & Saling, M. (1986). On psychoanalysis and neuroscience: Freud's attitude to the localist tradition. *International Journal of Psycho-Analysis, 67,* 397–416.

Spence, D. (1982). *Narrative Truth and Historical Truth: Meaning and Interpretation in Psychoanalysis.* New York: W. W. Norton.

Spencer, H. (1864). *The Principles of Biology.* London: Williams & Norgate.

Stechler, G., & Halton, A. (1987). The emergence of assertion and aggression during infancy: A psychoanalytic systems approach. *Journal of the American Psychoanalytic Association, 35*(4), 821–839.

Stern, D. N. (1977). *The First Relationship: Infants and Mothers.* Cambridge, MA: Harvard University Press.

Stern, D. N. (1983). The early development of schemas of self, other, and "self with other." In J. D. Lichtenberg & S. Kaplan (Eds.), *Reflections on Self Psychology* (pp. 49–84). Hillsdale, NJ: Analytic Press.

Stern, D. (1985). *The Interpersonal World of the Infant*. New York: Basic Books.

Stevens, W. K. (1990). New eye on nature: The real constant is eternal turmoil. *New York Times*, July 31, C1–C2.

Stolorow, R. (1983). Self psychology—A structural psychology. In J. D. Lichtenberg & S. Kaplan (Eds.), *Reflections on Self Psychology*. Hillsdale, NJ: Analytic Press.

Stolorow, R. (1985). Toward a pure psychology of inner conflict. In A. Goldberg (Ed.), *Progress in Self Psychology* (Vol. 1, pp. 193–201). New York: Guilford Press.

Stolorow, R. (1986). Beyond dogma in psychoanalysis. In A. Goldberg (Ed.), *Progress in Self Psychology* (Vol. 2, pp. 41–49). New York: Guilford Press.

Stolorow, R. & Atwood, G. (1989). The unconscious and unconscious fantasy: An intersubjective developmental perspective. *Psychoanalytic Inquiry, 9*(3), 364–374.

Stolorow, R., Brandchaft, B., & Atwood, G. (1987). *Psychoanalytic Treatment: An Intersubjective Approach*. Hillsdale, NJ: Analytic Press.

Stolorow, R., & Lachmann, F. (1980). *Psychoanalysis of Development Arrests: Theory and Treatment*. New York: International Universities Press.

Stolorow, R., & Lachmann, F. (1984). Transference: The future of an illusion. *The Annual of Psychoanalysis Vol. XII/XIII*. Madison, CT: International Universities Press.

Strachey, J. (1934). The nature of the therapeutic action of psychoanalysis. *International Journal of Psychoanalysis, 15*, 117–126.

Strachey, J. (1957). Editor's footnote to "Instincts and their Vicissitudes." *Standard Edition, 14*, 129.

Sullivan, H. S. (1953). *The Interpersonal Theory of Psychiatry*. New York: W. W. Norton.

Sulloway, F. (1979). *Freud, Biologist of the Mind*. New York: Basic Books.

Symons, D. (1989). A critique of Darwinian anthropology. *Ethology and Sociobiology, 10*(1–3), 131–144.

Szasz, T. (1963). The concept of transference. *International Journal of Psychoanalysis, 44*, 432–443.

Teicholz, J. (1988). Broadening the meaning of empathy for work with primitive disorders. Presented at Harvard Medical School Conference on Narcissism, Boston, MA, November 4.

Teicholz, J. (1989). Discussion of Slavin and Kriegman's paper, Beyond the Classical-Relational Dialectic in Psychoanalysis. Symposium, "Drive versus Relational Theories, Compatible or Incompatible?" Division 39 (APA), Spring Meeting, Boston, MA.

Tellegen, A., Lykken, D. T., Bouchard, T. J., Wilcox, K. J., Segal, N. L., & Rich, S. (1988). Personality similarity in twins reared apart and together. *Journal of Personality and Social Psychology, 54*, 1031–1039.

Tinbergen, N. (1951). *The Study of Instinct*. London: Oxford University Press.

Tomkins, S. (1962). *Affect, Imagery, Consciousness* (Vol. 1). New York: Springer.

Tomkins, S. (1963). *Affect, Imagery, Consciousness* (Vol. 2). New York: Springer.
Tomkins, S. (1987). Shame. In D. Nathanson (Ed.), *The Many Faces of Shame*. (pp. 133–161). New York: Guilford Press.
Tooby, J., & Cosmides, L. (1989). Evolutionary psychology and the generation of culture, Part I: Theoretical considerations. *Ethology and Sociobiology, 10*(1–3), 29–51.
Tooby, J., & Cosmides, L. (1990). On the universality of human nature and the uniqueness of the individual: The role of genetics and adaptation. *Journal of Personality, 58,* 17–67.
Tower, L. (1956). Countertransference. *Journal of the American Psychoanalytic Association, 4,* 224–255.
Trivers, R. L. (1971). The evolution of reciprocal altruism. *Quarterly Review of Biology, 46,* 35–37.
Trivers, R. L. (1974). Parent-offspring conflict. *American Zoologist, 14,* 249–264.
Trivers, R. L. (1976a). Foreword to *The Selfish Gene* by R. Dawkins. New York: Oxford University Press.
Trivers, R. L. (1976b). Haplodiploidy and the evolution of the social insects. *Science, 191,* 249–263.
Trivers, R. L. (1985). *Social Evolution*. Boston: Addison-Wesley.
Turner, M. B. (1967). *Psychology and the Philosophy of Science*. New York: Appleton-Century-Crofts.
Wallerstein, R. (1981). The bipolar self: Discussion of alternative perspectives. *Journal of the American Psychoanalytic Association, 29,* 377–394.
Wallerstein, R. (1983). Self psychology and "classical" psychoanalytic psychology: The nature of their relationship. In A. Goldberg (Ed.), *The Future of Psychoanalysis* (pp. 19–64). New York: International Universities Press.
Watts, A. (1964). Myself, a case of mistaken identity. Talk presented at the Symposium for Human Development, Dallas, TX. (Audiotape published 1991. Boston: Shambhala Publications.)
Weiss, J., & Sampson, H. (1986). *The Psychoanalytic Process: Theory, Clinical Observation, and Empirical Research*. New York: Guilford Press.
Wenegrat, B. (1984). *Sociobiology and Mental Disorder: A New View*. Menlo Park, CA: Addison-Wesley.
White, E. (Ed.). (1981). *Sociobiology and Human Politics*. Lexington, MA: Health.
Wilkinson, G. (1984). Reciprocal food sharing in vampire bats. *Nature, 308,* 181–184.
Williams, G. C. (1966). *Adaptation and Natural Selection: A Critique of Some Current Evolutionary Thought*. Princeton, NJ: Princeton University Press.
Williams, G. C., & Neese, R. (1991). The dawn of Darwinian medicine. *Quarterly Review of Biology, 66,* 1–22.
Wilson, E. O. (1975). *Sociobiology: The New Synthesis*. Cambridge, MA: Belknap Press.
Winnicott, D. W. (1950). Hate in the countertransference. In *Through Paediatrics to Psychoanalysis*. New York: Basic Books (1975).
Winnicott, D. W. (1951). Transitional objects and transitional phenomena. In *Through Paediatrics to Psychoanalysis*. New York: Basic Books (1975).

Winnicott, D. W. (1952). Anxiety associated with insecurity. In *Through Paediatrics to Psychoanalysis.* New York: Basic Books (1975).
Winnicott, D. W. (1954). Withdrawal and Regression. In *Through Paediatrics to Psychoanalysis.* New York: Basic Books (1975).
Winnicott, D. W. (1958). The capacity to be alone. In *The Maturational Processes and the Facilitating Environment.* New York: International Universities Press (1965).
Winnicott, D. W. (1959). Classification: Is there a psycho-analytic contribution to psychiatric classification. In *The Maturational Processes and the Facilitating Environment.* New York: International Universities Press (1965).
Winnicott, D. W. (1960a). Ego distortion in terms of true and false self. In *The Maturational Processes and the Facilitating Environment.* New York: International Universities Press (1965).
Winnicott, D. W. (1960b). The theory of the parent-infant relationship. In *The Maturational Processes and the Facilitating Environment.* New York: International Universities Press (1965).
Winnicott, D. W. (1963). Psychiatric disorders in terms of infantile maturational processes. In *The Maturational Processes and the Facilitating Environment.* New York: International Universities Press (1965).
Winnicott, D. W. (1963b). Communicating and not communicating leading to a study of certain opposites. In *The Maturational Processes and the Facilitating Environment.* New York: International Universities Press (1965).
Winnicott, D. W. (1965). *The Maturational Process and the Facilitating Environment.* New York: International Universities Press.
Winnicott, D.W. (1969). The use of an object and relating through identifications. In *Playing and Reality.* Middlesex, England: Penguin (1971).
Winnicott, D.W. (1971). *Playing and Reality.* Middlesex, England: Penguin.
Wolf, E. (1980). On the developmental line of selfobject relations. In A. Goldberg (Ed.), *Advances in Self Psychology* (pp. 117–130). New York: International Universities Press.
Wolf, E. S. (1988). *Treating the Self: Elements of Clinical Self Psychology.* New York: Guilford Press.
Wolfson, R., & DeLuca, V. (1981). *Couples with Children.* New York: Dembner.
Winson, J. (1985). *Brain and Psyche: The Biology of the Unconscious.* Garden City, NY: Anchor Press.
Wrong, D. (1963). The oversocialized conception of man in modern sociology. In N. Smeltser, N., & W. Smeltser (Eds.), *Personality and Social Systems* (pp. 68–79). New York: Wiley.
Wynne-Edwards, V. C. (1962). *Animal Dispersion in Relation to Social Behavior.* Edinburgh: Oliver and Boyd.
Yankelovitch, D., & Barrett, W. (1971). *Ego and Instinct: The Psychoanalytic View of Human Nature Revisited.* New York: Vintage.
Zetzel, E. (1956). The concept of transference. In *The Capacity for Emotional Growth.* New York: International Universities Press (1970).

Index

Adaptation
  developmental strategy in, 138–140, 147–152, 157–161, 163–170, 176–178, 180–183, 187–190, 192–193, 200–206
  as ego function, 44
  endogenous drives in, 161–175, 200, 269, 275, 279, 283
  existential conflict in, 144–147
  experience and intuition in, 147–149
  innate characteristics affecting, 140–144
  natural selection in, 55
  and paradoxes of relatedness, 149–150, 157, 161, 178, 240–241, 270
  promotion of individual interests in, 78
  reciprocal altruism in, 99
  repression in, 160–161, 181, 182–183, 185
  strategies evolved from ancestral environments, 39
  true self in, 78–79, 175–178, 200, 203
  universal dilemma in, 150–152
Adaptationist hypothesis, limitations of, 58–59
Adolescence
  classical narrative of, 194–198
  in evolutionary theory, 200–204
  relational narrative of, 198–199
  revision of self in, 192–204
  viewed by Anna Freud, 47–48, 194, 197
  viewed by Blos, 48, 195–198
  viewed by Erikson, 198–199
Affect states, compared to drives, 172–175, 305
Aggression, origins of, 167
  self assertiveness and, 167
  interpersonal function of, 169–170
  as inherent vital capacity, 171–172
Alienation
  in adolescence, 195–196
  developmental, 181
  and existential conflict, 144–147
Altruism, 298–301
  among related individuals, 85–86. *See also* Kin altruism
  among unrelated individuals, 86. *See also* Reciprocal altruism
  Freud's views on, 94
  in nature, 84–86, 98–99
  rage as a defense against, 169–170
  as reaction formation, 94, 103
  repression of, 159–160
Analysts
  biases and identity, 216, 222n., 254
  capacity for transference in, 191
  countertransference disclosures of, 256–257
  expectations, 233–234
  experiences in therapeutic rela-

tionship, 234–235
failures of, 229
interpreted by patients, 256–257
non-kin status of, 237–240, 242
self-deceptive strategies of, 222n., 254, 256
Analytic process
 alliance of interests in, 13, 242–243
  as unnatural process, 244, 255n.
 conflicts in, 12, 220, 243–244, 255n.
 empathy in, 239–240
  communication of, 245–248
  inherent ambiguities of, 248–254
  and negative therapeutic reactions, 219n., 222n., 250, 252
  and selfobject needs, 96n., 251–252
 interactions of patient and analyst in, 255n., 257–260
 paradoxes in, 3, 12, 234, 242–243
 resistance to. See Resistance
Anger. See Rage
Animal studies
 altruistic behavior in, 84–86, 98–99
 and applications of data from other species, 67–68
 and convergent evolution, 295n.
 kinship detection in, 89
 parent–offspring conflict in, 109–110
 sexual rivalry in, 118
Archaic heritage, envisioned by Freud, 35, 38–39, 72–73
Aristotle, 292–295
Attachment theory, 45–46
 adaptive utility as seen by Mitchell, 49
 and parent–offspring conflict, 133–134
Atwood, G.
 on existential conflict, 145–147
 on interpersonal, familial conflict, 22, 126, 130–132
 on biology and drives, 50, 172

B

Basch, M. F.
 on drives replaced with affects, 172
 utilization of biological logic, 53
Behavior
 evolutionary explanation of, 64–66
 goals of altruism in, 83–86
Biological approach to human psychology
 concerns with, 6–7
 evolutionary theory in. See Evolutionary theory
Blos, P., views on adolescence, 48, 195–198
Blurton-Jones, N. G., on toddler night awakenings, 65–66
Bollas, C., on true self, 176, 177
Bowlby, J., attachment theory of, 45–46
Brain evolution, and formation of civilization, 102–103, 299
Brandschaft, B., subjectivist view of reality, 128, 130–132, 183, 188, 191n., 217–218

C

Chilhood. See Infants and children
Chomsky, N., 4
Classical narrative of human condition, 20–21
 adolescence in, 194–198
 basic dimensions in, 23–24
 compared to relational narratives, 26, 268, 275
 evolutionary perspective in, 267–283
 conflict and deceptions in, 122–126, 134–135
 in therapeutic relationship, 244
 defenses in, 272
 Freud's theories in, 33–42
 Hartmann's ego psychology in,

Classical narrative *(cont.)*
  42–45
  individualistic tradition in, 30
  oedipal theory in, 117–119, 128
  repetition in, 21, 24
  resistance in, 228
  subjective experience in, 24
  transference in, 188, 214
Communication
  deception in, 151
  symbolic, as encoded version of reality, 138–139, 150–151
Competition
  and collectivist altruism, 92–94
  in nature, 83–86
  in sibling rivalry, 114
Conflict
  in adolescence, 194
  affecting formation of self-structures, 248
  anticipation by innate motivational system, 165–172
  deemphasized in relational model, 23
  from disruptive environments, 24, 27
  and drive theory tradition, 271
  and ego development, 43
  existential, 144–147
  in Freudian view of relational reality, 71–72
  between individual and group aims, altruistic behavior in. *See* Altruism
  infantile, parental role in, 41
  inherent in classical narrative, 23–24
  intrapsychic struggles in, 277–282
  negotiated in families, 119, 124
  oedipal, critiques of, 117–120, 128
  parent–offspring, 107–120
  in therapeutic relationships, 243–244
Conscious life, shifting boundary with unconscious, 187
Cosmides, Leda, 4, 39, 77–78, 104

Cultural experience, influence of, 66–68

D

Da Costa, E., on toddler night awakenings, 65–66
Daly, Martin, critique of oedipal theory, 117–118
Danger, in Kleinian view of relational reality, 73
Death instinct, 215
Deception
  as adaptive strategy, 149
  in adolescence, 197
  anticipation by innate motivational system, 165–172
  in communication, 151
  in parent–offspring conflicts, 115, 122–123, 180
  in renegotiation of self, 214
  in repression, 157–158
  in transference, 28
Deep structure of psyche, 68–79
  and belief that knowledge cannot go beyond experience, 70, 74, 125
  and conflict during development, 114–116
  as evolved adaptive design, 5, 40, 148–149, 270, 276–277
  and Freud's archaic heritage, 72–73
  and hard-wired rules affecting adaptive behavior, 69, 70, 138
  and individual interactions with relational world, 142–144
  individual variations in, 76–79
  infant research in, 70–71
  Klein's version of, 73–74
  natural selection affecting, 57
  and preparation for relational world, 116, 281
  psychoanalytic assumptions about, 71–76
  relation to evolutionary adaptedness, 63–66

in relational models, 74–76
self-interest in, 269
superego as, 35
as system to organize experience and generate behavior, 64, 69
transferences rooted in, 185–186
universal nature of, 76
Defenses
in adolescence, 194
in classical and relational narratives, 272
Depression, as adaptive process, 224–225, 226
Depressive position of child, and negotiated solution to conflicts, 123, 165, 166
Determinism, as problem in evolutionary theory, 63–66
Development, adaptive strategy in, 138–140
Distortions in transference, 28, 183, 186, 188, 214
and renegotiation of self, 219–223
Drives, instinctual, 161–175, 271
in adolescence, 194, 196–197
compared to affect states, 172–175, 305
critique by Mitchell, 49
and individual autonomy, 143
as motivational system, 163–164
in anticipation of conflict and deception, 165–172
origins of, 35, 44
and parent–offspring conflict, 122, 135

E

Ego
adaptive functions of, 42–45
antirepressive action in transference, 184, 216
Ego psychology, 42–45
conflicts in therapeutic relationship, 244
resistance in, 228
transference in, 214
Empathy in therapeutic relationship
communication of, 245–248
inherent ambiguities of, 248–254
Environment
affecting development of self, 142–143, 206
disruptive, conflicts from, 24, 27
pathology from failure of, 129, 131–132, 136, 144–146
relational. *See* Relational reality
Erikson, E.
on adolescence, 198–199, 202
identity foreclosures of, 177, 199
psychosocial model of, 46–47, 176, 199
Evolutionary theory
adaptive mechanisms in, 6–7
adolescence in, 200–204
clinical implications of, 209–212
compared to classical and relational narratives, 267–283
and distal causes of proximal mechanisms, 60, 302–304
Freud's concepts of, 33–42
reactions to, 37–38
and influence of cultural experience, 66–68
on parent–offspring conflict, 135–136
perspectives in, 55–80, 274–283
problem of determinism in, 63–66
problem of reductionism in, 59–62
repression in, 272–274
transference in, 218–219
Existential conflict, and intersubjectivity, 144–147
Experience
organized by deep structure of psyche, 64, 69
responses to, in relational theory, 74–75
subjective
as deceptive distortion in classical narrative, 24

Experience *(cont.)*
    as personal reality in relational narrative, 24

F

Fairbairn, W. R. D., on libido as object-seeking, 133, 134, 162
Fitness
    inclusive, 86–90. *See also* Inclusive fitness
    personal, 86–87
Freud, A.
    on adolescence, 194, 197
    on hostility between ego and instincts, 44, 47–48
Freud, S.
    on archaic heritage, 38–39, 72–73
    confusion of proximal and distal causes, 302–304
    on death instinct, 215
    drive theory of, 271
    evolutionism of, 33–42
        reactions to, 37–38
    and Haeckel's law, 296–297
    on innate individual characteristics, 142
    on parent–offspring conflict, 122
    on parental motivation, 94–95
    on phylogenetic memories, 38, 40, 72, 237
    on relational reality, 72–73
    shift from biological thinking to historical approach, 61–62
    on sibling rivalry, 114
    speculation on deep structure of psyche, 71–73
    on transference, 183, 188
    use of Lamarckian theory, 36, 290–296
Fuss, Diana, 3

G

Game theory, reciprocal strategy in, 100–101
Gene's-eye view
    of balance of interests in parents and children, 111
    of self and relatedness, 88, 90–91, 93, 198
Genetics
    and altruism, 85–86, 108
    and degrees of relatedness in families, 99
    and determinism as problem in evolutionary theory, 63– 66
    and variability in deep structure of psyche, 76–77
Genotype, definition of, 55–56
Glantz, K., 240–241
Goldberg, A., utilization of biological logic, 53
Good-enough environment. *See* Relational reality
Greenberg, J. R., comparison of psychoanalytic theories, 23, 266, 270
    transference relationship, 197, 201
Group selectionist theory, altruistic behavior in, 84–86
Guilt
    and reciprocal altruism, 102–103, 299–300
    and repression, 158

H

Haeckel's law, 296–297
Haldane, J.B.S., 87
Hamilton, W.D., 85, 86–87, 89, 92
Hartmann, H.
    ego psychology of, 42–45
    reality principle of, 44, 92
Heidegger, M., on existential conflict, 144–146
Hoffman, I.
    patient as interpreter of analyst, 255, 256
    psychoanalytic perspectivism, 210, 259
    social constructivism of, 28, 264

I

Id
  as motivational apparatus, 162
  origins of, 35, 44
Inclusive fitness, 86–90
  and concept of true self, 176
  and kin altruism, 91–97
  and reciprocal altruism, 98–104
Individual needs
  and adaptation to environments, 76–79
  and conflict between parent and child, 111
  and shared group goals, 84
  and true self concept, 175–178
  and unique aspects of self, 140–144
Infants and children
  abuse by related and unrelated adults, 108
  developmental strategy of, 138–140
  inherited motives of, 41
  innate strategies of toddlers, 65–66
  parent–offspring conflict theory, 107–120
  research in innate capacities of, 70–71
Innate capacities
  and concept of evolved deep structure. *See* Deep structure of psyche
  concern for related others, 91–97
  concern for unrelated others, 98–107
  infant research in, 70–71
  and influence of cultural experience, 67, 69
  for linguistic ability, 4, 68–69
  and strategies of toddlers, 65–66
Instinctual drives. *See* Drives, instinctual
Intelligence
  development through reciprocal exchange, 103
  as hypothetical construct, 62

Interpersonalist theories, 27–28
  on conflicts in therapeutic relationship, 244
  on resistance, 228–229
  transference in, 214
Intersubjectivist theories, 51
  on existential conflict, 144–147
  on parent–offspring (familial) conflict, 130–132
  on resistance, 229
  on therapeutic relationship, 243
  on transference, 217–218, 221

K

Kin altruism, 85–86, 192, 298
  and inclusive fitness, 86–90, 91–97
  indirect benefits of, 97
  and relations with siblings, 114
  survival of genetic material in, 87
Kitcher, P., criticism of sociobiology, 58, 63
Klein, M.
  on deep structure of psyche, 73–74
  on innate instinctual motivation, 165–166
  on parent–offspring conflict, 73, 123–126
  psychoanalytic theories of, 27
  on relational reality, 73–74
Kohut, H.
  on empathy, 246
  on intergenerational strife, 128–130
  on narcissistic rage, 168–169
  on parental devotion as innate satisfaction, 51
  on parental motivation, 95–96
  utilization of biological logic, 51, 53
Kohut, T., on empathy, 246

L

Lacanians, views on parent–offspring conflict, 127
Lachmann, F., theories of, 51
Lamarckian theory, Freud's use of, 36, 290–296

Language
- innate capacity for, 4, 68–69
- selective advantage of, 56

Lichtenberg, J.
- motivational systems of, 173–174
- utilization of biological logic, 53

Literalists, views of, 37

Locke, John, belief that knowledge cannot go beyond experience, 70, 74, 125

M

Mayr, Ernst, 4
- on adaptationist hypothesis, 58

Mitchell, S.
- on attachment and relating, 48–50
- comparison of psychoanalytic theories, 23
- criticism of developmental tilt, 306
- on need for relatedness, 143
- relational conflict model of, 133

Motivation, endogenous, 163–165. *See also* Drives, instinctual

Mourning, developmental, 203, 257

Munder-Ross, J., on parent–offspring conflict, 134–135

Mutuality
- and altruistic behavior. *See* Altruism
- in families, 107–117
- inherent in relational narrative, 23, 75, 83–106, 271

N

Narcissism of parents, 94–95

Narcissistic rage, 168–169

Natural selection
- and adaptation, 55
- affecting deep structure of psyche, 57
- definition of, 56
- in distal causes of proximal mechanisms, 60
- and group selectionist model, 84–86
- shaping of phenotypes in, 60
- in social environment, 35, 166, 287–289

Negotiation
- in families, 119, 124, 135, 139
- of self, 179–206
- in solution of conflicts, 165, 166
- in therapeutic alliance, 244–245

Nesse, R., 304

Neubauer, P. B., studies of twins reared separately, 76–77, 140–142

O

Oedipal conflict
- critiques of, 117–120, 128, 129, 134–135
- motivations in, 168
- paternal motives in, 41

Oral imagery, in parent–offspring conflict, 124

Ornstein, Paul, utilization of biological logic, 53

P

Paranoid suspicions of child, and detection of conflict, 73, 123, 165, 166

Parents
- conflict with offspring, 41, 107–120, 170
- in animals, 109–110
- and attachment theories, 133–134
- and biases in parent's view of reality, 116, 133, 148, 151, 180
- classical psychoanalytic model of, 122–126, 134–135
- competing interests in, 115–116, 132–134, 149, 180
- in evolutionary narrative, 135
- fate of siblings in, 111–113
- intersubjectivist views on, 130–132

negotiation in, 119, 124, 135, 139
and oedipal theory, 41, 117–120, 128, 129, 134–135
parental or environmental failure in, 129, 131–132, 136
parental point of view in, 198
relational conditions affecting, 182
in relational narrative, 126–134
and socialization process, 113
devotion as innate satisfaction, 51
inclusive fitness and personal fitness of, 86–87
motivation of
Freud's views on, 94–95
inherited, 41
Kohut's views on, 95–96
narcissism of, 94–95
Pathology
and conflicts in therapeutic relationship, 244
from environmental failure, 129, 131–132, 136, 144–146
and expectations about reciprocity, 240–241
Pearce, J., 240–241
Pizer, S., 234
Pleasure principle, as motivational apparatus, 162
Prisoner's Dilemma game, modified version of, 100–101
Projective identification
concept of, 73–74, 125, 191
repetitions in, 216
Psychic structure, viewed in classical and relational narratives, 23
Psychoanalysis. *See* Analytic process
Psychoanalytic theories
assumptions about deep structure of psyche, 71–76
basic dimensions in, 23–25
biological approach to, 59–60
classical narrative in, 20–21
contrasting beliefs in, 8–9, 28–32
dialectical tension between, 263–267

early Darwinian versions of
by Bowlby, 45–46
by Erikson, 46–47
by Freud, 33–42
by Hartmann, 42–45
evolutionary perspective in, 209–212
hypothetical constructs in, 62
reductionist barrier in, 60–61
relational narrative in, 21–22
repression in, 22
Psychosocial model of Erikson, 46–47, 176, 199
Psychotherapy. *See* Analytic process, and psychoanalysis, 211

R

Rage
as defense against altruism, 159–160
narcissistic, 168–169
self-interest in, 167–169, 171–172
Reaction formation, and altruism, 94, 103
Reality
biased parental view of, 116, 133, 148, 151
relational. *See* Relational reality
Reciprocal altruism, 86, 298–301
human examples of, 99–102
and inclusive fitness, 98–104
as strategy in game theory, 100–101
in transference, 191
Reductionism, as problem in evolutionary theory, 59–62
Regression
in adolescence, 193, 194, 196, 197, 199
in renegotiation of self, 187, 189, 214, 221–223, 308–309
Relatedness
gene's-eye view of, 88, 90–91, 93
paradoxes of, 149–150
reciprocity in, affecting resistance, 231–232

Relational narrative of human condition, 21–22
  adolescence in, 198–199
  and conflicts in therapeutic relationship, 244
  basic dimensions in, 23–25
  Bowlby's attachment theory in, 45–46
  collectivist tradition in, 30
  compared to classical narratives, 26, 268, 275
  evolutionary perspective in, 267–283
  defenses in, 272
  Erikson's psychosocial model in, 46–47
  innate knowledge and expectations in, 74–76
  Mitchell's views on attachment and relating in, 48–50
  mutuality in, 23, 75, 83–106, 271
    and promotion of self-interest, 128
  normal environment in, 126–134
  repetition in, 25
  subjective experience in, 24
  transference in, 217
Relational reality
  adolescent probing of, 201
  anticipation by innate motivation, 165–172
  childhood testing of, 125–126
  and concepts of deep structure of psyche, 281
    by Freud, 72–73
    by Klein, 73–74
    by relational theorists, 74–76
  conflict and mutuality in, 83–106
  deception in, 115
  individual interactions with, 142–144
  operating on prepared individual organisms, 57
  preparation by deep structural design in child, 116
  probed during transference, 191
  true self in, 175–178

Repetition
  as adaptive process, 188
  in adolescence, 196
  in classical narratives, 21, 24
  compulsion for, 214, 216
  creative, in renegotiation of self, 223–227
  and death instinct, 215
  in relational narratives, 25, 28
  in transferences, 214–219
Repression, 155–157
  as adaptive strategy, 160–161, 181, 182–183, 185
  and capacity for future repetition, 226, 227
  in classical narrative, 21, 22
  of conflicted inner experience in childhood, 132
  as deceptive tactic, 157–158
  evolutionary view of, 272–274
  and identity elements held in reserve, 157, 160–161, 182–183
  in adolescence, 203
  of innate altruism, 159–160
  in relational narrative, 22
  and retrieval of repressed versions of self, 187, 189
  in adolescence, 196, 197, 200
Resistance, 228–241
  clinical expressions of, 235–241
  comparative viewpoints of, 228–229
  and innate skepticism of patients, 229–231
  interpreted as pathology, 13, 236
  kin relationships affecting, 231–232
  unnaturalness of analytic situation in, 232–235

S

Sartre, J. P., on existential conflict, 144–147
Selection pressures. *See* Natural selection
Self
  divided character of, 204–206

evolutionary perspective of, 269
gene's-eye view of, 90–91
hidden from environment, 22
inherent internal conflict in, 277–282
negotiation and renegotiation of, 179–206
  in adolescence, 192–204
  and capacity for creative repetition, 223–227
  coordination of repression, regression, and transference in, 180–192
  provisional organization in, 179–180, 206
  regression in, 187, 189, 214, 221–223, 308–309
  in transferences, 186
as semisocial entity, 91
threatened in parent–offspring conflict, 124–125
"true self," concept of, 175–178, 182
unique aspects of, 140–144
Self-interest
  in adolescence, 203–204
  protected by rage, 167–169, 171–172
  and transference-based distortions, 219–223
Self psychology
  competing dominances in, 282
  critique by Mitchell, 50
  narcissistic rage in, 168–169
  therapeutic relationships in, 243
  transference in, 217, 221
  utilization of biological logic, 53
  views on intergenerational conflict, 129–130
Sexuality, conflicts in, 118–120
Shame, as innate affect, 305, 306–307
Sibling rivalry
  factors affecting, 114
  motivations in, 168
Slavin, J., on transference, 184
Social behavior. *See* Behavior
Social constructivism, 28

Sociobiology, criticism of, 58, 63
Stolorow, R.
  biology and drives, 50, 51, 172–173
  on drives as affect states, 172–173
  on existential conflict, 145–147
  on interpersonal (parent–offspring) conflict, 126, 130–132
  on intersubjective view of reality, 28, 183, 188, 191n.
Subjective experience, 217, 218
  as deceptive distortion in classical narrative, 24
  as personal reality in relational narrative, 24
Subjectivity. *See* Intersubjectivist theories
Superego, evolutionary origin of, 35
Symbolic communication, as encoded version of reality, 138–139, 150–151
Symbolist tradition, 37

T

Tension
  in classical narrative, 21
  in relational narrative, 21–22
Therapeutic relationship. *See* Analytic process
Therapists. *See* Analysts
Tomkins, S., concept of affective motivational systems, 173
Tooby, John, 4, 39, 77–78, 104
Transference, 183–187
  in analysts, 191
  bargaining function of, 221
  comparative concepts of, 214–219
  contrasting views of, 29
  deception and distortion in, 28
  developmental, in adolescence, 200–202
  distortions in, 28, 183, 186, 188, 214
  and renegotiation of self, 219–223
  learned expectations in, 186

Transference *(cont.)*
  and probing of relational reality, 191
  reciprocal altruism in, 191
  reorganization of self in, 186
  repetition in, 214–219
    and renegotiation of self, 223–227
  universal capacity for, 183–185
Trivers, Robert
  concept of reciprocal altruism, 98, 102–103
  on parent–offspring conflict, 109, 113
"True self," concept of, 10, 78–79, 145, 175–178, 182, 190, 192, 200, 203, 248n., 254n., 256, 270, 308–309
Twins reared separately, studies of, 76–77, 140–142

U

Unconscious life, shifting boundary with conscious, 187

W

Wilson, Margo, critique of oedipal theory, 117–118
Winnicott, D. W.
  on individuality, 90–91, 147
  on interpretation, 253
  on paradoxes of relatedness, 149–150
  on parental hate and ruthlessness, 135
  on regression, 213
  on separation anxiety, 127
  on true self, 175–176, 182
Wolf, E.
  on interpersonal conflict, 130
  utilization of biological logic, 53